W9-CAL-218

Darwin and the Novelists

DARWIN and the Novelists

Patterns of Science in Victorian Fiction

George Levine

Harvard University Press
Cambridge, Massachusetts
London, England

Printed in the United States of America
10 9 8 7 6 5 4 3 2

Publication of this book has been aided by a grant from the
Andrew W. Mellon Foundation.

This book is printed on acid-free paper, and its binding materials have
been chosen for strength and durability.

Library of Congress Cataloging-in-Publication Data
Levine, George, 1931–
Darwin and the novelists.
Bibliography: p.
Includes index.
1. English fiction—19th century—History and
criticism. 2. Science in literature. 3. Darwin, Charles,
1809–1882—Influence. 4. Evolution in literature.
5. Literature and science—Great Britain—History—19th
century. I. Title.
PR878.S34L4 1988 823'.8'09356 87-36201
ISBN 0-674-19285-0 (alk. paper)

For Marge
a natural selection

Preface

THIS PROJECT was originally conceived out of G. H. Lewes's remark that "science is penetrating everywhere." Intending first to write directly about the scientists' discourse and the Victorian Sages' responses, I turned toward the novelists when, in 1983, I was invited to be Helen Cam Visiting Fellow at Girton College, Cambridge. There, working at the University of Darwin, Herschel, and Whewell, in a little undergraduate library of a college to which George Eliot had contributed £50, I had the privilege of talking with and learning from Gillian Beer, whose extraordinary *Darwin's Plots* was arriving in galleys the week I was to lecture on Ruskin and Darwin to the Faculty of the English Department. My reading of those galleys was humbling and exhilarating. Oddly, while it suggested to me that I was lucky to have spent my time on the nonfiction prose rather than the novelists, it opened a new direction for me.

Beer's suggestion that unread ideas often become assumptions registered in my own experience, my sense of having absorbed without full consciousness the vision of the world created by our contemporary science; and I began to look for analogous assumptions in Victorian fiction. The project of the Victorian novel increasingly appeared to me as a cultural twin to the project of Victorian science; even the great aesthetic ideals of fiction writers—truth, detachment, self-abnegation—echoed with the ideals of contemporary science. Seeing the kinship everywhere, I often found myself wondering whether the juxtapositions I was coming to insist on were not too extravagant or too easy because too general. This book will at last put that possibility to the test.

I am confident, however, that shifting the angle of approach to Victorian novelists throws light on them as well as on the scientists and their culture. The Victorian novel clearly joins with science in the pervasive secularizing of nature and society and in the exploration of the consequences of secularization that characterized mid-Victorian England. The tradition of natural theology was threatened and largely dismantled by Darwinian science, and in the process nature, society, narrative, and language itself were desacralized, severed from the inherent significance, value, and meaning of a divinely created and designed world. The Darwinian quest for origins was the signal and the authoritatively scientific means by which fact was severed from meaning and value, "presence" became absence, and the world had to be reconstituted not from divine inheritance but from arbitrary acts of human will. The absorption and response to Darwinism played out in narrative can suggest the resistance and evasiveness and submission of nonscientists to this critical development in Western thought and imagination.

The immersion in Darwin has been not the least of the pleasures of this enterprise for me. The greatness and originality of his mind, despite the unhappy uses to which he has often been put, place him among the sagest of the Victorians, and the most delightful. The subtle craftsmanship of his prose, despite his Trollopean self-denigration, makes it very much worth reading and teaching, even for the layperson. And the web of connections he spins with all the significant writers of the century makes *The Origin of Species* and parts of *The Descent of Man* as worthy of literary consideration as *Sartor Resartus, Culture and Anarchy,* or Mill's *On Liberty*.

There are moments in working on such subjects that make it all worth while, that somehow redeem all the petty Casaubon-like anxieties of scholarship—the tedium of reading too much yet knowing one should have read twice as much; the frightening suspicion that one is merely repeating what has already been said better; the questions about whether there is any point in doing it at all; the boredom of checking citations; the fear that now at last one is going to be found out. Every scholar might add to this list ad nauseam. Yet I would do it all again for the sake of a moment that could not be turned into a chapter or a footnote or an idea, but that transformed for me all this airy speculation and these words: the moment when Peter Gautrey, the generous curator of the Darwin collection at the Cambridge University Library, put into my hands the very copy of Lyell's *Principles of Geology* that Darwin took with him on the *Beagle*.

Parts of this book were written with the support of an Independent Study Fellowship from the Rockefeller Foundation. Chapter 4 appeared in a much different version and with another object in *Raritan Review,* 3 (Winter 1984): 30–61; Chapter 5 appeared in an earlier version in *Texas Studies in Literature and Language,* 28 (1986): 250–280; and Chapter 6 appeared in an earlier version in *Dickens and Other Victorians,* ed. Joanne Shattock (London: Macmillan, 1987). I thank the publishers for permission to make use of the material.

Rutgers University has been generous and supportive, not least so in allowing me a year of leave to complete the manuscript and dozens of ancillary quasi-scientific projects, and in appointing me director of its new interdisciplinary Center for the Critical Analysis of Contemporary Culture.

I am particularly indebted to Girton College, Cambridge, for inviting me for two terms as Helen Cam Visiting Fellow, in 1983. This book simply would not exist (although I am not convinced she would approve of it) if it were not for Gillian Beer, of Girton. I want to thank, in addition, Barry Qualls, of Rutgers, who read several sections in early versions, and whose extraordinary generosity of mind and spirit have put an army of colleagues and graduate students in his debt. Various colleagues spurred me to work with invitations to visit and lecture, and I would like to note among them Jack Farrell of the University of Texas, Austin; William Thesing of the University of South Carolina; John Stasny of the University of West Virginia; Rob Polhemus of Stanford University; Millicent Bell of Boston University; Sally Shuttleworth of the University of Leeds; and Barbara Hardy of Birkbeck College, London. My thanks go, as well, to David Kohn of Drew University, an editor of the Darwin letters, a man who likes to talk about Darwin, and does it wonderfully. Alan Rauch began working with me as a graduate student who knew a great deal about biology and Darwin, and has since become not only a scientific resource, but a friend.

It is difficult to know, with close friends, where warmth of feeling becomes intellectual support. But I know this book would have been the worse without Matthew Baigell and Renée Baigell, who frequently pulled me out of my Darwinian confusions on our drives to see the New York City Ballet, and without Michael Moffatt, who, particularly during our outings looking for odd species of birds at odd hours in New Jersey and New York, thrashed out Darwinian problems with great patience. I owe very special thanks, of a kind no preface could encompass, to Isobel Armstrong, of Southampton University, one of those rarely generous spirits who make intellectual discussions gifts of friendship and knowledge, and to Michael Wolff, who almost thirty years ago welcomed me to *Victorian Studies,* and who has made studying the Victorians for me not only an intellectual adventure but an act of love. Speaking of which, there is no language for what I owe Marge Levine, an artist who has—pace Darwin—made my life meaningful.

Contents

Darwin and the Novelists

1

Darwin among the Novelists

OF THE GREAT nineteenth-century scientists, Darwin is the one whose impact on nonscientific culture is best known, or at least most widely discussed. *The Origin of Species* and *The Descent of Man* have revolutionized the ways we imagine ourselves within the natural world and have raised fundamental questions about the nature of self, society, history, and religion; and it did not take a scientist to know that this was happening. Obviously, Darwin's revolution was not single-handed, but he can be taken as the figure through whom the full implications of the developing authority of scientific thought began to be felt by modern nonscientific culture. Darwin's theory thrust the human into nature and time, and subjected it to the same dispassionate and material investigations hitherto reserved for rocks and stars. His history of the development of species gave authoritative form to a new narrative—or set of narratives—that has permanently reshaped the Western imagination. Darwin has transformed reality for us and, as Gillian Beer has put it, "We pay Darwin the homage of our assumptions."[1]

He presented himself as the genial empiricist, eager for the smallest minutiae of fact, and willing to submit any idea to the test of experience. Ostensibly a propagandist for the Baconian ideal, he seemed to piece his theory together, like a detective in a literary genre that owes much to science, through fragments and traces, building vast structures from seeds and spores and insects and fossils. Whatever his public presentation, the theory was only possible because Darwin had seized it imaginatively before he could prove it inductively. He had the power to imagine what wasn't there and what could never be seen, and he used analogies and metaphors with subtlety and profusion as his imagination actually defied the experience that Baconian theory privileged. As we follow him through his long argument, he seems to be counterintuitive as often as he is the merely sensible ordinary man he took such pains to be. He had a way to organize the facts—with sets of assumptions that belonged both to him and to the scientists who preceded him as well as to the culture at large, and with a theory that could not, in Baconian terms, be proved.

In this book I will be talking about that theory not as an expert in evolutionary biology but as a student of nineteenth-century narrative and culture particularly interested in the complex interweavings of science and literature. The power of Darwin's story is self-evident, and that it is a "story," or a cluster of "stories," should be clear, as well. Like other great stories—Milton's, for example, or Freud's—it did not emerge from nowhere and its emergence is clearly related to the special historical conditions of its time; moreover, it was quickly absorbed into the narratives by which the culture defined itself and its sense of the real. Coming from a mode of discourse self-confidently representational and nonfictional it enters into the dubiously representational realms of narrative and fiction; the boundaries between the two kinds of narrative, the two kinds of representation, blur.

So I am concerned with Darwinian theory as a historically locatable response to questions of particular urgency among the Victorians: questions about the sources of authority (religious, political, and epistemological), about the relations of the personal and the social to the natural, about origins, about progress, about endings, about biological and social organicism. And my focus will be on how Darwin's real or reputed response to these questions interacted with the responses and forms of nineteenth-century English narrative.

There have, of course, been many studies of the impact of Darwinian thought on the literature of the nineteenth century, particularly on the work of writers obviously familiar with Darwin.[2] Beer has usefully discussed Kingsley, George Eliot, and Hardy, for example, all of whom were close readers of Darwin. Sally Shuttleworth's book on George Eliot and science explores in some detail the uses to which Eliot put—and did not put—evolutionary theory.[3] Roger Ebbatson has tried to link the "texture" of the work of Hardy, Forster, and Lawrence—writers self-conscious about their philosophical and scientific sources—to Darwinian theory.[4] And most recently, Redmond O'Hanlon has read Conrad, particularly *Lord Jim*, in view of the connections with Darwin and Darwinian ideas.[5] But I want to take up another hint of Beer's: "Ideas," she says, "pass more rapidly into the state of assumptions when they are *unread.*" Whereas she then focuses on writers who did read Darwin because "reading is an essentially question-raising" activity,[6] I am more interested here in the "assumptions" than in the self-consciousness of questioning, and thus in writers who probably did not know any science first hand, who could have been "influenced" by Darwin only indirectly. This is not, then, a study of influences.

Science, as I discuss it here, is a shared, cultural discourse, "a cultural formation," as Michel Serres believes, "equivalent to any other."[7] Although it certainly has been privileged, both in Victorian times and in some quarters even in our own, as an activity somehow exempt from the skepticism to which almost all other cultural discourse is subject, it works within the culture and responds to its exigencies. Internalist histories of science are surely necessary, but they go only part way in elucidating how science develops and, in particular, what sort of interplay between scientific and nonscientific discourses characterizes their mutual developments. Katherine Hayles seems to me exactly right when she argues that "*both* literature and science are cultural products, at once expressing and helping to form the cultural matrix from which they emerge."[8] There were many evolutionisms before Darwin and there have been many since. His theory found ostensibly scientific form for the ideologies that dominated Victorian society and was received by that society, despite some ruckus, with remarkable speed; its language pervaded its literature—had already found some voice in it before he ever published the *Origin*.

It is because of this peculiarly complex and even counterchronological interpenetration between science and literature that I concentrate on writers who were probably not directly "influenced" by scientific writing.[9] My concern is not "influence" but the absorption and testing of Darwinian ideas and attitudes (even when the writers are not thinking of them as Darwinian) in the imagination of Victorian novelists. Theirs, after all, was a time when science was most forcefully extending its authority in the realm of knowledge and even beyond, into religion and morals, and when it really did seem for a while that apparently insoluble problems could be solved, that the limits imposed on human society by material conditions could be broken, and that knowledge was an aspect of morality, so that the highest Victorian virtue was "Truth." Science, particularly through technology, was visibly reshaping Victorian life—transportation, lighting, sewage disposal, communications, medical treatment, mass production; but it was also an important part of middle-class and working-class entertainment. Popularizations of science were filling lecture halls, journals, and workingmen's institutes; "lay sermons" were displacing religious ones; amateur fossil hunting, insect collecting, seashell study were holiday diversions and potential contributions to rapidly expanding scientific knowledge. "Evenings at a Microscope" were pleasant hours of quasi-educational amusement.[10] The ideas of science were helping to form the general view of the nature of "reality" itself, and Darwin's vision, his great myth of origins, was both shaping the limits of the Victorian imagination of the real and being tested in the laboratories of fiction as well as in scientific argument. Moreover, the relationship was certainly two-way: how the culture tells stories, that is, imagines its life, subtly informs the way science asks questions, arrives at the theories that reshape the culture that formed them. Milton, as Gillian Beer has shown, not only accompanied Darwin on the voyage of the *Beagle*, but may have helped shape his *scientific* vision.[11]

Darwin read Milton; George Eliot, Darwin. But does it, after all, make sense and is it useful to consider how far, say, Dickens's imagination is consonant with Darwin's, whether the curious correspondences and inevitable divergences of their ways of telling stories of growth and change can reveal anything about their work that more conventional and internalist criticism could not provide, and can suggest something about the assumptions of Victorian culture that were powerful enough to infiltrate both science and fiction? My experience as a nonscientist in so scientifically oriented a world as our own suggests that there are, indeed, connections between what we assume about nature, self, and society and what (at least popular) science works on and argues. Stephen Brush talks of several ways in which science and culture may be related:

> An idea from culture may enter science, where it can stimulate certain lines of theorizing and (perhaps) suggest new experiments and lead to new discoveries. This was what happened with the romantic concept of the unity of all natural forces. Conversely, scientific facts and theories may have a direct influence on those who construct philosophical systems, write novels, or criticize society. Thus the mechanistic materialism of mid-nineteenth-century physics and biology was reflected by "realism" in philosophy and literature, and by "positivism" in the social sciences. A third possibility is that the same notion may appear at about the same time in both science and culture without any apparent causal influence one way or the other. Such was the case with the principle of dissipation of energy in physics, and the corresponding theory of degeneration in biology, both of which

> flourished in the pessimistic atmosphere of the latter part of the nineteenth century.[12]

Obviously, the last possibility precludes the study of direct influences. And yet it is the most striking and interesting of the connections. The coincidences are such that one is almost driven to posit a *Zeitgeist*, but a close look at particular instances tends to make these coincidences less mysterious. Much of what follows in this book will be concerned, for instance, with Victorian gradualism, an idea that popped up in geology (on a Newtonian model), fought its way into biology, and was the groundwork of nineteenth-century "realism." But there is no great mystery about the intellectual convergence, given the political and social temper of the time. The relation between the scientific, literary, and even political uses of gradualism manifests how deeply ostensibly "disciplinary" ideas are embedded in the whole culture and suggests that there is a great deal to learn about the separate disciplines in attending to their mutual (and yet divergent) uses of such ideas.

The determination to see the impurity of scientific and literary ideas and the recognition that there is two-way traffic between them may seem, presumptuously, to challenge the authority of science by questioning its rationality. That challenge has been a major subject of debate among philosophers of science since the publication of Thomas Kuhn's crucial work, *The Structure of Scientific Revolutions* (1962). A significant minority of philosophers of science have followed Kuhn in challenging Baconian, and Victorian, and Popperian notions of science as progressive, empirical, verifiable and falsifiable, objective, and uncontaminated by the winds of interest or subjectivity.[13] For the purposes of this volume, I, like Serres, consider science as an unprivileged form of cultural discourse, "a cultural formation equivalent to any other," but one that happens to have been privileged for much of its modern history.

Without plunging into the wars among philosophers and historians of science, I want to borrow a notion from the work of Gerald Holton, who, in his attempt to account for the ostensibly irrational elements in scientific work and particularly in acts of discovery, talks about the "thematic origins of scientific thought." He argues that while normal scientific discussion is concerned primarily with the empirical and analytical validity of propositions, even the decision to regard this as "normal" is arbitrary. But his major point is that the concern with these internal elements of scientific activity is inadequate to account for the way science generates new ideas, the way it moves and changes;[14] nor, more important yet, can it account for the "irrational" elements in all great scientific work. The positivist preoccupation with the empirical and the analytic, whatever the success, does not, says Holton, "hide the puzzling fact that contingency analysis excludes an active and necessary component that is effective in scientific work, both on the personal and the institutional level; that is, it neglects the existence of preconceptions that appear to be unavoidable for scientific thought, but are themselves not verifiable or falsifiable." These "preconceptions" are "themata." In Holton's definition themata are preconceptions "of a stable and widely diffused kind, that are not resolvable into or derivable from observation and analytic ratiocination."[15] Obviously, these are the elements in scientific thought with which a nonscientist would feel most comfortable. Holton himself is not suggesting that scientific discourse is ultimately irrational, or that it is not "objective." He is allowing for the fact that, as is clear in the work of Darwin, the very impetus to ask the questions and to find a solution and the assumptions that underpin the methodology depend on preconceptions established without scientific verification. It is possible on this argument, then, for science to be rational and coherent internally while still being fully implicated in significant extrascientific concerns of the culture.[16] The "constraints of culture" can still be invoked to help understand the development of scientific ideas.[17]

Whereas Holton suggests that the source of these themata is best studied through the nature of perception itself and the psychological development of children, and thus seems to imply that the themata have an existence beyond history, I consider themata primarily as social and historical constructs rather than as inevitable and permanent parts of the human intellectual constitution. Gradualism, for example, was not an inference from empirical data, but an explanatory desideratum that made the evidence of geology consonant with the idea of a designed and stable universe.[18] Since the early nineteenth century, following Hutton's geological theories and Lyell's then definitive consideration of the history and methods of geology, and following Darwin's powerful theoretical and practical application of Lyell's strategies, nature seemed to become uniform in its operations and gradual in its transformations. This kind of nature still for the most part governs biological research, as it has long governed narrative; and we can watch a parallel in modern thought between a new philosophical emphasis on discontinuities and the arguments of Stephen Jay Gould, Niles Eldredge and Ian Tattersall. The last two discuss the "myth," deriving obviously from Darwin, "that evolution is a process of constant change." They argue that "the history of the world is rather one of fits and starts."[19] Michel Foucault and Edward Said also emphasize discontinuity and seriality rather than slow and constant change.[20] And the two ostensibly unrelated theories can be seen to converge in the area of narrative, where traditional "realistic" fictions are being displaced by fictions of discontinuity and fragmentation. Said's project, like Foucault's, entails the rejection of the assumptions about continuity and order that determine the substance and form both of Darwin's argument and of the classic nineteenth-century novel. Sally Shuttleworth shows that uniformitarian thought was allied to a conception of the self as unified and continuous, and thus to important elements of characterization and narrative form; she shows also that the psychology based on evolutionary biology introduces elements of discontinuity that in turn led to revised notions of the self, and consequently of character in fiction, and of course of the way narratives move and are resolved. But the Foucauldian method, says Said, is "postnarrative"; "the novelistic model of successive continuity is rejected as somehow inappropriate to the reality of contemporary knowledge and experience."[21] I believe that this "model," though clearly applicable to most Victorian fiction and to Darwin, is actually threatened by much that Darwin's theory entails. There are deep contradictions within the Darwinian project that parallel contradictions within nineteenth-century realism, itself. And in the final chapter I will be arguing that Conrad's use and rejection of Darwinian gradualism brings together narrative experimentation and the possibility of revolutionary change.

The nineteenth-century revolution into continuity, which was most effectively formulated in the work of James Hutton (through John Playfair's *Illustrations of the Huttonian Theory of the Earth,* 1802) and Charles Lyell, and which belongs to the secularizing and rationalist tradition of Enlightenment thought, entailed a radical rethinking (and refeeling) of history, religion, teleology; that it too, however, belongs to history and is not inevitable either in narrative or science is almost a commonplace now. Looking back, we too often tend to associate uniformitarianism with the myth of progress, and to see how easily Darwinism, which seemed so disruptive and threatening, could be assimilated to traditional teleology and to its new form, progressivism. Yet it is worth remembering, first, that catastrophism was closely associated with teleological progress, while Lyell's uniformitarianism was conceived to affirm a cyclical and steady-state view of history;[22] and second, that Darwin employed his uniformitarianism to disrupt the very ideas of design and intentionality that it was originally used to affirm. That is, Darwin played—for a moment, for some readers—the role in relation to traditional views of history and the creation that Foucault plays now in displacing the Darwinian vision and disrupting conventions of coherence and continuity.

There is no separating, then, the directions of scientific thought from the directions of cultural analysis or literary forms. Gould's "punctuated equilibrium" assimilates Darwin to another model of change; and Gould self-consciously recognizes the way conventions of thought, thick with history, ideology, and politics, constrain scientific arguments. "Gradualism," he says, "is more a product of Western thought than a fact of nature." He argues, with Foucauldian resistance to ideal and unifying conceptions, for "pluralism in guiding philosophies." And he asks "for the recognition that such philosophies, however hidden and unarticulated, constrain all our thought. The dialectical laws express an ideology quite openly; our Western preference for gradualism does the same thing more subtly."[23] Kuhn's paradigms feel rather like Gould's punctuated equilibrium. Both entail a recognition of the sociocultural constraints on scientific thought, both challenge, in different ways, that "product of Western thought," gradualism.

But my central concern here is that the gradualist model and sociocultural constraints were manifest even where the direct influence of scientific argument was absent. There is a striking mutuality of assumptions about continuity among scientists and writers. Darwin's uniformitarianism seems to be echoed in the works of the self-conscious realists of high Victorian fiction, even when they did not need Darwin to confirm their own unlabeled uniformitarianism. George Eliot knew the assumptions upon which uniformitarianism was grounded, and theorized about them. Trollope *may* have known but did not talk about them. His narratives, however, are programatically bound by gradualist principles as well, and he abjured surprises in his novels.

Like uniformitarianism itself, many of the ideas to be discussed here are not strictly "Darwinian." What Darwin said was part of a much broader sweep of historical change and was implicated in major nonscientific developments. James Moore has argued, for example, that "the significance we now attach to Darwin's book and its religious implications . . . causes us to over-dramatize its effects," and that in fact the developments that seem to issue from Darwinism had begun at least twenty to thirty years before. We do not need a particular scientist named Darwin and his particular book to account for the changes in attitudes toward religion. Moore tries to show that Darwinism tended to confirm established social and epistemological positions by replacing with secular forms social and religious doctrines that differed from the Darwinian primarily in placing authority in God rather than in nature.[24] Concerned with the social bases of Darwinian thought, or with the way the scientific and social projects interfuse, he shows that although the "authority" changes, the social-political program and consequences remain similar.[25] Moore is almost a pure "externalist" historian of science, and his particular business with Darwinism is not my concern here. Moreover, although Darwinism clearly was used to reaffirm, even harden established positions, it included disruptive elements that were potentially revolutionary, and that were used increasingly into the twentieth century.[26] These are points that I will argue more fully later, but Moore is surely correct that however distinct Darwin's scientific arguments were, they were part of a whole movement of which Darwin can be taken as the most powerful codifier.

Another complication in this book's argument is that Darwin's theory often led to divergent, even contradictory interpretations, so that, for example, it was quickly adopted by many clergy as compatible with orthodox religion while often condemned as irrevocably hostile to religion. Peter Morton has carefully studied the extraordinary variety of response and interpretation to which Darwinian ideas were subjected, and argues that it is impossible to disentangle "authentic" Darwinism from what Morse Peckham has called "Darwinisticism."[27] Asa Gray, for example, could assimilate Darwin's evolution to the tradition of design in natural theology, while the Duke of Argyll, accepting much of evolutionary explanation, bitterly denied that it excluded divine intention, claiming that God works *through* the mediation of natural law.[28] Again, it was possible to argue that Darwin's notion of "species" paradoxically denied the reality of species at the same time that his theory itself attempted to explain how they came to exist. The theory even raises the chicken and the egg problem. Darwin insists that all generation is of like by like, but he never tries to account for a first organism, which would either have to have been created spontaneously or by the divine fiat that Darwin needed to deny for the rest of his argument; gradualism is pushed back to a point at which only a discontinuous and "catastrophic" change can account for beginnings. Moreover, Darwin himself was not consistent on the question of the inheritance of acquired characteristics. The most original element of his theory was in his explanation of the mechanism of evolution, "natural selection." But, as he reminds us nervously in the later editions of the *Origin*, his theory always left some leeway for inheritance of acquired characteristics, and he felt obliged to give that inheritance stronger weight as his confidence was shaken by attacks on "natural selection" (particularly from Lord Kelvin and the physicists who refused to allow that the earth could be old enough for the process).

To take more briefly some other complicating factors that make it difficult to identify purely Darwinian ideas: Darwin's position changed in many respects during his career, not only in the six, increasingly conservative editions of *The Origin of Species*, but in the effects of his direct consideration of "Man" as object, as in *The Descent of Man*. Moreover, it is inevitably difficult to disentangle his ideas from the already strong traditions, in science, political economy, and philosophy, to which they are akin. And, finally, consideration of the relation between a scientist's ideas and their emergence in literature must depend at least in part on the way they were understood by nonscientists like novelists. Showing what Darwin actually said might be largely irrelevant to "Darwinian" elements in a novel by a writer who did not follow in great detail the Darwinian arguments and controversies.

His name, with "ian" as suffix, has been invoked in the name of progress, competition, individualism, eugenics, and the whole assortment of social developments associated with the rise of industrial and monopoly capitalism. There is obviously enough in Darwin to justify almost any of these, and *The Descent of Man*, in which Darwin deliberately moves into anthropology, borrows heavily from writers who had already been influenced by him and advanced ideas that justify the charge that he was a "social evolutionist." John C. Greene argues that "those who view Darwin as a 'social Darwinist' have no difficulty in finding passages that outSpencer Spencer in proclaiming the necessity of competitive struggle between individuals, tribes, nations, and races as a requisite for social progress . . . On the other side of the argument, there are equally striking passages in which Darwin seems to recognize the role of education, public opinion, religion, humanitarian sentiments, and social institutions generally in social evolution, especially in civilized societies."[29] *The Origin of the Species*, however, does not directly sanction "social Darwinism" (whereas *The Descent of Man* does), although social Darwinism grows fairly naturally out of it.[30] While many contemporaries regarded Darwin's as a world in which nature can be best described as red in tooth and claw (Tennyson got this from Lyell, not from Darwin), and where the individual has ontological priority over the group, and where competition is a condition for survival (all these points would need careful qualification in a detailed study of Darwin's thought), this is only one of many possible Darwins, and not all Victorians saw his world as so bleakly competitive or individualistic. Here, George Eliot and G. H. Lewes, who assimilated Darwin's thought into their altruistic ideal of morality, are the most obvious counterexamples. The point, however, is that even in identifying certain dominant Darwinian ideas and motifs that emerge in narrative, I need to insist on possible qualifications, complications, alternatives. And the *precise* identification with Darwin will be difficult to make.

One further caveat, which will in fact become an important subject of this book: although scientific ideas have thematic and ideological roots, the move from a specifically scientific to a social meaning of an idea has nothing inevitable about it. For the sake of argument, let us assume, for example, that Darwin describes a natural world in which evolution works through individualism and competition: there would still be no logical necessity to the argument that *therefore*, society must be individualistic, run on competitive principles, and sanction social hierarchy according to degrees of "fitness." To say that nature *sanctions* any particular form of social organization or any set of moral or political values is to make an enormous metaphorical leap: what *is* true about, say, baboons, *should* be true about humans. Sociobiology seems to be Darwinian, even in its analogical leaps, yet it makes precisely the assumption that biological "fact" can become social prescription (even assuming that there are ideologically uncontaminated "facts"), and is reductionist in politically potent ways.[31] The other form of the argument, that nature rules human behavior whatever humans may will or intend, is a metonymic leap, and it effectively yields to a determinist amoralism in which the categories of will and morality are mere fantasies.

Connection between the behavior of other organisms and human morality and organization is often both illogical and reductionist. Nevertheless, it is a relatively normal connection to make, and it was apparently regarded by most Victorians as inevitable. It seemed to follow as the night the day that a nature that worked gradually was sanction for a society that rejected revolutionary change, and Darwin's theory was immediately pressed into the service of politics, while much recent literature of the "externalist" kind emphasizes how Darwin's theory, as Marx believed, implanted laissez-faire economics in nature.

But there is another side to the story. If Darwinian theory was domesticated and "bourgeoisified" very quickly, it also carried within it threats against traditions of stability and value that the society was eager to contain. In fiction, Darwinian theory did seem to sanction traditional narrative forms, particularly the growing dominance of "realism." But the complications that develop in the move from biological to social theory are enormous, and in fictional explorations Darwinian ideas often exposed disruptive and self-contradictory elements—challenges, as it were, to the empiricist and gradualist assumptions that seemed to guide most narratives. If there were alternative interpretations on scientific points, there were far more diverse interpretations on social matters.

The metaphorical and illogical leap exists on both sides of Darwin's theory. That is, a strong case has been made that the theory drew upon the basic themata of modern commercial and industrial capitalist society and that it was quickly absorbed into the ideology of that society by many thinkers. At the same time, it is important for a study like this, which depends a great deal on analogy and metaphor for its critical method, to resist the kind of metaphorical reductionism I have been attributing, for example, to sociobiology. Darwin's arguments can be read in the context of ideology but cannot be reduced to it, and in fact they contain elements hostile to the ideology with which they are normally accused of being complicit.

There are other aspects of the complications I have been describing. If Darwin's thought is itself multivalent and subject to diverse interpretation, the fiction that absorbs it raises other problems. What relation can there be between a literature of ideas, making truth-claims, aspiring to the condition of falsifiability, and a literature that uses ideas within a structure of fictions but that is not strictly about ideas or concerned with their rational or empirical validity? Obviously, however much we might want to accept contemporary reading of nonfictions as disguised fictions, the novels that are the main subject of this book belong to a different sort of discourse from that to which Darwin's books belong. In its simplest form, my premise is that for the most part, Victorian fiction, although sophisticated about the impossibilities of a naive realism, aspired to represent the "real," that is, a nonverbal reality,[32] and worked within the imaginative possibilities constructed by the culture. Perhaps more intensely than in any prior period, those possibilities were conditioned by the discourse of science, which had begun to assume almost exclusive responsibility for reporting on that real; and self-conscious speakers for narrative art frequently invoked science as a model or analogy for their own work. There was no precise analogue in England to Zola's epical "scientific" enterprise, but on the model of Serres's analogical method it is possible to locate in fictions, which are free to construct their realities with no verification but the common sense of their readers, the same sorts of structures as science seemed to be building in its own discourse. Science enters most Victorian fiction not so much in the shape of ideas, as, quite literally, in the shape of its shape, its form, as well as in the patterns it exploits and develops, the relationships it allows. The novel assumes serial, causal continuity of the kind that Foucault, in his "postnarrative" theory, was to reject. In such continuities fictions expose the culture's deepest assumptions (or desires), and it is in such quite fictional elements that the "nonfiction" of science makes its presence felt. Novels are not science; but both incorporate the fundamental notions of the real that dominate the culture.

It would be a mistake to assume that "the fundamental notions of the real" were monolithic. Just as interpretations of Darwin were multiple, so narrative embodiments of Darwinian notions of the real took many forms. It will be impossible even to suggest all variations, but I want to emphasize two fundamental directions—the one which absorbs Darwinian patterns to confirm the culture's dominant ideological and epistemological positions, and the other which resists that absorption and reveals fundamental contradictions in the realist project (and its alliance with a political status quo).

Given these complications—the innumerable possible interpretations of his arguments, biological and social, and the elusiveness of the science in fictions—it might well be possible to find Darwin anywhere. One of my problems, indeed, is that the argument is not, to cop a word from Popper's scientific theory, falsifiable. In a certain sense, I am free to play the game any way I like, to draw on fluctuating notions of "Darwinian" whenever I want to argue for his presence—metaphorically, at least—in a text. To an extent, this will be unavoidable. My foray into thermodynamics midway through the book is an attempt to keep me honest, for the fact is that I went hunting in *Little Dorrit* for Darwin and kept finding William Thomson, Lord Kelvin, instead. (Dickens, I have no doubt, would have been surprised to find either of them lurking in his pages.)

What I am after is a sort of gestalt of the Darwinian imagination, a gestalt detectable in novels as well as in science; and no simple list of "Darwinian" ideas will quite suffice to evoke it. Nevertheless, the Darwinian gestalt includes several clearly identifiable ideas, whose presence might be recognized anywhere, and certain fundamental attitudes toward science and toward the study of life that, if not exclusively Darwinian, were essential to Darwin's project. These ideas recur throughout the arguments of this book, and it will be useful here briefly to intimate what they are, how they work within Darwin's argument, and how they manifest themselves in the fiction.

The human subject. Part of the Darwinian enterprise was to create a theory that would be recognized as "scientific" within already acceptable terms for science, which Darwin had found most attractively formulated in John Herschel's *Preliminary Discourse on the Study of Natural Philosophy* (1830). What made Darwin's work problematic both for lay and scientific culture at the time was the attempt to apply scientific procedures appropriate to stars and chemicals to biological phenomena, and particularly to the "human": "Precisely because he was extending science into an area that his contemporaries thought unsuitable," writes Peter Bowler, "he was determined to minimize the risk of being criticized on grounds of inadequate methodology."[33] That is, the very attempt to be scientifically conservative was a radical act, and this doubleness is characteristic of the Darwinian imagination, and is implicit in even the most conservative nineteenth-century narrative. The patient, ostensibly detached registration of human character and behavior is an aspect of the Darwinian ethos central to the experience of the Victorian novel; it is part of a movement describing a new place for man in nature and tends to imply an ultimately material explanation for human behavior. As is evident in George Eliot's self-conscious commitment to the "natural history" of agrarian life, it is potentially disruptive of established social and moral categories.[34] Although we take for granted the strategies of representation within a third-person realist novel, there is nothing inevitable about those strategies. The classics of eighteenth-century fiction, for example, *Robinson Crusoe*, *Pamela*, *Tristram Shandy*, or even *Tom Jones*, all take a different view of the human experiences they describe. Even the third-person stance of the narrator of *Tom Jones* is deeply personalized, and the voice at its most solemn is not that of a scientist but of a moralist. Within Victorian fiction, novels that seem to resist the conventions of "realism"—like *Wuthering Heights*—reject also that stance of third-person detachment through which the Victorian novelist seeks the authority of science in the recording of human life.

Observation. The authority of science and its extension from natural phenomena to human was both a condition of Darwin's enterprise, and its consequence. The Baconian shift from traditional authority to the authority of experience (qualified by the self-conscious purgation of the idols that distort experience) was almost official dogma in the early nineteenth century, not least for Darwin himself. While recent study of Darwin makes clear that he was anything but a true Baconian in practice—"his entire scientific accomplishment must be attributed not to the collection of facts, but to the development of theory"[35]—Darwin expressly insisted on the accumulation of facts, most notoriously in his *Autobiography*. His work, with its sometimes disingenuous style of patient and plodding detail, helped foster the illusion that the power of science, and hence its authority, lay in its self-denying surrender to observed fact.

Only the establishment of an authority alternative to religious tradition made it rhetorically possible to extend the rule of science to the human. And that authority was a rigorously defined "experience" to be achieved through disinterested observation and experiment. Observation is the power that opens up the fact and subdues it into knowledge, and the disinterested observer is the true scientist. In nineteenth-century realist narrative not only is observation the primary source of the materials of the story, but the observer and the act of observation become increasingly the focus as much as the means of attention. The omniscient author convention—with its apparently unself-conscious directness of representation—does not inevitably treat the novelist's and narrator's activity of observation as unproblematic, and even when it seems to, it raises the problems of observation by filling narratives with unreliable spectators. It has become a commonplace of modern thought that the capacity to know is a form of power, as is evident in Fanny's story in *Mansfield Park* and in figures like Dickens's Jaggers and Tulkinghorn. The trick, as Darwin's own self-effacing strategies attest, is to avoid the exposure and thus the vulnerability that the act of observing normally if ironically entails. The peculiar Darwinian wrinkle in the scientific preoccupation with observation is that the observer becomes vulnerable, particularly because—as Darwin extends the rule of science from inorganic to organic phenomena—the observer also becomes the observed.

Uniformitarianism. I have already discussed briefly some aspects of this idea in its crossover from science to fiction. Novels as much as geology depended on the apparent plausibility conferred by the idea that all events can be explained causally, and by causes now in operation, and that extremes are to be regarded as the consequence of the gradual accumulation of the ordinary. Lyell's uniformitarianism was meant as a sanction for secular scientific explanation against biblical authority. In its purest form, Darwinism broke from Lyell's essentially antihistorical position and implied development but without teleology. The central tradition of Victorian realism—as we can see it in such different writers as Eliot and Trollope—adopts that form, although the pressure of teleological thinking can be detected in that tradition, just as it can in Darwin's own writing. Dickens tended to find thoroughgoing gradualism inadequate and often implied through his narratives the possibility of causes outside the secular. His complex relation to this idea is an important register of the culture's ambivalence about Darwinism and about the extension of scientific study to human history; and it suggests some of the limitations and contradictions within the realist project.

The scientifically conservative affirmations of authority implicit in these first three ideas were essential to Darwinism in part because his theory itself implied radical disruptions of epistemological, religious, and moral traditions. Again, it is not that Darwin's theory introduced such notions for the first time—far from it; nor is it that the disruption of tradition always worked against social, political, even religious stability. It was easy to *use* Darwinism to serve a multiplicity of antithetical purposes. But for the lay public, as well as for many scientists, Darwinism could be deeply threatening, and if, as I argue in the next chapter, the lay model for understanding the natural and human world before Darwinism was natural theology, in any of its various forms, then Darwinism could indeed be seen as a radical dislocater of the culture's understanding of nature and of the self. It is certainly the case that most science before Darwin—Lyell's even in its resistance to biblical authority was no exception—could have been assimilated to natural theology. And in England, far more than on the continent, science and religion were allies (if sometimes nervously), many of the best known scientists themselves clergy. But the accumulated secular emphases of Darwin entailed other forms of compromise if religion were to avoid hostility to science.

Change and history. Obviously, the theme of change did not need Darwin to invent it. But in his world *everything* is always or potentially changing, and nothing can be understood without its history. Species, which had been conceived as permanent, transform into other species or are extinguished. The earth and all of its local ecological conditions are shifting. Traditionally, the more things change, the less "real"—that is, ideal—they are, the more corrupt and corrupting. But in submitting all things to time, Darwin challenged the ideals of a permanent substratum of nature and of permanent categories of thought. Categories become fictions, historical and conventional constructions, mere stopgaps subject to the empirical. In realist narrative change and development become both subject and moral necessity, and they tend to be as well a condition of plausibility; character can only be understood fully if its history is known because character, as George Eliot wrote, is not "cut in marble," and it is intricately embedded in "plot." Moreover, closure is perceived as artificial and inadequate because it implies an end to history and is incapable of resolving the problems raised by the narrative. Conventional comic marriages are subjected to ironies of time and are often explicitly treated, as by Thackeray, as mere conveniences that allow books to end. The alternative tradition, as in a novel like *Jane Eyre* or in some of Dickens, provides closure and appropriate resolution to what has preceded. Here again, the most obvious and "natural" aspects of fiction turn out to belong to a particular historical formation, and one that operates with great force theoretically and substantively in Darwin's theory.

Blurring of boundaries. The continuum of time is, in Darwin's world, an aspect of the continuum of life itself and of all other sharply defined categories. The boundaries between species and varieties blur, and the further Darwin carries his investigations the more this is the case. All living things in Darwin's world are quite literally related, and, as he will say in a variety of ways, graduate into each other. Isolated perfection is impossible, and science and fiction both concern themselves with mixed conditions. Fiction's emphasis on the ordinary and the everyday, its aversion to traditional forms of heroism and to earlier traditions of character "types," all reflect the tendency obvious in Darwin's world to deny permanent identities or sharply defined categories—even of good and evil. Note how rarely in Trollope or, more programatically, in Eliot, genuinely evil characters appear. Typical stories are of decline or of development; the case in Dickens, of course, is quite different. Character tends increasingly to be a condition of time and circumstance rather than of "nature." In Dickens, the tension between these two ways of imagining is reflected in his attempts to move in his later novels from characters whose natures are fixed to characters who, like Pip, appear to develop. Change, in Dickens's world, nevertheless tends to be radical and "catastrophic," rather than gradual; like Dombey, rather than like Pip.

Connections—ecological and geneaolgical. Darwin's world is, as the famous last extended metaphor of the *Origin* puts it, a "tangled bank." All living things are related in intricate and often subtle patterns of inheritance, cousinship, mutual dependence. Adaptation of organism to environment is not, as in natural theology, a consequence of a divine fiat, but a result of history—of organic and environmental changes. To discuss the life and nature of any organism requires discussion of the many others with whom it struggles, on whom it depends, in seemingly endless chains of connection. Victorian realist narratives equally entail complex and intricately inwoven stories of many figures so that it is often difficult to determine which characters are the true protagonists, which the subordinate ones. The Victorian multiplot novel is a fictional manifestation of the attitudes implicit in the metaphor of entanglement in Darwin. Such entanglement is an aspect of the gradualism discussed earlier and reflects a distrust of abrupt intrusions from outside the system such as one might find in "metaphysical" fictions like, *Jane Eyre* or *Wuthering Heights*, whose narratives are also sharply focused on a small, defined set of characters.

Abundance. The ecological vision is connected with a view of a world bursting with life, always threatening overpopulation. In Darwin's world survival ultimately depends on variation and diversification, multiplicity of life and of kinds, some of which, from the vast and continuing waste and competition, will survive. Absence of diversity means vulnerability to change, and change is similarly a condition of life. The overpopulated worlds of the Victorian novel, those "large, loose baggy monsters," as James called them, are narrative equivalents of Darwin's "endless forms, most beautiful, most wonderful."[36] Like Darwin's theory itself, they reflect the Victorian taste for excess in ornamentation, the Victorian sense of a newly crowded and complicated life in which there were new opportunities for variety in possessions, art, relationships.

Denial of design and teleology. The Darwinian narrative unfolds "naturally," that is, without external intrusion. It is, as it were, self-propelled, unfolding according to laws of nature with no initiating intention and no ultimate objective. Adaptation, a key element in Darwinian as in natural-theological thought, seemed to imply design and intention, but Darwin had to show that it was merely "natural." His rejection of the natural-theological assumption of teleology fundamentally undercut the basis of most Western narrative. In the realist novel itself, certain conventional elements continue, willy-nilly, to imply teleology, but the movement is very clearly away from "plot"; and the Trollopean determination to focus on characters and to let the plot emerge from their encounters is a characteristically Darwinian way to deal with narrative and change. The characters, like Darwinian organisms, learn to adapt to their environments. The explanation of that adaptation is not metaphysical but "natural," and the emphasis on psychology is a means to explanation.[37] In character-oriented narrative, the events appear "natural"; they grow from the posited conditions of the fictional world and do not seem to be imposed by the author. This Jamesian ideal was implicit in the realistic narratives James often criticized, and it disguised well the romance or mythic elements that Northrop Frye suggests are the ultimate source of all literary narrative. The growing nineteenth-century dissatisfactions with closure—the most marked and inevitable feature of "plotting"—are further reflections of this Darwinian movement away from teleology and, as I have suggested, toward a new kind of emphasis on continuing change.

Mystery and order. In the multitudinous and entangled Darwinian world, order is not usually detectable on the surface, but the apparent disorder of nature is explicable to the keen observer in terms of general laws that can be inferred from phenomena. Similarly, the world of the realistic novel tends to be explanatory and analytic, showing that behavior is psychologically explicable and that events are "probable," that is, consistent with what might be regarded as empirical law, even if not strictly logical. Darwin's science aspired to the regularity of physics and astronomy, but in its biological preoccupation with individual differences could not achieve that. Nevertheless, Darwin demonstrates the regularity and comprehensibility of phenomena without reducing them to the strict form of logic and mathematics. On the contrary, "metaphysical" fiction, corresponding to the modes of natural theology, tends to be very strictly and rationally ordered. (The symmetries of *Wuthering Heights* are perhaps the most obvious example.) But no rational explanation can account for the order. Some force beyond nature is required. Full evidence for Darwin's theory is not immediately available, but the mysteries can be filled in by induction and extrapolation from the observable. In realistic fiction, similarly, mystery is merely a temporary gap in knowledge (despised by Trollope), but in metaphysical fiction, as in natural theology, mystery is the effect of a spiritual and inexplicable intrusion or initiation from outside of nature.

Chance. Darwin abjured chance but required it for his argument. Minute chance transformations are the source of all variations (Darwin could not explain the mechanism although he ventured a distinctly unsuccessful theory of "gemmules" later in his career), which are the first steps in speciation. Realism is programatically antagonistic to chance, but like Darwin almost inevitably must use it to resolve its narrative problems. The complications of chance in Darwin's theory and in narrative will require extensive discussion (and speculation), but it is important to note that like Darwinian theory, realism tends to depend on the smallest of events and on psychological minutiae for its stories and for change within those stories. Moreover, chance encounters seem like intrusions from another mode when they occur in realistic narratives. By contrast, in "metaphysical" fiction chance and coincidence play important roles, though almost invariably they seem not an intrusion from another mode but evidence of design and meaning in the world.

To some extent, all that follows in this book will expand this list and explain in some detail the complexities hidden in it. The interplay of Darwinian and literary form and thought is too complex and rich to be captured in a list, yet it should suggest the major elements and make clear the kinds of ideas I take as "Darwinian" in the novelists I will be discussing: Jane Austen, Walter Scott, Dickens, and Trollope, as well as Conrad and Hardy. Most of these, I should repeat, did not know Darwin's work well, and, at certain stages of their careers, *could* not have known it.

I make no claim that the Darwinian elements I will be identifying are the only ones that account for the nature and structure of these writers' work. There are almost always alternative sources, in earlier literature, philosophy, contemporary debate, social and economic developments; and my arguments are not intended to preclude the more traditionally received alternatives. The likely multiplicity of "sources" is part of the point. The novels Darwin read so enthusiastically in order to escape from the pain of his illness and of too much work helped strengthen the tendencies that led him to his theory. The happy endings may have seemed to him a relief from the pain of much of the world he was describing. But they could have provided relief only if they concluded narratives that were rich with the variety, particularity, complexity, and toughness of his own world. Conventions of literary form are assimilated and adapted both by the novelists and by those whose job it is to discover what the world is "really" like—the scientists. The overlap of scientific thought and literary convention is one of the points I want to emphasize and part of the excitement of this enterprise.

Darwin's narrative itself has a "happy ending," is comic in form. That form has sources which go deep into the history of narrative, but it can be detected clearly both in the eighteenth-century rationalist tradition, with its faith in perfectibility and the explicability of nature, and in romantic tradition, with its preoccupation with change, history, evanescence, and "cloudiness," as Ruskin called it. Darwinian thought was deeply implicated in the same culture that was producing Scott's historical novels, Dickens's large episodic and multitudinous fictions, Thackeray's complex social panoramas. Darwin's theory emerged midway in the life of Victorian fiction, with its preoccupation with multiple and complex social relations, with growth and change, with uniform and minute and inexorable sequences. In the earlier novels we can, as it were, watch Darwin coming; in the later we can feel his presence.

I will begin by looking at some pre-Darwinian documents in the tradition of natural theology to suggest what a "pre-Darwinian" narrative would probably look like. Natural theology, with its roots in an eighteenth-century rationalist tradition, with its bias toward sciences that treat of regularities, closed systems, recurrences, and clearly defined relationships, and which assumes that explanations will always be assimilable to the forms of logical discourse, might seem to provide a perhaps unfairly extreme antithesis to the Darwinian model. Part of my point, however, is that they are not at all extreme, since Darwin grew from the traditions of natural theology, and its language echoes in his. Moreover, by looking at William Whewell's Bridgewater Treatise and then more briefly at his *History of the Inductive Sciences* and *Philosophy of the Inductive Sciences*, it should be clear that the natural-theological position was itself more mixed than, in its most obvious arguments, it seems to be. And if Darwin reflects something of the tradition of natural theology, it should be clear, too, that Whewell's insistently historical way of reading scientific achievement (if not nature itself) implies "themata" that will come to seem after Darwin rather "Darwinian."

The discussion of natural theology requires consideration of some significant fiction written in the years when natural theology was flourishing and evolutionary theory—via Erasmus Darwin and Lamarck—had not acquired respectability and had associations with revolutionary and materialist theory. For that purpose, I consider Austen's *Mansfield Park*, which seems to me to reflect many of the same kinds of attitudes that went into the resurgence of natural theology, with all its conservative political implications. *Mansfield Park* beautifully embodies (even sometimes plays with) certain dominant conventions of fictional structure and perspective, which will help serve as a control in the study of the very different sorts of fictions with which this book will be primarily concerned. It will make one standard against which to measure the distance nineteenth-century fiction traveled both in realism and in the "metaphysical" tradition when it absorbed Darwinian notions.

Before proceeding to direct study of Dickens and Trollope, I will expand and explain some of the elements I have described as "Darwinian." As preliminary to the strictly narrative enterprise, I will consider some aspects of Darwin's thought, its rhetorical embodiment, and even its place in nineteenth-century debates about science. Such exposition should help suggest some of the limits of the speculations that follow. Darwin was more complex, more prodigiously inventive as a thinker than could ever be detected in a discussion that isolates some of his ideas and transplants them into other forms. It is my aim to shadow forth a Darwin more disruptive, perhaps, than even the greatest of his literary followers can suggest, a Darwin who, if fully absorbed by his contemporary novelists, might well have led to other kinds of narratives. In any case, the discussion of Darwin should make it possible to sustain in the readings that follow something of a double vision—of the Darwin whose imagination could not be contained in conventional narrative form, and of the Darwin who helped shape late-century narrative form and became a conservative political and literary force.

The final two chapters develop some aspects of the disruptive implications of Darwin's work. Here, I will have occasion to discuss Hardy, if only briefly, in considering the importance to nineteenth-century science of the theme of the authority of "observation and experiment" in the determination of truth, the location of adequate materials for scientific study. It will be useful to juxtapose some of the many Hardy characters who strive for positions of quiet observation with Fanny Price, the "quiet thing" of *Mansfield Park*, whose very quietness allows her to achieve a reliable perception of what goes on around her. Hardy dramatizes Darwin's double attitude toward observation in such a way that the respectable Darwin becomes a dangerous man. In Conrad, whose public politics were notoriously conservative, Darwin has revolutionary implications. While Darwin's kind of organicism and gradualism has been implicated in deep resistance to rapid change, the disruptions implicit in the chancy nature of Darwinian change explode into such fictions as *Under Western Eyes* and *Nostromo*, not only in the revolutionary subject, but also in the very artful forms by which Conrad attempts to resist the danger.

Either way, within the more comforting terms of mid-Victorian realism or the increasingly disruptive forms of the later century, the two Darwins have, whether we like it or not, transformed the way we can think about reality, about the way we exist in time, about what, after all, is possible. However bleakly Darwin's vision settled on a culture unprepared for the supremacy of natural explanation, he offered us—still offers—a vision of abundance, possibility, and life in which, as he says, there is "grandeur." Darwin provides the framework for a way of thinking about life in time that is still the primary antagonist to modern and Foucauldian views that argue for discontinuity and disruption (although, as will become clear, Darwinian thought is partly responsible for the development of those views).

2

Natural Theology: Whewell and Darwin

THROUGHOUT THIS BOOK, I call the scientific view that Darwin displaced "natural theology." Although this risks oversimplification, it accurately indicates that English science was intimately connected with its religion, and that its "themata" were frequently religious assumptions—as, for instance, that the universe is unified, coherent, and rational. Religion and science alike were concerned to describe a cosmos all of whose phenomena made sense, manifested intelligence and design. That very assumption, Walter Cannon has argued, "turned into a powerful tool of scientific research." Cuvier, Buckland, and Richard Owen, "by reasoning on the necessary construction of an animal designed to live and flourish in a particular environment . . . were able to carry out their masterpieces of reconstruction from fossil bone fragments."[1]

There were large variations within natural theology, of course, but the most serious scientists within it (most of whom were not "scriptural geologists," who made up only a very small minority of practicing geologists)[2] agreed on the importance of maintaining the integrity, coherence, and independence of scientific discourse and on the inevitable confirmation by science of religious faith. There were important disagreements on the relation of science to scripture, on the possibility of divine intrusion into the sequence of secondary operations that made the subject of science, on whether uniformitarian or catastrophist explanation best described the world's history, and so on. But the fundamental attitude of the scientists was concisely summarized by the most distinguished and respected scientist of his time, John Herschel, in his *Preliminary Discourse on the Study of Natural Philosophy* (1830): "Nothing . . . can be more unfounded than the objection [that science] leads [its cultivators] to doubt the immortality of the soul, and to scoff at revealed religion. Its natural effect, we may confidently assert, on every well constituted mind, is and must be the direct contrary."[3]

William Buckland, whose impressive and controversial contribution to the Bridgewater Treatises (*Geology and Mineralogy Considered with Reference to Natural Theology*) was catastrophist in its theoretical orientation, accepted, like Charles Lyell, the geological evidences of a deep time that belie the traditional interpretation that the world is six thousand years old; and like Lyell, he accepted the fact of innumerable and continuing extinctions of species. Nevertheless, his treatise typically reconciles geology with scripture. "No reasonable man," he writes, "can doubt that all the phenomena of the physical world derive their origin from God." And "no one who believes the Bible to be the word of God, has cause to fear any discrepancy between this, his word, and the results of any discoveries respecting the nature of his works."[4] Any story scientists told would, on this account, be governed by the rules of a rational science, and any complication or possible contradiction of established faith could be resolved by interpretation based on assumptions about the world that underlay and even initiated the investigation in the first place. Science was concerned with secondary causes, religion with primary.

Natural theology was the lens through which the natural world was seen and understood, and secular fiction, no more than science, could see with other eyes. The central conventions of narrative—its teleological unfolding; its providential use of coincidence; its implicit faith in the ultimate coherence, rationality, and intelligibility of the world being described; its movement to closure—all are consonant with the natural-theological view of things. The convention of coincidence resolving and explaining all may lead us back at least as far as Oedipus, but natural theology endorsed that kind of narrative convention. Coincidence is the ultimate confirmation of design. Within a Darwinian frame, however, where the themata of natural theology are used and disrupted, coincidence is only coincidence, reflecting, perhaps, the design of the author, but not of nature.

Darwin's influence on narrative can only be fully understood in the context of his theory's relation to natural theology's teleological creationism, which, because it seemed to him to threaten the very practice of science and to foreclose close empirical investigation of natural phenomena, he deliberately sought to displace. One of the most interesting examples of the natural theology Darwin resisted—after Paley, of course—is William Whewell's Bridgewater Treatise, *On Astronomy and General*

Physics Considered with Reference to Natural Theology (1836). Darwin, in fact, quotes Whewell, in an epigraph to the *Origin*: "But with regard to the material world, we can at least go so far as this—we can perceive that events are brought about not by insulated interpositions of Divine power, exerted in each particular case, but by the establishment of general laws" (p. 50). The quotation suggests that Whewell shared with Darwin a commitment to the scientific project, a rejection of any explanation that seems to entail looking beyond laws of nature. And Whewell's treatise fairly represents the way a serious working scientist (whose influence on Darwin's thinking has recently been made clear)[5] would articulate his sense of the close relations between science and religion.

Darwin read the treatise at least twice, even though or perhaps because he needed to reject its fundamental teleological orientation.[6] Thus, he alludes to it contemptuously in his 1838 notebook: "Mayo (Philosophy of Living) quotes Whewell as profound because he says length of days adapted to duration of sleep in man!! & not man to Planets.—instance of arrogance!!"[7] In the midst of developing a theory that would decenter man, Darwin finds natural theology "arrogant." He challenges the idea not that one phenomenon is adapted to another, but that the adaptation is *to* "man." Darwin's revolution is a consequence of his challenging the assumptions underlying the interpretations of natural theology; on the "facts" and even on the methods of investigation, they often surprisingly agree. A look at Whewell's treatise will allow sustained attention to the kinds of discriminations necessary to distinguish Darwinian from natural-theological positions.

The relations between Darwin and Whewell, between evolutionary theory and natural theology, are too complex to be reduced to the ultimate, insuperable divergence over natural selection, which led Whewell to bar the *Origin* from the Trinity College library.[8] It is partly because Whewell occupies a position apparently antithetical to Darwin's but at the same time holds many attitudes entirely compatible with Darwin's method and achievement that I find it particularly useful to consider his version of natural theology.[9] His most famous works, the *History of the Inductive Sciences* (1837) and *Philosophy of the Inductive Sciences* (1840), are ventures in the philosophy of science and not natural theology—brilliant and voluminous blendings of the British empiricist tradition with Kantianism, and, according to Walter Cannon, bridges between uniformitarian and catastrophist theory that enabled Darwin's own secular compromise. They are of particular interest here because they represent, in contradistinction to Mill's *System of Logic* (1844), which can be seen as a Herschelian response to Whewell, an argument against the extension of scientific method to the study of human morals and society.[10]

The obvious differences between Darwin and Whewell should not obscure the fact explicit in the epigraph to the *Origin*—that Whewell was a very serious scientist who resisted the closing off of scientific discourse for religious dogma as much as Darwin would. Moreover, by the time he came to write his *Plurality of Worlds* (1854), Whewell's antitransmutation teleology had been transformed into at least sympathy with developmental positions. Darwin, for his part, may well have adopted self-consciously Whewell's theory of the *vera causa*, his well-known idea of the "consilience of inductions." The exact nature of this influence will not concern us here, but "consilience" is the "jumping together," or agreeing, of inductions from very different classes of facts. Great scientific theories, Whewell claimed, tend to be confirmed by their application to areas they were not originally intended to explain. Darwin consistently argued for his own theory that it not only explained biological development, but many other matters as well, including such complex ones as the geographical distribution of species, and geological change.[11]

The Darwin-Whewell connection suggests that natural theology was closer in many respects to Darwinian argument than it would on the surface seem. Its anthropocentrism and teleological progressivism may have been utterly inimical to Darwin's developing theory; yet that theory was easily assimilable to both an anthropocentric—indeed Eurocentric—view of nature and a teleology whose final cause was the fullest development of mankind.[12] Thus, formally, natural theology actually implies a world structured rather like Darwin's; and the narrative analogues to natural theology and to Darwinian theory are also similar. To detect the important differences, we must consider not only such large issues as anthropocentrism and teleology, but the texture of the prose and of the arguments in which these writers found expression for their ideas. In order to understand what was to change in English narrative under the pressure of Darwinian "reality," we must examine both the nature of Whewell's position and the implications of that position for narrative.

Whewell, whose contribution to the Bridgewater Treatises was the most popular of all (going through seven editions in six years),[13] obviously was unusually sophisticated in the philosophy of science. Science, he argues, cannot begin by assuming what natural theology will draw on science to prove. That is, whereas teleological explanation was important in some scientific work at the time, and Darwin tries to justify his theory by denying the validity of teleology, Whewell, even in a treatise dedicated to reconciling science to religion and demonstrating that nature was indeed designed by God, took—*as scientist*—the antiteleological position. "Final causes," he says, "are to be excluded from *physical inquiry;* that is, we are not to assume that we know the objects of the Creator's design, and put this assumed purpose in the place of a physical cause."[14] (He goes on to argue that *after* the general laws have been discovered it is legitimate to speculate on their relation to divine intention.) Whewell's natural theology is "scientific" in that he maintains a strong professional commitment to the purity of the activities of science, which is defined by its exclusive concern with secondary causes and is not to be contaminated even by the religion that it will be enlisted to affirm.

The scientific bias of Whewell's view entails at least one important difference from Paley's natural theology. For Paley, evidence of design is empirical. We can infer God's creative presence by examining the details of his creation directly. Whewell occasionally adopts precisely that position, but when he argues, for example, about the nebular hypothesis, his position shifts. He refuses to commit himself on the validity of the hypothesis, addressing only the question of what consequences it would have for the argument from design if it turned out to be true. Good scientist that he is, Whewell won't prejudge the science. But he can comfortably argue that a naturalistic explanation of the origins of the universe does not contradict the idea of God's creative presence. It only means that God placed all the potential for designed development in the primitive nebular murk, and that He relied on His laws to develop the design we see about us now. Darwinian evolution, on this argument, would also be entirely conformable to religious explanation. The burden of design is on "law," not on intention (which natural theologians infer from law) or on divine intrusion.

Whewell's justification for eliminating teleological explanation from science is itself pious (Darwin would adopt the argument himself, later): we must not presume to know what God intends; we must rather work to find it out. Darwin's antiteleology was more secular: we must not presume to know the object of a phenomenon because assuming that it *has* an object, and defining what that object is, closes out the possibility of scientific investigation, of gathering new knowledge.

The difference between the secular and the religious justification should not disguise the coincidence of the conclusion. Both Whewell (who was also a Kantian) and Darwin are empiricists, both are committed to the idea that science is directed at discovering the laws that govern all natural phenomena. Both, too, are inheritors of that eighteenth-century rationalist tradition of which Paley's natural theology is so prominent a development. Their explanation of natural phenomena, in the tradition of natural theology, tended to be mechanistic; and, ironically, a mechanistic analysis was encouraging to the inference of a divine intelligence, for "mechanisms"—like the famous watch or clock of the design arguments—imply regular, lawful behavior, and, consequently, an intelligence that created the laws. Until the romantic revival and the explosion of concern about faith and religion that accompanied it, these secular-utilitarian tendencies of the tradition constituted the main stream of English thought. But while it may have hastened the secularization of the culture, it tended to be comfortable with a nondogmatic religion, with good, hearty, rough and ready English common sense. Thus, although we might infer from Darwin's comments on Whewell and from the revolutionary character of Darwin's thinking that nothing could be more un-Darwinian than Paley, Darwin noted in a letter of 1859, "I do not think I hardly ever admired a book more than Paley's 'Natural Theology.' I could almost formerly have said it by heart" (*Life and Letters,* II, 219).

Darwin admired the mode of Paley's argument; he made the idea of "adaptation" as central to his theory as it was to Paley's; he even admitted, in rethinking his arguments in the *Origin*, that in first working out his theory, "I was not able to annul the influence of my former belief, then widely prevalent, that each species had been purposely created; and this led to my tacitly assuming that every detail of structure, excepting rudiments, was of some special, though unrecognised, service."[15] The *form* of Darwin's arguments and, consequently, their implications for writers of narrative are thus in many respects close to the form and implications of natural-theological arguments.[16] Every detail has its function. Natural theology found its explanations in divine intelligence and intention. Darwin, seeking with similar intensity an intelligible world, finds the explanations in the apparently law-bound operations of nature itself. Darwin issued no public disclaimer of Asa Gray's praise of him for reintroducing teleology into biology, although he did deny his commitment to design in a letter to Gray,[17] and in the conclusion to *Variations of Plants and Animals Under Domestication* (1868), he wrote,

> However much we may wish it, we can hardly follow Professor Asa Gray in his belief "that variation has been led along certain beneficial lines," like a stream "along definite and useful lines of irrigation." If we assume that each particular variation was from the beginning of all time preordained, then that plasticity of organisation, which leads to many injurious deviations of structure, as well as the redundant power of reproduction which inevitably leads to a struggle for existence, and, as a consequence, to the natural selection or survival of the fittest, must appear to us superfluous laws of nature. On the other hand, an omnipotent and omniscient Creator ordains everything and foresees everything. Thus we are brought face to face with a difficulty as insoluble as is that of free will and predestination.[18]

What he had done, as Edward Manier points out, was to naturalize final causes.[19] This difference, which would seem to have no effect on the *form* of the arguments, has very large consequences. The form, indeed, remained similar, sometimes taking the shape of a direct inversion (for example, the comparison of artificial with natural, as in breeding, is used to prove the absence of will and intention). But the inversions are accompanied by tendencies not in Paley, an intense interest in aberrations, for example. Darwin constantly treats the exception, not, as in Paley, to assimilate it to regularity, but because the individual, in Darwinian theory, is fundamental. As I shall be discussing later, one of the major problems for Darwin in relation to Herschel's theory of science is that as a biologist his subject was inevitably singular, not generalizable. Moreover, there can be no role for the random in natural theology. Paley saw chance as only an "appearance" resulting from the ignorance of the observer, or as an occasional consequence of multiple laws intersecting, and where he found it, he immediately justified or compensated for it. Darwin, too, seemed to want to make chance disappear, but for science not for God.[20]

Whewell, like all of the Bridgewater writers, wanted to make strong logical arguments that would demonstrate the "power, wisdom, and goodness of God as manifested in the creation." But austerely logical arguments cannot lead from the empirical to the numinous, from nature to God. The "logic" of natural theology is built first of all on analogy—if we infer a designer from a watch, then we must infer one from a perfectly running machine; and of course it is interfused with appeals to feeling, strategies of rhetoric that make it, from a philosophical perspective, something less than deductively binding. Newman, we remember, entirely rejected natural theology as *proof* of God's existence because such a proof depended on giving the evidence of nature prior authority on matters of faith. And that move gives the whole game away. Bishop Butler, in the eighteenth century, working through his "analogy of religion," was careful to argue that the evidence of the natural merely helped confirm what had already been revealed spiritually.

Whewell makes a similar claim: "We are very far from believing that our philosophy alone can give us such assurance of these important truths [about the reality and nature of God] as is requisite for our guidance and support; but we think that even our physical philosophy will point out to us the necessity of proceeding far beyond that conception of God, which represents him merely as the mind in which reside all the contrivance, law, and energy of the material world" (*Astronomy,* p. 315).

Whewell saw, perhaps as clearly as Newman, that the experience of faith depended on an intuited immediacy of experience outside the realms of logic, analysis, or induction. Newman begins with that stunning assertion of belief in "two and two only absolute and luminously self-evident beings, myself and my Creator."[21] Whewell, too, was a "realist," that is, he believed intuitively in the reality of what he perceived. And the evidence he accumulates in his natural theology is ultimately designed to create a third luminousness, however much the methods of logical argument are also entailed. For it is part of Whewell's fundamental assumption about the nature of reality that it *is* rational, that the world of feeling can be confirmed by the activity of rational intelligence. So, near the conclusion of his chapter on final causes, Whewell asserts, "when we examine attentively the adjustment of the parts of the human frame to each other and to the elements, the relation of the properties of the earth to those of its inhabitants, or of the physical to the moral nature of man, the thought must arise and cling to our perceptions, however little it be encouraged, that this system, everywhere so full of wonderful combinations, suited to the preservation, and well-being of living creatures, is also the expression of the intention, wisdom, and goodness of a personal Creator and Governor" (*Astronomy,* pp. 297–298). Notice the rhetoric: "the thought *must* arise and *cling* to our perceptions"! No question of logic here, and the passage is not a bad summary of the overall argument of the book. The study of nature forces faith, the reality of God clings to our perceptions. It is a by-product of secular study, but all the more powerful for coming thus, unbidden.

It might even appear that Whewell denies the importance of logical argument. "However strong and solemn," he says, "be the conviction which may be derived from a contemplation of nature, concerning the existence, the power, the wisdom, the goodness of our Divine Governor, we cannot expect that this conviction, as resulting from the extremely complex spectacle of the material world should be capable of being irresistibly conveyed by a few steps of reasoning, like the conclusion of a geometrical proposition, or the result of an arithmetical calculation" (*Astronomy,* p. 11). Of course, like Newman, Whewell does resort to "logical" argument. But even the argument that design implies a designer, law, a lawgiver, is more intuitive than logical. The usual argument in natural theology takes a simple, ostensibly logical form: *a* is adapted to *b*; if *b* were different, *a* could not survive; there is no reason for *b* to be as it is except that it allows *a* to survive; therefore, *b* and *a* were designed for each other; the design is thus benevolent and implies a benevolent and forethoughtful designer. So Whewell tells us that the cyclical movement of the earth around the sun is adapted to the life of the plants and animals; the spinning of the earth is adapted to the needs for human sleep—eight hours asleep, sixteen hours awake, which is much better than, say, four-eight, four-eight. And so on. But it is no single argument that carries the weight Whewell wants his treatise to bear. It is not even this kind of "logic" that makes the case. At the heart of Whewell's theory is a Kantian view that refuses the most rigid rationalistic conventions of natural theology, but is compatible with its irrationalist religious position. He develops this view at great length in *The Philosophy of the Inductive Sciences*, a few years later, and in long public debates with J. S. Mill. For Whewell, there is a "fundamental antithesis of philosophy," the antithesis of "thoughts and things," "theories and facts," "ideas and sensations," the subjective and the objective—that is, mind and its objects.[22] The objects are experienced as sensation, made meaningful by mind. But they *can* be made meaningful, are intrinsically compatible with mind. As it was for the not so philosophical reading public (and most novelists, for that matter), his world, even in *Astronomy and General Physics*, has a firm, solid, material reality; but it also is informed by the structures of the mind.

For Whewell, the greatest scientists do not merely work deductively. Deduction depends on some prior "discovery"—one might even call it "revelation"—and can only educe the implications already implicit in that discovery. Induction, which is the primary means to discovery, includes both fact and idea. And facts, as Whewell argued in his famous controversy with Mill, imply ideas: "There is no definite and stable distinction between Facts and Theories; Facts and Laws; Facts and Inductions."[23] Mind is implicated everywhere. The structuring activity of mind is innate, and any strictly empiricist account that treats experience as the sole source of knowledge is, for Whewell, misguided. The mind is no blank slate.

Thus, when Whewell discusses Newton's achievement, he can emphasize Newton's extraordinary powers of mind; but Newton's greatness is beyond logical accounting: "When Newton conceived and established the law itself, he added to our knowledge something which was not contained in any truth previously known, nor deducible from it by any course of mere reasoning" (*Astronomy,* p. 282). Here Whewell is arguing on the empiricist side for the importance of Baconian experience, but beyond this, he wants to argue that the scientist is no passive receiver and organizer of fact; he is an inventive and active participant. Newton "conceives" and "establishes" the law of gravity. Whewell marks the history of science by the achievements of genius; all else builds to as Prelude, or infers from in Sequel. "However rare the mathematical talent, in its highest excellence, may be," he says, "it is far more common, if we are to judge from the history of science, than the genius which divines the general laws of nature" (*Astronomy,* p. 283).

Finally, Whewell does believe in the absolute rationality of the creation, and thus of the activity of both genius and natural theologian. The question is how we attain the "data" on which the rational process is to work, and then, secondarily, what "rational" processes one is to use. Whewell points out the weakness of deduction without the priority of data. So, he concludes, the principle that a design must have had a designer "can be of no avail to one whom the contemplation or the description of the world does not impress with the perception of design. It is not therefore at the end, but at the beginning of our syllogisms, not among remote conclusions, but among original principles, that we must place the truth, that such arrangements, manifestations, and proceedings as we behold about us imply a Being endowed with consciousness, design, and will, from whom they proceed" (*Astronomy,* p. 296). Note the burden this places upon Whewell to drive home the *experience* of design. And the rhetorical thrust of his book—consistent with the traditions of natural theology—is precisely there. *Astronomy and General Physics* is as much a compendium as an argument. Its strength is in the almost encyclopedic notation of examples from every aspect of his subject—from the weather to cosmic movements—that, cumulatively, make the inference of design automatic, intuitive.

A similar point could be made about the *Origin*. Huxley talked of it admiringly as an encyclopedia of facts. And certainly, as Gillian Beer points out, Darwin attempts to make his argument from "a plethora of instances." The difference is not in the degree to which each needs to be encyclopedic, but in how each allows the encyclopedia to be constructed. Whewell's bias is toward regularity and immediate intelligibility, and the detail of his argument seems always under control; more often than not the subject is a "law" rather than an instance of that law. He does not give a "plethora," but a precisely moderated accumulation. There is no chance of Whewell's argument being outstripped by the materials used to make it, just as there is no chance in Whewell's world that its details will ever escape the regulation of law, the benevolence of divine intention.

Like Darwin's, however, Whewell's argumment is highly rhetorical, having qualities akin to those John Holloway has classically ascribed to the Victorian sages.[24] These qualities are, moreover, peculiarly relevant to narrative (two of Holloway's sages are, after all, novelists), most obviously in their experiential bias. Like novelists, Whewell must create the experience of belief before he can persuade discursively.

In what follows, the question of rhetoric will always be latent, but Whewell's arguments build on certain reiterated motifs that coincide interestingly with Darwin's, and it is these that I want now to discuss, comparing them, when it is appropriate, with Darwin's, and suggesting what assumptions underlay them.

The rule of law. The obstacle to complete understanding in Whewell's world is never the world's irrationality, but always the infinitude of God and his creation, and the littleness of human consciousness. "Nature," says Whewell, "so far as it is an object of scientific research, is a collection of facts governed by *laws*: our knowledge of nature is our knowledge of laws" (*Astronomy*, p. 3). Thus far, however, in only a very few cases have we achieved the condition of science, that is, the capacity "to trace a multitude of known facts to causes which appear to be the ultimate material causes, or to discern the laws which seem to be the most general laws" (p. 9). Our failure should be no deterrent to belief in the universality of law. Instead, Whewell proceeds—as did most Victorian scientists—by taking astronomy as a model for all science, all knowledge.

The choice was a crucial one, for science and for culture. Ernst Mayr has argued that assuming the unity of science and the validity of the model of physics for all scientific activity has led to important errors.[25] Most important, from his point of view, is the inapplicability of the model of physics to the phenomena of biology. Whewell's Newtonian astronomy seemed to confirm not only the universality of law, but the perfect regularity of movements, the mathematical precision of relationships, the superiority for knowledge of the general to the particular. Mayr argues that biology has as its subject phenomena which are always unique. In the first half of the nineteenth century this provoked considerable discussion among biologists about how it was possible to make biology into a science. W. B. Carpenter's review of Whewell's *History* includes a long attack on Whewell's reluctance to consider physiology a science; Carpenter insists that only the difficulty of acquiring data impedes physiology. Living phenomena are not so easily observed as the stars. To observe organisms in their particularities was, in effect, to kill them. And, of course, once dead, they were no longer "biological."[26]

The model of astronomy had important consequences for Whewell's whole theory of science. For one thing, it led to his reading all actions in nature as mechanical, on the model of Paley's watch, or, to be more fair, on the model of the movements of the planets and stars as traced by the Herschels, William and John. For another, it allowed Whewell to risk developing his argument through extensive references to particular phenomena because the singularity of particulars is not their essence. In effect, despite science's Baconian focus on the data of experience as the source of all knowledge, Whewell's science, and theology, implied a world not only ordered, but ideal and timeless, in which each element is fully and permanently what it is. I don't mean to suggest that Whewell was unhistorical. In fact, as Yehuda Elkana points out, the most striking thing about his *History* and *Philosophy* of the sciences is that they are committed to the view that science develops historically and, in keeping with his catastrophist bias, progressively.[27] But the nature whose reality is being unfolded operates regularly, within laws always and everywhere the same.

The immediate preoccupation with law makes the style of *Astronomy and General Physics* far more abstract than Darwin's. Whereas Darwin dwells on particular instances beyond their apparent usefulness for the most general argument, and he injects himself, as observer, into the exemplary material, Whewell sustains the dignity of generality and abstraction. His instances are reiterative, that is, whatever various phenomena are discussed, they tend to affirm the same point about nature, and rarely does he convey the sense of intense, individual, empirical investigation. He is far more interested in averages than in the individual units used to compile the averages. Thus, he minimizes the fluctuation of weather in England to demonstrate that, for example, despite the severe frost of 1788 the mean temperature for the year was 50 degrees, and despite the terribly hot summer of 1808, "the mean heat of the year was 50 ½, which is about that of the standard" (p. 59). The difference is of texture, for we should not forget that "law" was of equal importance to Darwin.

Whewell seems, however, to posit an unproblematic natural world. Whatever his theory about the mind's activity in nature, his rhetoric implies a clear narrative of a stable reality to be discovered, of the progressive unfolding of the laws exemplified; and the history of science is the history of gradual progress toward complete understanding of the "ultimate material causes," or of "the most general laws." Particulars have precisely defined characteristics that do not change through time (until castastrophe obliterates and replaces them) and whose relationship to larger classes is firmly established. Time moves regularly, predictably. Every story reflects other stories, retells them with new data. The narrator can be wise about the narrated experience because he or she knows the generality of what the characters experience as particular. When reality is understood, time effectively ends; the story closes. Questions of identity imply the limits of the inquirer's consciousness, not uncertainty about the reality of classification. Although we can never attain or even approximate to God's omniscience, the natural world must be regarded as comprehensible.

This is not entirely true about Whewell's "moral" world, which is governed by different sets of "laws" still impenetrable to human intelligence. It was crucial for Whewell that the "rule of law"—that is, natural law—*not* be applicable to human consciousness and will, for he knew full well what the later materialistically oriented scientists like T. H. Huxley, John Tyndall, and W. K. Clifford were to affirm, that the extension of physical law to moral action implies a reduction of the purely human to the animal, the animal to the mechanical, and thus makes for a literally demoralizing determinism. "There can be no wider interval in philosophy than the separation which must exist between the laws of mechanical force and motion, and the laws of free moral action" (p. 374). The adoption of astronomy as the model science had enormous implications not only for biology, but for ethics, as well. In a sense Whewell was anticipating and rejecting the metaphorical extension, implicit already in Hobbes and Locke, of scientific law to social and moral activities.

The self-conscious extension of science to morality and society was given major impetus by Comte and Mill. Working from essentially positivist perspectives, they argued that scientific law applied to the entire world, inorganic, organic, human. But most pre-Darwinian speculation about science drew an absolute line between animal and human life, and the detected *physical* similarities between humans and beasts was not taken to reflect any consanguinity. A moral and spiritual nature distinguished the human from all other living creatures, and Darwin's apparent refusal to recognize the distinction kept his work off the Trinity College shelves.[28] Whewell must argue for the primacy of law while resisting its positivist extension because he needs to believe that the divinely created world makes sense in all its aspects. If, as he says, God's nature is incomprehensible so that we cannot know the "laws by which God governs his moral creatures," we know that there *are* such laws, and "natural reason" allows us to "trace laws that imply a personal relation to our Creator" (p. 324).

Insofar as it is reasonable to say that scientific thought displaced religion in the nineteenth century, "law" provided the principle of order that made the displacement tolerable and provided the surest evidence of a presiding intelligence. Nature as bearer of value is also dependent on the presence of law, and its evidence of "selection, design, and goodness" (p. 8). Faith in law makes the new science possible and makes the novel's shift to domestic and "realistic" subjects compatible with the possibility of its seriousness, even though such subjects had traditionally been the province of comedy. Realists working within a culture where the system was understood to be lawful and intentional will have a very different relation to particulars than will those post-Darwinian writers who can accept the idea of system, but not that of intention. The center of value must shift from the divine creator to the human sharer.

Transformation of the ordinary. The forces at work in the development of realism, as Erich Auerbach has described it, were at work in science as well. The social and political transformations within European culture touched not only imaginative literature, but also the literature of natural philosophy. The shift from high to comic subjects, from formal to colloquial styles, reflects and is reflected by the famous Baconian arguments and by the whole process of the secularization of knowledge, of which science is the most important element. Despite the religious sanctions argued for science, from Bacon through Newton to John Herschel, science represented an institutionalized shift of authority, from the revealed book, from the church, and ultimately from God himself, who once severed from his human vicars became almost too abstract to allow many intellectuals to accept his authority when it seemed to conflict with scientific discourse.

The authority was to be nature itself, and secular (and even religious) theorizing concentrated on the evidence nature, in all its multitudinousness and variety, provided. True knowledge is rooted in an intimate acquaintance with the details of the here and now, and not in a theological beyond, a Platonic world of timeless ideas. But the new interest in the particulars of nature entailed a new ascription of value to those particulars. In fiction Victorian authorial intrusions declared the importance of trying to understand and sympathize with the very ordinary protagonists. Science itself implies that all the details of ordinary life are manifestations of those general laws, some of which, we have seen, are traced to the model science, astronomy. Those laws are accessible through careful observation of details, but law, in the scientific view, gives the value to particulars that in religion derives from their creation by God, in morality from their relation to others. So Herschel, for example, explaining the nature of scientific knowledge, insisting on the need to get beyond "insulated independent facts," at the same time finds those seemingly independent facts intimations of larger truths: "The colors which glitter on a soap-bubble are the immediate consequence of a principle the most important from the variety of phenomena it explains, and the most beautiful, from its simplicity and compendious neatness, in the whole science of optics."[29] The soap bubble matters. If we attend to it closely we see not only that it is beautiful, but that it reflects, in its beauty, great principles.

The strategy of science is, thus, to make the most ordinary particulars meaningful. The shadows of religion and of Platonic idealism linger in the theory that identifies science with the activity of ascertaining the laws of nature and thus implies that the real inheres in the laws rather than in the particulars. They linger, too, in the realistic novelist's equation of art with the activity of extending sympathies from traditional heroic and romance subjects to the ordinary details of ordinary life. Natural theology (which builds on the scientific preoccupation with law), scientific theory, and realistic fiction all need to insist on the value of the ordinary.

Paley himself is concerned to reenforce the emphasis on the ordinary that emerges so clearly in early romantic poetry. He not only wants to make the intellectual argument, that every detail of experience reflects large general laws. He points out, too, that the very abundance and variety of life leads us to underestimate the significance of the most trivial seeming aspects of daily life: "One great cause of our insensibility to the goodness of the Creator, is the very *extensiveness* of his bounty." We tend, in our greed, to value only those things that others cannot have. But "nightly rest and daily bread, the ordinary use of our limbs, and senses, and understandings, are gifts which admit of no comparison with any other."[30] Thus, natural theology builds on a rhetoric that will develop new habits of perception. For the natural theologian nature is a continuing testimony to the divine, its sands marked with footprints of the creator. He is committed to bringing out what might be called the romantic side of familiar things. Or, to put it in a less Dickensian way, natural theologians like Whewell must constantly make us aware of how the most ordinary aspects of our lives, the things we take most for granted, emerge from an almost miraculous conjunction of natural laws, an intricate interdependence of forces and objects and organisms. As, in chapter after chapter, Whewell demonstrates these miraculous relationships, a sense of design comes to "cling" to the most ordinary perceptions, and a sense of a designer must grow. The world means intensely, and it means good.

Within the context of a work of discursive and argumentative prose Whewell cannot use the devices of particularity common to the great realists like Thackeray and George Eliot. But he does not undertake the discussion of astronomy, the movements of the planets and the stars, before he undertakes a more humanly recognizable study. His subject, he says, "may be treated as Cosmical Arrangements and Terrestrial Adaptations," and he chooses to begin with the "Terrestrial Adaptations." His reasons are illuminating and consonant with the biases of realistic fiction: "In treating of these the facts are more familiar and tangible, and the reasonings less abstract and technical, than in the other division of the subject. Moreover, in this case, men have no difficulty in recognising as desirable the end which is answered by such adaptations, and they therefore the more readily consider it *as an end*" (pp. 13–14). The larger view will come with a vision of the vast, yet mathematically ordered, movement of the stars; but to begin with, Whewell notices the blossoming times of honeysuckle, gooseberry, currant, and elder; he considers the times of the rising of sap, or the various ways in which flowers open or shut. We are to *feel* that "every step we take" and "every breath we draw" is dependent on the weight of the core of the mysterious earth beneath us.

It is, finally, the recognition that every "ordinary" element of our experience is dependent on a myriad of relationships, marvelously coordinated, that makes reality resonant with the presence of the mysterious and divine. The weight of our planet, the time it takes to circle the sun, the action of heat and cold in producing clouds and rain and wind, the alternations of light and darkness—all of these, working according to laws the natural philosopher can discern—affect our eating, breathing, sleeping, working, walking. The complicated interplay makes the most ordinary activity mysterious and yet satisfyingly part of some great master plan. Ironically, then, as Whewell unfolds the laws by which the most trivial elements of the material world are governed, he appeals rhetorically, like Dickens, or Thackeray, or Eliot, to feelings of awe.

Plenitude. "Abundance," I have suggested, is a dominant motif of Darwin's argument; Malthus figures importantly in Darwin's thinking because overpopulation entailed "selection," and Darwin found his mechanism in a nature that was constantly expanding beyond the limits of the space it could occupy. Beer talks of the "unruly superfluity of Darwin's material" which "at first gives an impression of superfecundity without design."[31]

But the importance of abundance for Darwin is foreshadowed in natural theology itself. For the rhetoric of natural theology had always depended on strategies for conveying the wonder of the abundance, fecundity, and variety of the natural world. That old metaphor, "the great chain of being," included within its meaning a nature filled to plenitude, and Renaissance writers praised and glorified the variety of nature that filled every point in the order of things. Plenitude does not, however, imply overcrowding, waste, and potential disorder, which Darwin's abundance does. In Darwin, every niche is immediately filled by adaptive organisms; in the world of natural theology, there are no gaps to be filled. There are no vacuums, no empty spaces. The world is a plenum.

The laws of nature are multiple; the nature they govern is prodigiously fertile and various. The miracle celebrated by natural theology is the order that coordinates, relates, adapts the multitude of natural forms and natural laws. The more overwhelming the number, the more convincing the argument from design. Random connections could not make the world work. Only a designer, all-knowing, with intellectual powers beyond the imagination of humankind, could have organized so bounteous and various a world.

Darwin needed to explain the multiplicity and abundance of life, which Whewell used rhetorically to enforce the experience of God's supervising presence, without recourse to traditional religion. His laws without a lawgiver make the particularities in his work seem more aberrant, more dangerous and less adapted to human needs than the particularities of Whewell's work. Darwin focuses with almost obsessive concern on the minutiae of nature, whereas Whewell's "particulars" have an eighteenth-century generality about them. His emphasis is on how the details are ordered, but, as Beer puts it, "only gradually and retrospectively does the force of [Darwin's] argument emerge from the profusion of example."[32]

Whewell lays out one of the major stategies of the natural-theological argument in his discussion of the "adaptation of laws":

> The number and variety of the laws which we find established in the universe is so great, that it would be idle to endeavour to enumerate them. In their operation they are combined and intermixed in incalculable and endless complexity, influencing and modifying each other's effects in every direction. If we attempt to comprehend at once the whole of this complex system, we find ourselves utterly baffled and overwhelmed by its extent and multiplicity. Yet, in so far as we consider the bearing of one part upon another, we receive an impression of adaptation, of mutual fitness, of conspiring means, of preparation and completion, of purpose and provision. (p. 11)

Natural theology is an effort of containment. Multiplicity, in this argument, is not itself a confirmation of the presence of a creator; the wonder of Whewell's world is in the way the multiplicity is always joined with order. The multiple laws are adapted to each other and to their objects.

Whewell is cautious about arguing that abundance in itself is admirable or evidence of a wise creator, although he finds it further confirmation of the power and wisdom of the creator. He notes, in his discussion of the remarkable ways vegetables and animals are adapted to the large cosmic movements of the earth, the sun, the planets, that there are many people who "will probably see something admirable in this vast variety of created things . . . in these apparently inexhaustible stores of new forms of being and modes of existence." These are, indeed, "well-fitted to produce and confirm a reverential wonder" (p. 63).

Here, again, persuasion is not strictly logical. Whewell knows that beyond the initial Paleyan assertion that design implies a designer, he has no particular argument to make here. He needs to make the sense of a designer "cling" to our perceptions, and one of the ways to do this is to remark—or exclaim—about the inexhaustible variety of the creation, the stunning way in which, with all this variety, each species is adapted to its place in nature. But for the most part, Whewell's arguments are relatively austere. They are not as abundant in illustration as those of some of the other Bridgewater writers, and they lack the particularity of Darwin's prose. Whewell can only claim that the abundance and variety of life is beyond human imagination: "The number and variety of animals, the exquisite skill displayed in their structure, the comprehensive and profound relations by which they are connected, far exceed any thing which we could have beforehand imagined" (p. 232). He opens before us a vast universe, beyond the planets, beyond our solar system, beyond the stars we can see with naked eye. He reaches for a vision that will awe us into submission to the overwhelming power and wisdom of God:

> If we take the whole range of created objects in our system, from the sun down to the smallest animalcule, and suppose such a system or something in some way analogous to it, to be repeated for each of the millions of stars which the telescope reveals to us, we obtain a representation of the material universe; at least a representation which to many persons appears the most probable one. And if we contemplate the aggregate of systems as the work of a Creator, which in our own system we have found ourselves so irresistibly led to do, we obtain a sort of estimate of the extent through which his creative energy may be traced, by taking the widest view of the universe which our faculties have attained. (p. 234)

He talks of the difficulties of bringing to the "common apprehension" the scale of the universe.

At last Whewell completes the argument quite traditionally: whereas some people feel that the sheer scope of the universe makes the idea of divine concern for each individual seem impossible, in fact numbers and size make no difference. "*Large numbers* have no peculiar attributes which distinguish them from small ones" (p. 242). If God can attend to a hundred, he can attend to a billion. All the evidence suggests that God's care extends everywhere in this enormous, and microscopic, universe. The limits of Whewell's rhetoric in creating a sense of abundance and vastness contrasts markedly with the range of Darwin's. The difference has to do primarily with particularity. Darwin's universe, too, is superfecund, almost incomprehensibly various. But the effect of his rhetoric is to move in the opposite direction from that of the natural theologians, even though, in his commitment to law, plenitude, and particularity, he takes very similar positions. What Darwin includes that Whewell, with his far more abstract and general rhetoric omits, is the unique, the individual, the aberrant, the grotesque. The experience of Darwin's prose raises the question of how it is possible to believe that this entangled, abundant, aberration-filled world is "designed." Plenitude, so crucial to Darwin's theory of natural selection, implies not an all-provident creator, but waste, loss, trial and error, the absence of design.

Darwin, in fact, seems to be closer to Paley's version of natural theology in his treatment of fecundity. While Paley too treats fertility and variety as "imperfections," he argues (in a way that anticipates Darwin's emphasis on the importance of crossing in breeding for survival and development) that "it is probable that creation may be better replenished, by sensitive beings of different sorts, than by sensitive beings all of one sort."[33] He recognizes that plants produce far more seeds than will survive, that death ultimately contributes to life. But he sets all these limits within a Panglossian world, in which what seems bad ultimately works for good because all things come under God's design. Darwin simply refuses—or tries to refuse—the last step, of seeing all phenomena as designed and teleological, and alternatively refuses the kind of selectivity in examples that allows Whewell so total a commitment to "law," and that allows Paley to assimilate all pain to a view of the world as perfectly adapted and benevolent.

Adaptation. Modern science has taught us to look to Darwin when we think of adaptation. Popular essays about evolution are full of the industrial moth that has turned black to adapt to the soot and smoke of mining country; and this is almost always taken as evidence for Darwin's theory.[34] But Darwin learned the language and many of the strategies of adaptation from natural theology. Adaptation is Paley's argument and Whewell's, and it provides evidence not for natural selection, but for God's prevision and care. Whether we look at the phenomenon as a secular or divine one, adaptation exercises a peculiar fascination.

It concerns itself with relations, with how things fit. How is it that fish survive in water, humans cannot? How is it that the Coryanthes orchid is shaped so as to collect water, that bees "gnaw off the ridges within the chamber above the bucket" and push each other into the water so that they have to crawl out in such a way that the pollen is glued to their back? And how is it that when the bee repeats the experience in another Coryanthes, the pollen on its back comes into contact with "the viscid stigma, and adheres to it, and the flower is fertilised"? (*Origin*, pp. 179–180). The example is Darwin's, but could easily, to this point, be part of an argument from natural theology. What extraordinary, minute contrivances are here used for the sake of propagating the flower, what clearer evidence of a designer?

Darwin describes this "contrivance" (a word he consistently used despite its natural-theological implications) for quite other purposes. In the preceding paragraph, I have clinched the natural theologians' case with one of their typical rhetorical devices—the question that seems to answer itself. But the rhetorical question can have another answer; the argument from adaptation works in more than one way. For example, the adaptation may be evidence of intelligent design, or the bees and orchids may only exist because they are adapted to each other. Darwin uses this peculiar adaptation as evidence of the *absence* of design. Where the natural theologian would say that it is impossible that so subtle a set of relationships could have happened without deliberate contrivance, Darwin sees this particular contrivance within the context of a multitude of other related ones. In nature there are, he says, innumerable variations for the achievement of the same ends, and in every case he shows there are "transitional grades" from one form to another. And he asks, staying close to the rhetorical formulas of natural theology, but inverting them,

> Why, on the theory of Creation, should there be so much variety and so little real novelty? Why should all the parts and organs of many independent beings, each supposed to have been separately created for its proper place in nature, be so commonly linked together by graduated steps? Why should not Nature take a sudden leap from structure to structure? On the theory of natural selection, we can clearly understand why she should not; for natural selection acts only by taking advantage of slight successive variations; she can never take a great and sudden leap, but must advance by short and sure, though slow steps.[35]

One of the reasons he can turn the natural theological argument on its head is that he refuses to be as selective about his examples as Whewell and Paley; more important, he asks the rhetorical questions as a clincher to the argument only after he has examined the particular phenomenon and as many related phenomena as he can manage.

Whewell's handling of a similar argument implies a very different attitude toward "contrivance" and "adaptation." In discussing the length of the year, Whewell actually anticipates and denies Darwin's kind of interpretation of adaptation. "Why," he asks, in concluding his argument,

> should the solar year be so long and no longer? or, this being of such a length, why should the vegetable cycle be exactly of the same length? Can this be chance? And this occurs, it is to be observed, not in one, or in a few species of plants, but in thousands. Take a small portion only of known species, as the most obviously endowed with this adjustment, and say ten thousand. How should all these organised bodies be constructed for the same period of a year? How should all these machines be wound up so as to go for the same time? Even allowing that they could bear a year of a month longer or shorter, how do they all come within such limits? No chance could produce such a result. And if not by chance, how otherwise could such a coincidence occur, than by an intentional adjustment of these two things to one another? by a selection of such an organisation in plants, as would fit them to the earth on which they were to grow; by an adaptation of construction to conditions; of the scale of the construction to the scale of conditions. (*Astronomy*, pp. 24–25)

More than half this passage is composed of questions. Alternative possibilities are not considered. The details are rather general, and one can bet that Darwin would have contributed here a great many examples of vegetables and animals for which the cycles of the year and of day and night (given, too, that they vary so greatly in different parts of the world and at different times) do not appear so convenient.

Such a response would have disrupted the major assumption of the kind of natural theology we find in the Bridgewater Treatises—that all adaptations are perfect. Whewell can then go on to infer an intelligent overseer. The selectivity of the natural theologian's argument is of particular importance here for one did not need Darwin's acute powers of observation to note that astonishing as the eye is, for example, it is not a perfect instrument. Nor, would common perception fail to note that there are many organs, human and animal, that serve no function, but are what Darwin was to call vestigial. In fact, although many biologists had abandoned the idea that adaptation could be used as an argument for intelligent design (and in this respect Whewell was retrograde, and consequently chastised by W. B. Carpenter), almost all scientists agreed with theologians that adaptation was perfect.[36] Darwin would continue, like the theologians, to marvel at adaptations, but perfection was no part of his scheme.

Whewell did recognize that there is an alternative—naturalistic—

interpretation of adaptation: "that no plants could possibly have subsisted, and come down to us, except those which were thus suited to their place on the earth." This was, in fact, to be a part of Darwin's own argument, the major point being that the "system" of nature is self-sufficient, self-adjusting, and does not require an intelligent superviser. And Darwin would have found much earlier support, from Owen, Carpenter, and others, for this position.[37] But Whewell, while agreeing that his alternative explanation is true, argues that "this does not at all remove the necessity of recurring to design as the origin of the construction by which the existence and continuance of plants is made possible" (*Astronomy,* p. 25). Darwin could not have complained much about this retreat. He himself was to be accused of not having treated the ultimate "origins." This was correct, but beside the point. As long as Darwin could explain present adaptation naturalistically, he had no problem accepting a divine originator of the secular system. In other words, both interpretations of adaptation might be true without contradiction of each other. Darwin's theory of evolution by natural selection might be regarded as an elaborate development of a natural- theological theory of adaptation, which simply extends the area in which naturalistic explanation is acceptable.

Although intellectually plausible, this interpretation misses the felt difference. In fact, Robert Chambers's *Vestiges of Creation* was precisely natural theology extended to evolution, and this was possible by using Whewell's view that God is the originator of laws which operate naturalistically from the time of their invention. Chambers is quite explicit: "Those who would object to the hypothesis of a creation by the intervention of law, do not perhaps consider how powerful an argument in favour of the existence of God is lost by rejecting this doctrine. When all is seen to be the result of law, the idea of an Almighty Author becomes irresistible."[38] Whewell's explanation implies the perfection of adaptation, the pervasiveness of order and intelligence in all aspects of nature. Darwin's explanation implies imperfection, allows for waste, error, clumsiness—even, and this was the most intolerable possibility, chance.

Paley, for his part, had also entertained and rejected emphatically purely naturalistic explanations. His reasoning is a little surprising. Some people, he says, have argued that the wonderful adaptations and contrivances of nature result from "a *principle of order* in nature." But, he points out, not all of nature is ordered. "Where order is wanted, there we find it; where order is not wanted, i.e., where, if it prevailed, it would be useless, there we do not find it." We don't find it, he says, "in the forms of rocks and mountains, in the lines which bound the coasts of continents and islands," and so on. But even if there is disorder, there is no chance: "I desire no greater certainty in reasoning, than that by which chance is excluded from the present disposition of the natural world." "What," he asks, as though the answer would logically exclude chance, "does chance ever do for us? In the human body, for instance, chance, i.e., the operation of causes without design, may produce a wen, a wart, a mole, a pimple, but never an eye."[39] Without plunging into Paley's extensive and unlikely discussion of chance at the end of *Natural Theology*, we can see that wherever there is organization he excludes chance. Order exists where order is wanted. Darwin's intimation that chance might play a role in what appears to be organized and intentional would be, on Paley's account, atheistic.

Adaptation, then, trickily carries within it the whole struggle over the order, intelligibility, teleological directions, and divine presence in the world. For Darwin, all adaptation implies history, a gradually adjusting set of relationships between organism and environment, vestiges of which will be detectable in the structure of the organism. Current conditions imply, in the Darwinian world, a long past narrative. In natural theology each adaptation has been fixed permanently at the outset. Theological narrative closes; evolutionary narratives can only be closed artificially (or by death, which is, however, the condition of life, too). Even when individual lives end, evolutionary narrative implies further, generational change.

Ecology. Although Darwin is often thought of as the presiding spirit of individualism, he might equally well be taken as the father of ecology. No nineteenth-century book, scientific or fictional, more elaborately works out the delicate, manifold, intricate interaction of organisms with themselves and with their environments. The idea of adaptation also implies the idea of interdependence. The bees and the orchids depend on each other, for example, as they both depend on the rain that fills the orchid, the soil that nurtures it, the worms that enrich the soil. Darwin impresses us with the astonishing forms of interdependence that govern the natural world. In place of the great chain of being, modern thinkers put the food chain, which makes the strongest and most noble predators, eagles and tigers, for example, dependent on the fate of insects and vegetables; or we marvel at the extraordinary shapes and sizes of parasites which manage to develop camouflage that makes them look like their hosts.

But Darwin's ecology also has its roots in natural theology, which—

once again through the idea of adaptation—imagines a world intricately interconnected. The natural theologians' emphasis on adaptation and law, plenitude and benevolence, ultimately implies a unity that makes sense of it all. Whewell points to this unity as he indicates what his project in *Astronomy and General Physics* is to be:

> When we have illustrated the correspondencies which exist in every province of nature, between the qualities of brute matter and the constitution of living things, between the tendency to derangement and the conservative influences by which such a tendency is counteracted, between the office of the minutest speck and of the most general laws: it will, we trust, be difficult or impossible to exclude from our conception of this wonderful system, the idea of a harmonising, a preserving, a contriving, an intending Mind; of a Wisdom, Power, and Goodness far exceeding the limits of our thoughts. (pp. 11–12)

That extension beyond the limits of our thoughts reappears in Darwin, but this time with threatening implications. For the connections Darwin describes don't always make sense, don't always "compensate." This question of "connection" across diverse areas is a leitmotif of Victorian thought, penetrating the fiction and the great nonfiction prose writings as well. Narratives are built on the assumption that such questions—"What connection can there be?"—will be answered. Somebody must lose in the interconnections, fictional or Darwinian. The balances are beyond human control. Breeding for certain qualities produces unintended consequences. Extermination of pests affects other organisms in the food chain in so elaborate a way that yet other organisms necessary to man might also suffer. Whewell's theory implies an ultimate balance between "derangement" and "conservation," and again, therefore, implies stability. But Darwin's nature is continuous and irreversible, and there is always a loser.

The interconnections in Whewell point back to laws issuing from a central consciousness. Organisms were created for their niches, survive by virtue of divinely instituted natural laws, and are dependent only accidentally on each other, but essentially on the divine. Normally, Whewell tends to think in only two terms, of the adjusted organism and its inorganic environment. Although he affirms the wonderful complexity of the interaction of many laws, his examples tend to be relatively simple. Darwin's, however, deliberately proliferate and call attention to processes (as a third term).

Finally, the difference between Darwin and Whewell has to do with Darwin's willingness to risk the loss of any consolation from non-naturalistic explanation. Whereas Darwin's language "outgoes," as Beer says, all of his attempts to tame it to law and univocal meaning, Whewell rests from the abundance, intricacy, and complexity of the world he describes through his faith in an order not yet available to scientific consciousness, nor ever able to be. In his last paragraph, he puts it this way:

> And if, endeavouring to trace the plan of the vast labyrinth of laws by which the universe is governed, we are sometimes lost and bewildered, and can scarcely, or not at all, discern the lines by which pain, and sorrow, and vice fall in with a scheme directed to the strictest right and greatest good, we yet find no room to faint or falter; knowing that these are the darkest and most tangled recesses of our knowledge; that into them science has as yet cast no ray of light; that in them reason has as yet caught sight of no general law by which we may securely hold: while, in those regions where we can see clearly, where science has thrown her strongest illumination upon the scheme of creation; where we have displayed to us the general laws which give rise to all the multifarious variety of particular facts;—we find all full of wisdom, and harmony, and beauty: and all this wise selection of means, this harmonious combination of laws, this beautiful symmetry of relations, directed with no exception which human investigation has yet discovered, to the preservation, the diffusion, the well-being of those living things, which, though of their nature we know so little, we cannot doubt to be the worthiest objects of the Creator's care. (pp. 327–328)

We leave the argument with the crucial phrase: "we cannot doubt."

The overlap of ideas and attitudes in these five areas should suggest that the tendency to think of Darwin as in total conflict with natural theology leads to the obscuring of important similarities. What I want to emphasize, in what might be taken as a Darwinian way, is that Darwin's divergence from natural theology is a divergence by descent. Darwin's argument begins by using a model of design and intention, artificial selection; he makes his case by an encyclopedic profusion of examples; he describes a world abundant, interdependent, lawful, and apparently progressive (although this point needs heavy qualification). The very examples that a natural theologian might use to demonstrate intelligent design become, for Darwin, evidence of a natural process. And nature displaces God as designer. The assumptions and attitudes underlying these similar patterns make the critical difference.

Perhaps the most important assumption, implicit in the language Whewell uses to describe natural process, is that the natural world can be described in terms of mechanical relationships. Kin to the romantics though he was, Whewell as naturalist inherited an eighteenth-century tradition of seeing nature as a complex machine. He had adopted Kant's view that "any objective explanation must be a mechanical one, and since organisms cannot be explained mechanically, they can never be given an objective explanation."[40] For that reason, biology could never be a science, and the tradition of absolute separation between the inorganic and the organic was reenforced. Our failure to understand all of nature is partly a result of the enormous complexity of this machine, largely a result of our own limitations. At the point when the mechanical model no longer works, Whewell automatically invokes the divine. That is, his world, though he insists on the inseparable nature of mind and matter, is fundamentally dualistic. His thinking about nature is almost exclusively materialist. The model is Paley's watch. And the importance of this is that mechanisms do not grow; they do not grade minutely into each other. They are what they were designed to be, work according to the design, allow modifications only by direct interpositions of the designer. His mode of explanation normally implies an either/or structure. Either the phenomenon is explicable in the mechanical terms of, say, astronomy, or it is a naturalistically inexplicable divine creation.

Although Whewell rarely spells out the full implications of this metaphorical model, they suggest where a major divergence from Darwin, whose model is the tree, not the machine, asserts itself. Darwin, though an inheritor of eighteenth-century materialist thought, had nevertheless absorbed the organicist assumptions of the romantic poets, and is what Beer calls a "romantic materialist." His modes of explanation rarely employ the sharp implicit logical alternatives of Whewell. Laws that govern the branching of trees are different in kind from laws governing the movements of the planets or the spinning of the earth, for, among other things, the units that operate within laws relating to organic growth are peculiarly difficult to discriminate. The dominance of law in Whewell's scheme of things implies also the dominance of the idea that all the elements in nature can be perceived as separate units, with clearly demarcated identities. History, on this mechanical model, develops by virtue of a designer's objectives. It is progressive, but the progress is, again, imposed from the outside.

The view is atomistic. And it assumes that large material structures (or organisms) can be adequately understood by the mere multiplication of microstructures. Whewell's discussion of the way the constitution of matter implies mind suggests, fairly, the general view of nature and relationships assumed by natural theology, and this view is utterly at odds with the view implied by Darwin's kind of biological science:

> Every particle of matter possesses an almost endless train of properties, each acting according to its peculiar and fixed laws. For every atom of the same kind of matter these laws are invariable and perpetually the same, while for different kinds of matter the difference of these properties is equally constant. This constant and precise resemblance, this variation equally constant and equally regular, suggest irresistibly the conception of some cause, independent of the atoms themselves, by which their similarity and dissimilarity, the agreeement and difference of their deportment under the same circumstances, have been determined. Such a view of the constitution of matter, as is observed by an eminent writer of our own time, effectually destroys the idea of its eternal and self-existent nature, "by giving to each of its atoms the essential characters, at once, of a *manufactured article* and a *subordinate agent*." (*Astronomy,* p. 260)

The quotation is, significantly, from Herschel's *Preliminary Discourse*, which is the crucial text for the establishment of astronomy as the model for scientific knowledge. It reaffirms the argument made by Paley at the beginning of *Natural Theology*, but more important for our purposes, it closes out individual variations, argues for regularity, permanence, repetition. Whewell's rhetoric, while evoking awe for the vastness and profusion of the material world, tends in style, metaphor, and argument, to emphasize moderation, limits, and repetition. Where Darwin's world knows no beginning, or end, Whewell emphasizes size but at the same time reassures us with limits. We will eventually achieve a telescope that will assure us that any more powerful one would fail to discover new stars (p. 286). The biological world is similarly limited: "In like manner, although the discovery of new species in some of the kingdoms of nature has gone on recently with enormous rapidity, and to an immense extent;—for instance in botany, where the species known in the time of Linnaeus were about 10,000, and are now above 100,000;—there can be no doubt that the number of species and genera is really limited; and though a great extension of our knowledge is required to reach these limits, it is our ignorance merely, and not their non-existence, which removes them from us" (p. 247). Here Whewell seems to make nature commensurate with man's power of understanding, even though the argument begins with real sensitivity to the traditional natural-theological argument about the vastness of the universe. This affirmation of limits is an attempt to keep the multitudinous world under control, and more important, to keep man at the center of it.

Whewell was in fact taking an unusual position, one that he developed in *The Plurality of Worlds* twenty years later. The scientific *and* theological assumption was that the vastness of the world almost certainly included other worlds with intelligent beings on them. Whewell says no:

> The mere aspect of the starry heavens, without taking into account the view of them to which science introduces us, tends strongly to force upon man the impression of his own insignificance. The vault of the sky arched at a vast and unknown distance over our heads; the stars, apparently infinite in number, each keeping its appointed place and course, and seeming to belong to a wide system of things which has no relation to the earth; while man is but one among many millions of the earth's inhabitants;—all this makes the contemplative spectator feel how exceedingly small a portion of the universe he is; how little he must be, in the eyes of an intelligence which can embrace the whole. (p. 240)

Whewell's chapter "Man's Place in the Universe" implies a kind of narrative, beginning here with the protagonist's recognition of littleness and isolation in a vast universe. The plot entails the recognition that if we look closely at the details of that universe, we will realize that the "intelligence which can embrace the whole" looks after each individual in it, as well.

The next step in the plot is intensification of vision, and the use of analogy. The protagonist discovers that the world is as minutely small as it is vastly large, and through the microscope the minuscule world turns out to be as well organized as our own, the organisms as well adapted. If design extends to such small matters, the analogical argument goes, then the mere fact that the world is vast does not at all imply that the individual self is untended, alone: "We find . . . that the Divine Providence is, in fact, capable of extending itself adequately to an immense succession of tribes of beings, surpassing what we can image or could previously have anticipated; and thus we may feel secure, so far as analogy can secure us, that the mere multitude of created objects cannot remove us from the government and superintendence of the Creator" (p. 244). Although Whewell begins by projecting vastness, the language suggests how much that projection depends upon unchallenged conventions. The "arched" "vault," the "appointed places" imply no unease or tension, but a rather lax absorption of bromidic formulations. Yet even against these comforts he finds strategies of reassurance, the primary one being the analysis of phenomena to assure us that however vast the world may seem, it is always under the control of caring intelligence, and that the intelligence, while not always comprehensible to man, has not only kept man at the center of the contrivances of creation, but as a consequence has made a world humanly intelligible—and satisfying.

The finitude of the universe, the centrality of man, combine in Whewell with a curious sense of permanence such that narrative can only mean the unfolding of what is given—divine superintendence—rather than changes in the nature of the given. The material units of the world of natural theology have their identities perfectly and permanently. Whereas Darwin finds the present intelligible only by recognizing in it vestiges of the past that will help account for the imperfections or apparent aberrations in nature, Whewell explains the aberrations as he explains the norms—as part of a divine plan that may, it is true, require God's further intervention, but that has not "developed" from past conditions. We need not explain what is by study of what was. Charles Gillispie has pointed out that "so long as natural philosophy was devoted to the construction of nature, natural theology emphasized design." "A sense of history," he argues, was "uncharacteristic" of the utilitarian tradition to which Paley belonged.[41] And though Whewell was surely not a utilitarian, his emphasis on design and on physics, as opposed to geology (the first science, Gillispie says, to be "concerned with the history of nature rather than its order"),[42] gives to his understanding of nature a characteristically ahistorical perspective.

The Whewellian narrative, unlike the Darwinian, tends toward repetition, resemblance, regularity, predictability. The story of natural theology closes at the point that consciousness discovers two things, the imperfection and the finitude of what is already there. All the elements in Whewell's crowded world are fully adapted to their positions, are unequivocally and permanently what they are, and operate within laws that relate them to each other and to the teleological object. They are defined not by their pasts, but by their places in the present system. We do not here have the story of Maggie Tulliver, who is shaped by her past, and who cannot fully work out her place in the present system. We are far closer to the story of Elizabeth Bennet.

A characteristic movement in Whewell's argument, as for example in the chapter "The Mass of the Earth," is to "illustrate one or two cases" that demonstrate the validity of the generalization. The argument then moves from "The first instance we shall take . . ." to "As another instance . . ." to "Another instance . . ." The instances do not constitute, to be sure, a Darwinian "plethora," but they suggest an ultimate similarity in the world. Any "instance" will do because all alike demonstrate the order of things. It might seem that so instance-full a book as Darwin's would work, rhetorically, in the same way, but when Darwin says that his book has been "one long argument," he is speaking accurately. Whewell's "argument" is made early and clearly. The rest of the book fills in some of the blank spaces.

Each instance Darwin uses is educed as part of complex and entangled strains of argument. The first four chapters of the *Origin* are too busy establishing the argument to settle into instances, and the language is rarely of Whewell's sort—"an instance," "another instance." Here are some examples (to use a Whewellian mode) of Darwin's introduction of instances: "Altogether at least a score of pigeons might be chosen" to make the point that varieties might be mistaken for species; "Some facts in regard to the colouring of pigeons well deserve consideration," and there follows a long set of examples, all of which are considered to discover what is the best mode of *interpretation* of them; or, "In favour of this view, I may add, firstly . . ." whereupon follows a sequence up through "fourthly," in which each detail is not only presented, but also argued. That is, Darwin's rhetoric is built around a problematic world, one that needs not only to be experienced, but to be interpreted. Oddly, the secular, materialist world of Darwin is more mysterious, more difficult, than the spiritually determined world of Whewell. The consequence of this in the reforming, or deforming of teleological, natural-theological narrative is obviously great. Post-Darwinian narrative will become much more oddly opaque, much more concerned with the nature of the interpreter of the experience, much more unlikely to allow *any* details to carry the weight of meaning we can expect, say, in the *apparently* obscure six paper bags out of which the editor constructs Teufelsdröckh's biography in *Sartor Resartus*, much less likely to tend in a single, intelligible direction.

Subscription to the catastrophist position in the great geological debate put Whewell in the progressivist camp but also allied him to teleology and to the possibility of divine intrusions into the sequence of history. The discontinuous evidence of the fossil rocks led geologists to believe not in transmutation, but in successive divine interpositions, which perfectly adjust the new creations to the changed environment. As late as *The Plurality of Worlds*, Whewell put it this way:

> The best geologists and natural historians have not been able to devise any hypothesis to account for the successive introduction of these new species into the earth's population; except the exercise of a series of acts of creation, by which they have been brought into being . . . It is true, that some speculators have held that by the agency of natural causes such as operate upon organic forms, one species might be transmuted into another; external conditions of climate, food, and the like, being supposed to conspire with internal impulses and tendencies, so as to produce this effect. This supposition is, however, on a more exact examination of the laws of animal life, found to be destitute of proof; and the doctrine of the successive creation of species remains firmly established among geologists.[43]

Each creation, paleontologists showed, produced more advanced species, and the last, of course, produced humans. Such a "history" implies a clear originary moment, the *Fiat* of God himself. Within each period there is no movement or change. Each successive creation produces another *Fiat*. Thus we have progress, but history in only the most skeletal form.

The plots implied by such positions could incorporate change and apparent coincidences, but these would all be resolved into patterns of meaning. A careful reader would find the ends implied in the beginnings, as God's creation was nomothetic—establishing the laws of nature that would without much further intervention lead to the telos, spiritual perfection. Since the structure, design, adaptations within the world are perfect, there is no place for the random, and God intrudes only for further creation. The chain of natural order is filled and will always remain filled, and always ordered and lawful. What seems accidental will turn out to be meaningful, what seems loss will be compensated for. As the work of natural theologians justified the ways of God to man, and the order of things as they are to society, so the Whewellian plot develops within a language and a structure that confirms the given, even as it raises questions about apparent mystery. At certain points naturalistic—or secular—explanation must stop and divine intervention moves the world one more step toward its ultimate timeless perfection.

3

Mansfield Park: Observation Rewarded

DISTANT as they are from science and theology, Jane Austen's worlds, neither timeless nor perfect, reflect the assumptions and possiblities also present in the natural-theological predispositions of pre-Darwinian science. Her obvious distaste for abstract theorizing and language is manifest in her letters as well as in her novels, but her imagination of her two inches of ivory is constrained by conceptions of order that constrained contemporary science as well. In *Mansfield Park*, more than in any other of her novels, she adopts a narrative mode that self-consciously resists change, almost as natural theology in the Bridgewater Treatises represented a last-ditch resistance to the secularization of science and knowledge and to its social and political consequences.

This reading of *Mansfield Park* is a kind of pre-Darwinian test of my overall argument about the relation between fiction and Darwin's thought. What would a novel outside the aura of Darwinism look like? What sorts of differences would there be from the Darwinian kind? *Mansfield Park* is particularly useful for these purposes because creatively and dialogically, it works out tensions between the stable vision of a world governed by the principles of natural theology and the destabilizing vision of Darwin's. In the process, it suggests how both ways of seeing are implicated in social and political struggles that will help determine how epistemological struggles will be resolved. By placing at its center a keen observer whose passive observations become a source of power, it dramatizes one of the key principles of scientific and novelistic practice.

Mansfield Park tends to confirm the kind of world Herschel was to describe in his *Preliminary Discourse*, a world rationally designed and governed, stable, ordered, under the control of disciplined will, a world of precisely defined categories, in which what is true is experienced through observation. Austen's, like the world of Newton, Lavoisier, and Dalton, whose atomic theory was emerging at about the same time as her novels, is constructed from the rational conglomeration of discrete, atomic entities. But within what we might call this Newtonian frame, an alternative imagination—of a world more opaque yet more fluid, multifarious, unstable, unsusceptible to clear rational regulation, and endlessly subject to change—challenges and threatens to destroy the frame.

As geology and biology emerged from natural history to the status of "science" (Lamarck first used the word "biology" in French in 1802), the dominance of Newtonian physics as the model for scientific method was itself challenged. The beauty of the Newtonian design almost *required* a theological interpretation, whereas Darwin's messier world had more difficulty assimilating God. But with the development of uniformitarian geology, history begins to disturb the universality of truth, and with the emergence of "biology," not so much through Lamarck's evolutionism, but through Cuvier's comparative anatomy, which makes structure "a function of [each creature's] way of life," the model for structure shifts from the mathematical ideal back to the empirical, away from reason, toward multiplicity and uniqueness.[1]

Jane Austen's fictions, where moral anarchy seems imminent among large groups of people, self-consciously resist this shift. Noise and crowding have no Dickensian warmth and exuberance about them, but are invariably disruptive and perilous, as in Fanny's return to her own family near the end of the novel. Such sharply defined limits distinguish *Mansfield Park* from otherwise similar but more cluttered Victorian novels. At the same time, it incorporates and confronts the new forces of disorder, forces that, as Marilyn Butler has suggested, echo Locke's empiricism and Holbach's materialism. These, in turn, had opened the way for projects like Darwin's in that they attempted to explain nature without reference to God, spirit, or intention.[2] Such systems emphasize the importance of the individual, the possibility of improvement, and the dominance in human action of the irrational, of feeling. This whole complex of ideas and attitudes, threatening to burst the Newtonian frame, is associated socially and politically with the French Revolution, and Jacobinism in general. Obviously, these attitudes are largely incompatible with natural theology, which became more vociferous in the latter half of the eighteenth century in part as a reaction to continental materialism and its apparent political consequences.

Mansfield Park, of course, is not directly concerned with any of these issues, although the Kotzebue play, *Lover's Vows*, brings them indirectly into the action of the novel, and they are implicit in the very neatness of its structure, and in its treatment of moral and social issues. The form resists the materialist views that one might associate with both the new science and the new society: the absence of design, of a telos, the emphasis on the unique as opposed to the typical, the possibility of randomness in nature, open-endedness in change, the history-bound conception of structure, species, and value, the emphasis on the irrational and on "sensibility." The novel's conclusion affirms stability and order, allows for individual reformation, but excludes important social change, and re-establishes a closed system.

Many readings of *Mansfield Park* deny its apparent social conservatism and find its real strengths in the disruptive forces—the Crawfords (Mary in particular), and its awareness of the severe limitations of the symbolic place, Mansfield Park, so idealized by Fanny herself. Moreover, the curious self-conscious refusal to sustain the realistic method in the last chapter, those determinedly novelistic and abrupt passages that give to Fanny everything she desires, release an irony subversive of the ostensibly dominant teleological "Newtonian" model.

But Austen's fiction (most self-consciously in *Mansfield Park* and with the most complicated results) bears the kind of relation to the sentimental novel, which it often parodies and uses, that the tradition of natural theology bears to the materialist science and philosophy that helped inform the ideology of the French Revolution. Like Whewell, who affirmed intentionality, intelligence, teleology, and the pervasiveness of order and design in the material world, while assimilating latently Darwinian material from contemporary science, Austen assimilates techniques, which sentimentalism had sponsored, into the natural-theological vision.

Of course, Austen is not talking about science or epistemology; but she was, I believe, working with Whewell's culturally shared model of meaning and explanation, derived from seventeenth-century science and absorbed into the discourse even of the moralists she loved. It was so firmly part of the way things were seen that it became the language of "common sense," and if the ideal of it was a mathematical symmetry, it purported to be derived from experience itself. Moreover, it clearly *worked*, and in the chemistry of her own day, the assimilation of Newtonian mechanics to the study of the material constituents of nature was producing yet another scientific triumph.[3] In *Mansfield Park* the clarity of outline and functioning that certain aspects of Whewell's vision entail (not, of course, the natural-theological tendency to optimism and complacency) is firmly, self-consciously sustained.

The novel constructs an ordered and essentially closed system. Change within that system depends on the repositioning of its components, not on changes in structure and procedure. In this sense it is like the world of Lavoisier's chemistry, built in part on the axiom, "in natural and artificial processes alike, nothing is lost and nothing is created."[4] Characters replace each other in their positions and relations; they may even acquire a stronger sense of what their positions require. But the end of *Mansfield Park* simply replaces inferior parts with superior ones, as the end of *Emma* confirms the order that Emma's various strategies had sentimentally aimed at disrupting.

Successful functioning of the system requires strong control and supervision. A nomothetic God made the world like a clock, to go by itself with lawful regularity, although creationist theories required divine intervention wherever something new appeared. But like eighteenth-century machinery, the system requires a "governor" to control the release and distribution of energy. All loss to the forces of disorder is compensated for by assimilation of new elements into the ordered system. All disorder is recuperated by the return of authority. The system can be threatened by the intrusion of disruptive forces, by the absence of its governor, by the selfish energy of its separate parts. The system extends metaphorically to the structure of *Mansfield Park*, where external discipline, of the sort the young people think will be exercised by Sir Thomas, must be supplemented by internal discipline. Fanny exercises just such discipline over her feelings (which threaten, by way of the sentimental novel, to overwhelm her and the narrative). There is no explosion of feeling to break down the lawful regularity of the Mansfield system. Even the love scene—the telos of the plot from the first pages—never happens.

Instead, the last chapter is concerned to explain why Sir Thomas's daughters were inadequate to their positions, why Henry acts in a way both disruptive to the Mansfield system and contrary to his own interests. Sir Thomas comes to fear that "principle, active principle, had been wanting, that they had never been properly taught to govern their inclinations and tempers, by that sense of duty which can alone suffice."[5] Henry was "ruined by early independence and bad domestic example" (p. 451). Against the power of independent feeling, Austen poses authority, discipline, self-control, and rationality. Henry loses in Fanny "the woman whom he had *rationally*, as well as passionately loved" (p. 453).

Reading Jane Austen into modernity badly oversimplifies her complex handling of her own conservatism. One commentator has suggested, for example, that Austen's "epistemological frame of reference . . . is inherently plural, and that plurality mocks the search for determinate meaning."[6] Austen's "epistemology," if it makes any sense to use the term in connection with her work, implies a knowable and a representable world. Language, for Austen, can "represent," but only when it is severely disciplined, when the user puts aside interest, and honors both the medium and experience tested against other's perceptions. Stuart Tave suggests that "in every sentence are the words of meaning which, if correctly used and correctly understood, are the adequate expression of the reality. There is a language in which she can move to arrive at its place and give it the proper names. If that seems simple, it imposes a complex moral discipline. If it seems to us that the word is inadequate to the complexities of the life, the point is rather that to come to the proper word is to do justice to the complexities." There is a social and natural world out there, firm as Dr. Johnson's stone, to be described, as the world of contemporary "natural history" required describing and classification. Truth and morality themselves are bound up in using the right word to mean the right thing. As Tave puts it, in language that reaffirms some of my earlier argument here: "There is a definable reality, not to be made or unmade, to which Jane Austen's men and women must bring themselves; and it is in proportion to their success that they make or unmake their own lives. What is difficult of definition is, characteristically, painful; if Anne Elliott or Elizabeth Bennet cannot define her own feelings she is in distress. But the means of definition are available to her if she has the sense and the moral will to use them."[7] The novel is obviously not exempt from deconstructive reading, but my point is that *Mansfield Park* is self-consciously representational, even where it calls attention to its participation in novelistic tradition. Firmness of definition is essential both to narrative and morality, as it was to the practice of science itself. Although her characters may not always be capable of defining the nature of their moral acts and conditions, the narrator can imply them clearly enough, and the plot will respond accordingly—as it responds to Fanny's virtues.

Austen was writing at a time when scientific discourse, in purging natural description of its tendency to see the sign as part of the thing signified, had long since, as Foucault states, opened "a gap between things and words."[8] That silent gap is filled, in Western tradition, by the idea of representation, the notion that things can be "named." What a thing is can be known, and knowledge is ordered—that is, created in scientific terms—by naming. The extension of the Newtonian tradition to chemistry, a science that achieved preeminence in Austen's lifetime, develops the tradition of the Linnaean model of "naming," which had dominated natural history. Lavoisier joins the procedures of observing and naming in ways that parallel Austen's method. "The sciences," he says, "have made progress, because philosophers have applied themselves with more attention to observe, and have communicated to their language that precision and accuracy which they have employed in their observation: *in correcting their language they reason better.*"[9] Calling things by their right names is the central moral (and aesthetic) project of Austen's fictions; the characters who earn her respect are those who can define even the most difficult "things"—like feelings—who can find the language to represent them fairly. The character who consistently demonstrates this capacity most effectively is the narrator.

The Darwinian project would be in this respect antithetical to Austen's, for, as I suggested in the first chapter, Darwin needed to break the hold of traditional classification, to expose it as a merely verbal activity. Accepting the utility of the Linnaean system as an "artificial means for enunciating, as briefly as possible, general propositions,—that is, by one sentence to give the characters common, for instance, to all mammals" (*Origin,* p. 399), he denies that this implies any essence, any eternal, divine idea. Classification moves from an ideal "natural system" to a genealogical one. By arguing that the perceived similarities between, say, mammals are a consequence of genealogical connection, he denies Aristotelian essentialism and makes any classification temporary and incomplete. A "dog" can be conceived in ideal doggy permanence only by giving language an ideal and atemporal relation to reality. "Dog," for Darwin, has to be a merely pragmatic label for an organism genealogically related to other transient forms and itself only temporarily what it is. Definition thus becomes arbitrary and inadequate.[10]

For Austen, however, the ambiguity of classification and definition is a consequence of individual limitation rather than of the nature of language and things. Thus, the first move toward definition, again in consonance with the developing tradition of Western science, is to observe clearly. One must, as Lavoisier implies, be able to "observe" reality. Observation itself becomes a privileged activity—at least according to Foucault—at the point when the gap between word and thing opens, when the sign is withdrawn from the thing and becomes, rather, a means to represent the thing.[11] From Bacon forward, observation became the privileged way to truth, and keeping observation pure became one of the primary aims of scientific method. The complications of observation, which became so important to fiction, were well understood by Herschel and Lyell. A scientist's own feelings, desires, needs, biases must not be allowed to influence perception of the "thing." Observation must control the "idea" of what is observed; hypotheses, on the Baconian and Newtonian principle, are not to be made.

Observation could only achieve authority with the help of objectivity, and "objectivity," in the modern sense of the word, does not seriously enter the language until the early nineteenth century (probably by way of Kant, but ultimately in the service of positive science).[12] Austen's heroines learn to see clearly by curbing their desires, and by so doing they can then see those desires more clearly. Fanny attempts to repress all desire except her safe affection for her brother, William, even her love of Edmund, because she will not imagine them as taking shape in reality. In consequence, she sees with great clarity from the first. One of the clearest indications of Austen's pre-Darwinian imagination is that she can imply and dramatize Fanny's (repressed) private desires and her very limited perspective and still allow her to achieve precision of observation and understanding—"objectivity," as it were. Part of the reason for Fanny's relative unpopularity among readers of Austen is that her repression of desire smells vaguely of hypocrisy, and her disinterested observation feels interested, especially to those who distrust the notion of objectivity in the first place. To the modern mind, the translation of a scientific ideal into a moral one produces reverberations that call the idea of "objectivity" into question.

Nevertheless, *Mansfield Park* belongs very clearly to the tradition of nineteenth-century realism, and that is partly because the weak and inadequate Fanny has these powers of clear perception. Pre-Darwinian science, and the epistemology that sustains it, are compatible with the kind of realism we find in, say, Richardson, as well as in Austen. But realism has many faces, and it is subtly a very different literary mode once the contest between design and chance, so much overbalanced toward the former in Richardson and Austen, begins to dip in the other direction, and once the categories of order that Austenian "realism" represented begin to dissolve.

The representational texture of *Mansfield Park* joins with a particularly emphatic symbolic drama; like her other novels, this is "realist," and yet it uses its precise observation of social movement and scene to evoke a world totally meaningful. No possible nuance of meaning is sacrificed for realist (and Darwinian) abundance, so that even the landscapes signify. The contrivance of the quasi-symbolic episodes does nothing to diminish the force of the narrative, and it is generally agreed that the finest sequences are those dealing with the amateur theatricals, so rich in double significances for all the participants, and with the outing to Sotherton, which enacts the ultimate fate and significances of the characters, and establishes Fanny as the reliable observer.

One scene will serve as an indication of how such symbolic moments work within the convincingly representational frame. Playing Speculation, Mary and Edmund have been discussing his commitment to be ordained and to take a place at Thornton Lacey, which Mary has plans to improve. Edmund tells her that he must be satisfied to live within his small income. Mary turns back to the game, and secures William Price's knave "at an exorbitant rate." "There," she says, "I will stake my last like a woman of spirit. No cold prudence for me. I am not born to sit still and do nothing. If I lose the game, it shall not be from not striving for it" (p. 251). There is nothing particularly subtle about this: Mary's play in the game obviously reflects the way she will try to maneuver. The passage works, however, because Mary knows that she is talking about more than cards. She is helping the novelist turn event into meaning, as she reads her way of playing as her way of living. And her language is carefully designed to set her off against her true antagonist, Fanny. The "cold prudence" might be a way to describe Fanny's apparent behavior. Certainly, Fanny seems to have been born to "sit still and do nothing." But while Mary wins the card game, she loses the larger struggle with Fanny.

Fanny wins by sitting still; Mary loses by striving. The casually domestic scene links knowledge and power through Fanny, the "scientific observer." In the sequel Mary's loss of Edmund results from her failure to have seen clearly, to have understood the ideal order that governs Edmund's aspirations and Fanny's life; not seeing this ideal, she has no idea what is entailed in violating it, or how apparently minor detail can expose a moral vacuity. Edmund responds in horror at Mary's use of words—at her satisfaction in calling what Maria and Henry did "folly" (p. 441). (Edmund himself, from Fanny's perspective, is similarly misusing words, saying that Mary has been "spoilt.")

In any case, Edmund rejects Mary's striving to hold on because she has not sufficiently understood. He tells his story to Fanny, who has long resisted telling him what she knows, that Mary "loves nobody but herself and her brother" (p. 414), and that she and her brother have been corrupting each other and their friends. But Fanny sits still and waits for the moment when it is "natural" for Edmund to stop loving Mary and start loving Fanny. It is the moment, moreover, toward which the audience knows the whole novel builds. As author, Austen depends on her reader's awareness of this teleological movement.

Such sequences, meticulously presented, coherent with the realist texture of the novel, are characteristic. Austen's world in *Mansfield Park* resonates with meaning. The tightness of the structure and of almost every scene in the novel are essential to Austen's strategy of representation. The ideal of design that lies behind the representation is present *in* it. Characters reveal themselves through minutiae; the direction of the plot is similarly intimated; the moral significance is constantly part of the descriptive surface of the text, not merely in Edmund and Fanny's moralizing. The word "spoilt" manifests Edmund's weakness as clearly as "folly" marks Mary. In such a world, clarity of perception is a moral necessity.

Just as there are surprising similarities between natural theology and Darwinian doctrine, natural theology sustains in Austen a realism recognizably—I am tempted to say "genealogically"—related to its mid-Victorian forms. And one of the points of convergence (and ultimate difference) is in the scientific ideal of disinterest, with its inverse side, repression of desire. Fanny Price is in this respect almost the perfect representative of the realist mode. Science and realism deny desire, with its anarchic tendencies, its assertion of an unstable self across the limits of social or intellectual community. The conjunction between realism and repression, which Leo Bersani has investigated, can be located, I would argue, in the cultural commitment to the idea of "objectivity," which derives its sanction from science itself. Deferring to the object requires submerging the self; the world appears to be designed, life and self seem to make sense only when desire is denied.[13] Like the ideal scientist, Fanny Price watches from a distance, and the accuracy of her vision is a consequence of her forced detachment. Like the realistic novel, the scientist attempts to avoid the consequences of knowledge by disengaging from the object. Fanny, like scientist and realist, keeps herself safe by sitting quietly and watching.

After Darwin the ideal of disinterested observation came under increasing pressure. Arnold's celebration of a "best self" that transcends the limits of the practical world was, ironically, consonant with the aims of the positivists, whose cold imagination of the world he sought to counter; and it was as much a last ditch dug against the armies of socially sanctioned selfishness as the Bridgewater Treatises were against the secularizing of knowledge. Austen and Fanny's struggle with knowledge are an early manifestation of the nineteenth-century novel's increasing experimentation with point of view, its preoccupation with what I will be calling the "perils of observation." Fanny's "interest" in Edmund may distort her perception after all, but her whole presence in the novel accentuates the importance and the price of knowing disinterestedly. It is a long way from Darwin's theory, which withdrew consciousness from the ideal by deriving it from the lowest forms of animal life, while it transformed the human into an object of scientific investigation, thus making disinterest impossible. Davydd Greenwood points out that, for Darwin, "even the observer's point of view is not absolute. Darwin argued that all perception is relative to the material structures that do the perceiving. Thus even our knowledge of evolution is conditioned by the structure of our own perceptual apparatus, itself a product of evolution. The special creationists, by contrast, must believe that perception of the absolutes in the design that lies behind the world of variable appearances is possible and that the human mind is capable of it."[14] In some respects Austen herself, partly through her vicar, Fanny, partly through her own narrative strategies, participates in what Greenwood is defining as a "creationist" view.

Although to a degree the question of observation is already problematized in Fanny, in that she is both extraordinarily weak and subjected to intense questioning by all the characters, in the end she is shown to be, as Tony Tanner tells us, never wrong.[15] Of course, she is "interested." But her condition does what can be done to keep that interest from playing an active role. She still implies the possibility of perceiving accurately an ultimately unproblematic reality, for the human mind, however limited, can make sense of the world. The narrative is written from the perspective of someone who judges the irrationalities of a natural/social world against a rational ideal.

Through Fanny, Austen associates true knowledge with power, for the narrative finally denies the impending scientific separation between knowledge and feeling. Fanny, knowing above all others how to observe, knowing how to discipline the self into disinterest, and learning how to call things by their right names, is rewarded with what the narrative has made us believe she has always desired and makes us desire for her. Virtue is rewarded because virtue is inherent in the design of the world. The power to see clearly is allied to a moral virtue, better here called selflessness than disinterest. "It is always Fanny," says Tave, "who sees the entire process, who sees what others are doing when they themselves do not understand their own actions, sees the whole drama of their interaction."[16]

Only such complete and careful observation—of the sort one would ask from a novelist (or scientist)—can develop adequate resistance to the threats of the romantic imagination. The scientific analogue to Fanny's achievement is astronomy itself, and the consequences of taking astronomy as the model science are manifest in her, for whom precise and wide knowledge is coextensive with value. Fanny Price allows no Darwinian blurring of margins or imprecision of language.

Given the richness—the slow unfolding and complex perspectives—of what has preceded Fanny's triumph, the abruptness, certainty, lightness of the final chapter as it breathlessly if self-consciously gives Fanny everything she had dared not allow herself to wish, make for disturbance, seem to require a rereading of all that has past. Passive observation gives power; like the novelist in accuracy of judgment, Fanny possesses the novel. Its narrative is finally determined by what she desires. This is, of course, characteristic of Pamela's story, the Cinderella myth, and the conventional happy ending. But Fanny's translation to power is more emphatic and disturbing than even Pamela's; her insipidity and weakness emphasize the extraordinary power the novel invests in disinterested observation. The strategies of the conclusion suggest that underlying Austen's realistic program is an ideal of order and design that owes a great deal to the cultural assumptions underlying natural theology, as well. And yet the facility with which everything is resolved into the telos of the story suggests that those cultural assumptions are now more self-consciously and thus ironically or tenuously held.

At the start of the last chapter, it is "My Fanny," of whom the narrator has the satisfaction of knowing that she "must have been happy in spite of every thing" (p. 446). The narrator dismisses guilt and misery at the moment when Fanny, who has always felt sorry for her oppressors, particularly the villainous Aunt Norris, has the greatest occasion to feel for their losses. Yet at just this moment the narrator is careful to suggest, parenthetically, that perhaps Fanny was not as concerned as we might have expected: "She must have been a happy creature in spite of all that she felt *or thought she felt*, for the distress of those around her." What does that clause mean? Is this the last intimation of Fanny's innocence and childishness, or are we seeing Fanny ironically, the desires, so deeply repressed that she herself did not fully know them, now at last exposed?

The nature of the novel is transformed, the tender-hearted Fanny suddenly exposed, her desires in fact all-potent, for they are manifest in what actually happens in the story. Repressed desire turns out, after all, to *be* desire, and not disqualifying. Fanny has occupied the correct position, not because she has no desires, but because she has been able to discipline them in case they were not to be satisfied. Like a good scientific natural theologian—like Buckland or Whewell, for example—Fanny investigates only to find that the world corresponds to her ideal of it, a world that emulates her desires (which are the novel's creator's desires).

The last chapter is a brilliant set of displacements through which Fanny moves into the positions formerly occupied by her superiors and oppressors. But the narrative is so rapid and evasive on the retributive implications of the reversals that it sounds like parody of the conventional distribution of rewards and punishments. One reader even claims that "Fanny and Edmund finally emerge as monsters, if only because they overpower the Crawfords so completely; by the close of the novel the high-spirited villains seem pitiful and defenseless by comparison."[17] But it is not merely the triumph over the Crawfords. The last chapter reaffirms the structure of the world into which Fanny moved when she was ten, except that that world has been purged of its (realistic?) weaknesses and diseases: the leader of that world, Sir Thomas Bertram, assumes his full authority by way of a new consciousness of his earlier failures in inculcating the principles of that world and in ensuring appropriate supervision of its activities. Fanny's "monstrosity," if that's what it is, is the affirmation of the meaning of the place, Mansfield Park, a meaning that had been threatened because of the weakness encouraged by lack of adequate supervision and failure of authority, which made it vulnerable to invasion and subversion.

The design revealed fully in the final chapter (though, through literary tradition, fully anticipated almost from the start) reveals a world in which, as Tanner says, *every* narrative wish is satisfied. Henry Crawford is expelled, but more than that, he is shown to be intelligent and sensitive enough to regret the loss of Fanny. Even at the "moment" of the elopement and in the exercise of his sexual dominance he feels it, and after: "A very few months had taught him, by the force of contrast, to place a yet higher value on the sweetness of her temper, the purity of her mind, and the excellence of her principles" (p. 452). What could be sweeter? Mary, too, is expelled, but with the same sort of narrative twist with which the desires of the protagonist are peculiarly satisfied. If Henry regrets Fanny, so, symmetrically, Mary regrets Edmund. She was long in finding, the narrator delightedly tells us, "among the dashing representatives, or idle heir apparents, who were at the command of her beauty, and her 20,000£ any one who could satisfy the better taste she had acquired at Mansfield, whose character and manners could authorise a hope of the domestic happiness she had there learnt to estimate, or put Edmund Bertram sufficiently out of her head" (pp. 453–454). The tables are not only turned, with the rogues in love and the protagonists indifferent (well, perhaps, almost), but most important, the values of Mansfield Park, embodied by Fanny, which the Crawfords did all in their power to violate, have now convinced them sufficiently that their pain is exacerbated.

Equally satisfying, Fanny takes the place formerly occupied by Aunt Norris. Not only does she become mistress of the parsonage, but she does so when Aunt Norris herself is expelled, to look after the Bertram daughter whom Fanny replaces. Again, almost gleefully, the narrator suggests Aunt Norris's fate with Maria: "Shut up together with little society, on one side no affection, on the other, no judgment, it may be reasonably supposed that their tempers became their mutual punishment" (p. 450). Fanny has never expressed hostility either to Aunt Norris or Maria, but the conclusion enacts it and the vindictive pleasure we are not explicitly encouraged to feel. The pleasure on Fanny's behalf is further intensified because Sir Thomas himself feels the "supplementary comfort" of Mrs. Norris's removal from Mansfield. The judgment, of course, vindicates Fanny, and places Mrs. Norris—both morally, and, as a consequence, physically—outside the paradise of Mansfield Park. The moral, in *Mansfield Park*, translates consistently into the physical.

Austen's recklessness with the qualifications and contingencies of what we take to be realist technique at this stage are emphasized by the abrupt, almost brutal pleasure implicit in all these turns. This can be measured by what it takes to get Fanny to occupy the parsonage: the novel must kill Mr. Grant, which earlier it had self-consciously refused to do for the wishful young Tom Bertram. Mr. Grant, a healthy forty-five years old, is living evidence of the consequences of Tom's profligacy, which has meant the loss to Edmund of at least half his living. But by the end of the novel, Mr. Grant can be expected to die suddenly, "just after [Edmund and Fanny] had been married long enough to begin to want an increase of income, and feel their distance from the paternal abode an inconvenience" (p. 457). How are we to take this? It is obviously funny and allusive, for it isn't possible to miss the further reminder that we are in a novel hastening to a totally happy ending. The reminder does not diminish the significance of Fanny's moral position, but blatantly emphasizes it. Mr. Grant will not die for Tom, as a consequence of expressed desire, but he will die for Fanny's unexpressed need, because Fanny controls the narrative.

The displacement of Mary is completed by Edmund in a passage in which the self-reflexive casualness of tone is most conspicuous. Edmund's dramatic last scene is perhaps the most "dialogical" in the novel, partly because it is narrated entirely by Edmund to Fanny so that the limits of point of view constantly imply possibilities Edmund does not register. We know the satisfaction to Fanny in Edmund's recognition of Mary's failings, and his decision not to turn back to her, but we can—almost must—consider the possibility of alternative readings, consider, that is, what the novel would seem like if Mary, not Fanny, were the presiding consciousness. This possibility is further developed because the scene toward which the whole novel has apparently been moving, when Edmund recognizes his love for Fanny, does not happen. It is, one might say, novelized, as the narrator turns directly to her readers: "I only intreat every body to believe that exactly at the time when it was quite natural that it should be so, and not a week earlier, Edmund did cease to care about Miss Crawford, and became as anxious to marry Fanny, as Fanny herself could desire" (p. 454). The perfunctory and comic quality here emphasizes Fanny's power over the narrative, and the novel's refusal to accept the alternative perspectives it has vividly dramatized. At the same time, the impossibly casual reliance on the reader's understanding of the telos deprives Fanny of the "scene" that all readers might have expected. The strategy both affirms Fanny's power and questions it, falls in line with teleology and intimates that the subversions of the Crawfords have a power beyond their literal place in the text.

All of the other displacements of the happy ending come under the cloud of this irony. They are almost too much. For Sir Thomas, "Fanny was indeed the daughter that he wanted" (p. 456), made literally so by the marriage to Edmund. Fanny replaces Maria, and Fanny's true sister, Susan, replaces Julia, so that the wicked sisters who oppress Cinderella at the start of the novel are expelled. Maria, meanwhile, gets (for a moment) the man Fanny has fled and the punishment she has escaped. Fanny even replaces Mrs. Norris as the woman with the greatest authority and she leaves her oppressors, Maria, Julia, Mrs. Norris, and Henry Crawford to face the punishment consequent on their violation of the standards of Mansfield Park.

The novel's peculiar comic insistence on its fictionality makes problematic what the story thematically affirms. While it fits neatly into the model of natural theology, it implies strong distrust of the too easy ordering of painful complications. If all pain is designed in the long run for the benefit of mankind, what are we to think of those who suffer that pain? The disruptive forces, so easily displaced by Fanny, remain outside the world of the novel as a threat requiring its formal control. The very strength and clarity of Austen's affirmation of order implies its perilous condition, its increasing vulnerability to the pluralist, irrational, amoral forces that will later take shape in Darwinian theory.

But with all of its latent questioning of its own affirmations, the conclusion does resolve the complications of narrative entailed by such forces: (1) Its many varying perspectives are essential not because multiplicity of perspective gets us closer to the truth but because it is so difficult to see with clarity, to avoid the blinding force of one's own interests and desires. (2) Seeing clearly depends upon repression of desire, and refusal of engagement with the world. (3) Precise observation not only reveals the true structure of the world, but is a means to power over it. (4) The world revealed to careful observation is ordered, regular, meaningful, designed, teleological. Knowledge, then, is consistent with feeling and value. (5) The rationality of the world is manifest in the clarity with which language can denote and define; language that obscures not only obstructs accurate observation but in sanctioning disorder implies moral weakness. It is dangerous enough to need perfunctory dismissal.

All of these ideas find focus in problems of classification, of definition of boundaries, of precise discriminations. Austen's novels are much concerned with boundaries, and they are preoccupied with the minutiae of social decorum, and with a more ideal propriety that is only roughly manifested in actual society. Violating the rules, or blurring the margins, is not merely a social but a moral mistake. Snobbery and false gentility are evidence of blurred boundaries, the business of characters like Mrs. Elton or Mrs. Norris, who aggressively affirm their new and shaky places in the social hierarchy. The novels are centrally concerned with sharp verbal and social definitions because these, like the natural theology of the Bridgewater Treatises, are the conditions of existence for the established moral and social order. Confused or confusing language misrepresents a natural category and becomes a moral as well as an epistemological failure. *Mansfield Park* can be seen in one sense as a novel about the significance of defining and understanding established categories.

Appropriately, it begins with a flurry of discriminations showing how social hierarchy and, more interestingly still, categories of moral action are being violated. These violations are the narrative cause of all that follows, and they entail a series of rewritings in the lives of the next generation, particularly, of course, in the life of Fanny. The discriminations are written into a system that remains closed, whose structure is to be explored and questioned. Fanny's narrative embraces and reworks the options taken by the three sisters. She enters the world of Mansfield Park, occupied by Maria Ward; she returns to Portsmouth, where Frances Ward has "fallen"; and she concludes in the parsonage, where Mrs. Norris had found her intermediate place. In each case Fanny's career illuminates and revises the choices of the three sisters.

The initial sentence announces immediately the preoccupation with social categories: Maria Ward, with "only seven thousand pounds, had the good luck to captivate Sir Thomas Bertram . . . and to be thereby raised to the rank of a baronet's lady." Maria's uncle, furthermore, "allowed her to be at least three thousand pounds short of any equitable claim to it." Here three categories are crossed: there is the simple fact of money, which becomes an index to the quality of the marriage for the community around; there is the matter of class, for Maria's shortage of three thousand pounds has to do with both financial parity and the power to buy into the aristocracy; finally, there is the indication of sexual attraction. Sir Thomas is "captivated." The collocation of these categories brings "good luck" to Maria. Irony emerges from the matter-of-factness. The abruptness of the prose and the absurd juxtaposition of pounds, luck, and a captivating appearance diminish each.

The irony of "good luck" is important because Austen's world has no room for chance. Maria aspires to the condition of "Lady," and in accepting her power to "captivate" Sir Thomas—in however vegetable a manner—she chooses an advantageous marriage quite coldly. The choices of the three sisters have moral significance as in a folk tale, and the judgment of the second sister is even more clear, for she makes a match which, "when it came to the point, was not contemptible" (p. 41). Lacking anyone as eminent, and rich, as her brother-in-law, she "found herself obliged to be attached to the Rev. Mr. Norris." That is, she begins her "career of conjugal felicity" with "very little less than a thousand a year." Here we have a similar juxtaposition of incongruous categories, a juxtaposition whose ironies are dependent on a subtext of calculating intentionality: Mrs. Norris's marriage was "not contemptible"—but by whose standards? "Conjugal felicity" announces itself as a cliché that answers a publicly necessary description. Yet Mrs. Norris does stay within the boundaries of respectability and can remain in communication with her richer sister. Having found their husbands by artifice, they quickly learn to take their conditions as "natural." The unnatural mixing of categories is quickly naturalized, the old class forms affirmed vigorously.

Bitterly funny as these opening sentences are, the world of careful discriminations and of prudential marriages is a thoroughly unpleasant disguise of inauthenticity. Yet the prose implies a moral authenticity that is the measure of the ironies. If there is justice in the argument that the novel idealizes the "place" that is to become the locus of the marriages, and of the way of life of the landed gentry that was quickly passing from dominance, it does so within the context of a strong critique of these things. The mixture of categories that allows for the comic ironies is characteristic of Austen's prose in almost all of the novels: it reflects the moral imperative of clear observation, precise classification, accurate naming. When categories are crossed or blurred, moral disaster is inevitable.

Fanny's idealization of the place in no way diminishes its real weaknesses, but suggests an ideal perspective from which it can be observed. The disruptive forces are "incorporated" into the novel by way of an implicit romantic critique of the ideals of civilization—the idea of Mansfield Park has been almost entirely corrupted. Sir Thomas's whole family is either silly or cruel, or both, and the two idealized men—Sir Thomas and Edmund—are not exempt. Most obviously, Sir Thomas sanctions the marriage of Maria to Rushworth, knowing that Rushworth is a hopeless fool but assuming that the financial rewards will make her happy. He continues with his daughter the pattern established by his wife and their sisters. Among many sharp and bitter sentences in the novels, there is none more biting than this, describing Maria's preparation for her marriage: "In all the important preparations of the mind she was complete; being prepared for matrimony by an hatred of home, restraint and tranquility; by the misery of disappointed affection, and contempt of the man she was to marry" (p. 216). The summary of Sir Thomas's "reasonings" on the subject are, in the context, brutally judgmental. Maria's feelings, he thought, "probably were not acute; he had never supposed them to be so; but her comforts might not be less on that account, and if she could dispense with seeing her husband a leading, shining character, there would certainly be every thing else in her favour. A well-disposed young woman, who did not marry for love, was in general but the more attached to her family, and the nearness of Sotherton to Mansfield must naturally hold out the greatest temptation, and would, in all probability, be a continual supply of the most amiable and innocent enjoyments" (p. 215). Despite such extraordinary human and paternal failure, Fanny's final valuing of Sir Thomas's authority and the system makes sense, not only because it empowers her, and because the alternatives are so obviously wretched, but also because she has managed in her own story to demonstrate the authentic values of the system when it is regarded properly.

Although almost all registration of feeling in the novel comes filtered through clichés and ironies that suggest the existence of a system both deeply engrained and humanly contemptible, the first alternative to the system comes ironically, to justify it after all. One might at first have thought that the young Miss Frances, in rebellion against her family, should have been rewarded for following feeling and for rejecting the system. "To disoblige her family" she marries a lieutenant of the marines, and "she did it very thoroughly." But she pays the price: a breach between the sisters develops, "the natural result of the conduct of each party, and such as a very imprudent marriage almost always produces." The third sister, too, makes her choice, and again the narrative voice places it in the context of the unspoken authenticity of a system in which categories do not mix.

The ironies and subversions implicit in the opening paragraphs do not, then, tend to the rejection of the system, but rather to its affirmation, since the system itself supplies the ironic perspective. At the same time, the ironies imply a tension between the necessity of the system and the often painful consequences of living by its rules. Frances Ward makes a mistake and pays the consequences; her rebellion earns her nothing but a difficult life and isolation from her family. The romantic thrust of feeling is no more authentic and virtuous than Maria Ward's emotionless marriage to Sir Thomas. The opening paragraphs set up the tension between feeling and prudence. The categories must not cross. And the ironies are far too complicated to allow the cutting attack on the systematic hypocrisy of the marriages to reflect on the system. Where the crossing of boundaries is only of the sort we have in Maria's marriage to Sir Thomas, the incongruities and fortuitousness are comic. The willingness to allow feeling to drive one's choice to "disoblige" and thus fall from one's class, however, must have immediate and serious consequences.

Jane Austen does not authorize the romantic ideal of the "natural." Civilization, warmed by feeling—a model of humane rationality—provides the principles of order. The very tough sequence in Portsmouth suggests how deeply unsentimental Austen can be, how much—against what she takes as a largely spurious ideal of spontaneous feeling—the principles of decorous order matter. They prevail in the reading of the family as well as in the narrative as a whole; as later eighteenth-century natural theology belongs to a rationalist and utilitarian tradition, so *Mansfield Park*, in its emphasis on prudence, common sense, and practicality, implicitly justifies Fanny's revulsion from her own family. There is no celebration of poverty, no indication that the marginal life is pleasant or that there is either dignity or worth in the crowded household. Fanny may seem like a prig, but the narrative voice endorses her: "Fanny could not conceal it from herself," but her family's home was "in almost every respect, the very reverse of what she could have wished. It was the abode of noise, disorder, and impropriety. Nobody was in their right place, nothing was done as it ought to be. She could not respect her parents . . ." (p. 381). Natural parents do not by virtue of their "naturalness" inspire or deserve affection. Having, through none of her own doing, crossed the boundaries of class and having understood the virtues of upper class values, Fanny is not interested in crossing back. Portsmouth simply does not inspire guilt; it is a mistake to be forgotten, and the narrator pauses no longer than Fanny in concern for the Prices: "After being nursed up at Mansfield, it was too late in the day to be hardened at Portsmouth" (p. 404). Frances Ward's choice is judged by Fanny's inability to accept it.

Fanny's position must be seen in the context of the discriminations suggested in the opening paragraphs. In keeping with the book's preoccupation with social definition, Fanny's entrance into Mansfield Park is determined by a preliminary discussion of the problem of her social status; where, in the subtle discriminations of class created by the situation of the three sisters, should she be located? As a *niece* of Sir Thomas, she is obviously entitled to comfortable treatment; but as a *poor* niece, she must be prevented from thinking of herself as entitled to the privileges due to Sir Thomas's daughters. On these matters Mrs. Norris is to watch closely to ensure that Fanny never forgets her "place." The preliminary discussion, like all major moments in the novel, is carefully articulated to emphasize the question of social status: in addressing the issue characters quickly reveal their worth. Sir Thomas takes the possibility of bringing Fanny to Mansfield gravely: "It was a serious charge;—a girl so brought up must be adequately provided for, or there would be cruelty instead of kindness in taking her from her family" (p. 43). But Mrs. Norris immediately assumes that the real question is how to prevent Fanny from attracting Sir Thomas's sons. Her theory is that growing up in the family would be precisely the thing to prevent this (which tells us, as experienced readers of novels, that it is what will happen). But Mrs. Norris is not wrong about Sir Thomas's fears. His place in the system requires that he be practically (and unsentimentally) concerned with his responsibility to the girl—to ensure that Fanny is secured "the provision of a gentlewoman" (p. 44). Yet, of course, to sustain the system he must see to it that Fanny understand her station. Sir Thomas is not exempt from the continuing ironic commentary on the ostensibly ideal Mansfield Park. "There will be some difficulty in our way," he says, "as to the distinction proper to be made beween the girls as they grow up; how to preserve in the minds of my *daughters* the consciousness of what they are, without making them think too lowly of their cousin; and how, without depressing her spirit too far, to make her remember that she is not a *Miss Bertram.* I should wish to see them very good friends, and would on no account, authorize in my girls the smallest degree of arrogance towards their relation; but still they cannot be equals. Their rank, fortune, rights, and expectations, will always be different" (p. 47). The attempt to sustain such delicate discriminations, without the most rigorous and intelligent supervision, proves chimerical.

When Fanny returns to Portsmouth, the arbitrariness of the social distinctions among the sisters is emphasized again:

> Of her two sisters, Mrs. Price very much more resembled Lady Bertram than Mrs. Norris. She was a manager by necessity, without any of Mrs. Norris's inclination for it, or any of her activity. Her disposition was naturally easy and indolent, like Lady Bertram's; and a situation of similar affluence and do-nothing-ness would have been much more suited to her capacity, than the exertions and self-denials of the one, which her imprudent marriage had placed her in. She

> might have made just as good a woman of consequence as Lady Bertram, but Mrs. Norris would have been a more respectable mother of nine children, on a small income. (pp. 382–383)

In complicating the categories and dramatizing the difficulty of keeping them pure, Austen suggests how fragily, delicately, artifically constructed is a world that answers to a natural-theological description. The romance pattern that determines the shape (and finally the texture) of the ostensibly realistic narrative keeps it alert to the complications it must finally refuse, but cannot really dismiss. Sir Thomas affirms the rigidity of social classifications and the nuanced sense of what is required to sustain them; ironically, he will eventually find himself making Fanny his daughter, partly in reparation for the inadequacies of his own children. The discriminations destroyed by corruption must be reinstated if the ideal of civilization is to be sustained.

The questions of category raised by the opening marriage are most elaborately explored in Fanny's relations with Henry and Edmund. Forced to choose between a marriage that will provide her wealth and station, and a love that might issue in nothing at all, Fanny rejects mere prudence. Although the categories must not cross, the situation cannot be reduced to that rejection. While Fanny loves Edmund, her resistance to Henry must be justified in its own terms, so that much depends on the case the novel builds against Henry, the evidence that his disruption of the conventions of love is sufficient ground for rejection. Most critics have found the case wanting.

They simply don't like Fanny and Edmund; they do like Mary and Henry Crawford.[18] And they rarely credit the novel for knowing what it is about. Here, for some reason, Austen has gone awry. But if we assume that *Mansfield Park* is seriously about the kinds of issues I have been intimating through the use of natural theology and Darwinism, Mary and Henry are critically disruptive forces, challenging the rule of law and the very ordering principles that construct civilization and give to the minutest details of life and language their meaning. Witty and charming as they can be, they represent forces that would reduce moral to biological issues, and they imply an irrationality in social relationships that reanimates the forces of the French Revolution itself.

Fanny is indeed a weak vessel against such force, and Austen herself builds the case against Fanny's priggishness. The narrator insists on the authenticity of Henry's feeling, his vitality, his respect for Fanny, and suggests that had Henry persisted Fanny would eventually have had to yield, and that the consequences of yielding probably would not have been bad. Henry alone visits Fanny in her desolation at Portsmouth, and Henry, not Edmund, endures discomfort for her, just as it was he who helped get William a commission. Fanny herself is aware that if she were to marry Henry, she could count on "the probability of his being very far from objecting" to her bringing Susan with her (p. 410). Yet Fanny will not concern herself with the practical benefits of marrying him, even when she is at Portsmouth. Her choice must test out the temptations that led her aunts to their marriages, but escape their taint. Fanny clings to her preference, perhaps "imprudently," perhaps—from a modern point of view—foolishly; but she loves Edmund, the one person in her first years at Mansfield Park who protected her and was sensitive to her needs.

In insisting on Henry's worthiness, modern critics have been far more sentimental than Jane Austen. The appropriate context in which to see Henry's relation to Fanny is more that of Laclos than of Richardson. Henry's relation to women, however sanitized and Anglicized, is straightforwardly a relation of power. Fanny has correctly observed this—to her pious horror—in his relation to Maria and Julia. The reader knows, as she cannot, that Henry has determined, like the Valmont of *Les Liaisons dangereuses* in relation to Madame de Tourvel, to demonstrate his power, to make desire win over piety and convention. Henry, like Valmont, acts out the same tradition of eighteenth-century commitment to the reductive materialism of science—to the despiritualizing of nature and the relativizing of morals—that we find in de Sade and Laclos. Henry's behavior through the first part of the novel represents the seductive disruptiveness of social and moral traditions, that celebration of desire and exercise of power that has its continental parallel in the French Revolution. When Henry seems to reverse himself and finds himself valuing those feelings he had spent the preceding years attempting to disrupt, the novel further suggests Austen's revulsion from the materialism and commitment to the categorical division of matter and morals that provides the shaping natural-theological order and stability of the narrative.

Henry's language as he formulates the idea of breaking down Fanny's defenses clearly belongs to the materialistic and revolutionary tradition of Laclos. It is, first of all, because she is physically worth the attack that Henry conceives of the idea at all: "I used to think she had neither complexion nor countenance; but in that soft skin of her's, so frequently tinged with a blush as it was yesterday, there is decided beauty; and from what I observed of her eyes and mouth, I do not despair of their being capable of expression enough when she has any thing to express" (p. 239). Her physical attractiveness is all the more seductive to Henry because she seems so impregnable, has been so difficult to charm: "Never met with a girl who looked so grave on me! I must try to get the better of this. Her looks say, 'I will not like you, I am determined not to like you,' and I say, she shall" (p. 240). And here is the wittier and yet more worldly Valmont, talking to the Marquise de Merteuil:

> The Love who is preparing my crown himself hesitates between myrtle and laurel, or rather he will unite them to honour my triumph . . .
>
> You know Madame de Tourvel, her religious devotion, her conjugal love, her austere principles. That is what I am attacking; that is the enemy worthy of me; that is the end I mean to reach . . .
>
> I have but one idea; I think of it by day and dream of it by night. I must have this woman, to save myself from the ridiculous position of being in love with her—for how far may not one be led by a thwarted desire.[19]

Moreover, with all Henry's charm, the assertion of power entails the ability to inflict pain, for pain is part of what gives pleasure to dominance. Henry's treatment of Maria Bertram had already reflected both the power and the pleasure. He claims he doesn't want to make Fanny unhappy, yet he laughingly finds himself saying: "No, I will not do her any harm, dear little soul! I only want her to look kindly on me, to give me smiles as well as blushes, to keep a chair for me by herself wherever we are, and be all animation when I take it and talk to her: to think as I think, be interested in all my possessions and pleasures, try to keep me longer at Mansfield, and feel when I go away that she shall be never happy again. I want nothing more" (pp. 240–241). Valmont will not be satisfied until Madame de Tourvel *gives* herself to him; and he tells the marquise with much pleasure that "the time will come only too soon when, degraded by her fall, [Madame de Tourvel] will be nothing but an ordinary woman to me."[20] Which, of course, is what happens to Maria.

Despite the enormous differences between the two books, their similarities here are striking: sexuality, wealth, and power are one. The blurring of these distinctions is intolerable to Fanny. Sex and money are dangerous to the order she so much values because both blur social boundaries and moral categories, and because with them, power has no basis in rational order. Within the tradition of enlightened self interest, money and sexuality become the primary sources of "interest" in society and therefore the primary obstructions to knowledge of the true order of things. Morality is reduced to biology, and Science is confused with moral law. Mere sensual desire threatens to explode into revolution or into something like Henry's elopement with Maria. A purely calculating Henry would never have eloped with Maria, but his action becomes entirely intelligible when viewed as a consequence of the irrational energy of sensual desire and lust for power. Henry's calculations are only the strategies that make it possible for him to take great risks, the condition for real power (at one point he envies William for having as a sailor taken so many real physical risks).

The rationalist-materialist dismissal of spiritual value elevates the irrational; and the "calculating" Henry is, as we are told at the end, a slave to his own pleasures, and to the overriding demands of his desire for power. He determines to make Fanny love him because she refuses to do so. When he sees Maria again, he confronts the same kind of resistance, and the language once again recalls the tradition of Laclos:

> He saw Mrs. Rushworth, was received by her with a coldness which ought to have been repulsive, and have established apparent indifference between them for ever; but he was mortified, he could not bear to be thrown off by the woman whose smiles had been so wholly at his command; he must exert himself to subdue so proud a display of resentment; it was anger on Fanny's account; he must get the better of it, and make Mrs. Rushworth Maria Bertram again in her treatment of himself.
>
> In this spirit he began the attack . . . (p. 452)

Thus Fanny's refusal of Henry's charms is a resistance to the power of sex and money, justified by the materialist cruelty that modern readers have become too sentimental to take seriously. That Henry is capable of good feelings, of generosity, vitality, charm, has nothing to do with Fanny's relation to him. (Valmont has these qualities, too.) She sees what he really is—a disruptive force in every respect, one who applies to human relationships the rule of biological law, denying the boundaries that define social and moral action. He exploits the material to gain power. It is not quite true to say that Fanny remains detached, since we are told that her power to hold off Henry derives from the previous engagement of her feelings to Edmund. Detachment, of a sort, saves Fanny, but *Mansfield*

Park comes close to positing a world totally subject to the rule of brute nature, of forces that disrupt the culture and civility through which humans transcend their material base. Only the devices of narrative, the return to the patterns of myth, allow for the preservation of the sharply defined, designed, and teleological world that the novel attempts to affirm.

Fanny is the figure endowed with the power to resist that threat. She mediates, like natural theology itself, between naturalistic observation and teleology. The natural is not, as the Bertrams imply, at odds with meaning and rationality, merely brute material, but totally consonant with intelligence, teleology, design. Yet to give Fanny such power, Austen must make her recognizably mythic. Thus, in her fairy-tale weakness and passivity, she becomes an observer of the disruptive forces that threaten to attract and engage. At Sotherton she watches the couples escape into the "wilderness"; in pain she observes Edmund and Mary; all too knowingly she observes Henry and Maria. Mary and Edmund each come to her to rehearse their love scene for the amateur theatrical as she tries to escape into quietness. In her very silence and stillness she acquires the opportunity and capacity to observe, and determines the novel's shape by denying her presence.[21] "We see Fanny literally 'at the still point of the turning world,' " says D. D. Devlin, "as, in a series of complicated movements, all the others move round her while she sits still."[22] Her stillness is implied even in her passive entrance into the novel by invitation from Sir Thomas.

Such narrative stillness does make it appear that through Fanny, *Mansfield Park*, as some critics have not unfairly noted, speaks for repression and negation, for fixity and enclosure, for caution and routine as opposed to the exhilaration of risk and change and liberated desire. But by incorporating the potentially revolutionary tradition of sentimental materialism into its very center, and by dramatizing the fundamental corruption of the place and the system it asks us to take as ideal, *Mansfield Park* makes the contention between objectivity and subjectivity, realism and imaginative license, repression and freedom, the focal point of its drama. Fanny resists her feelings, does quiet battle with the aggressive and world-making Mary. Imaginatively, that is to say, the novel makes clear the stakes in the contest over the scope and rights of the imagination within civilization.

It implies the possibility of keeping (or bringing) together the empirical validity of representation with the shaping power of imagination. Fanny's precise vision has to do with her power to read the present in terms of an invisible ideal—that authenticity that governs the narrator's ironies and sharply defined language. She is besieged by characters who have no such ideal frame, no natural theology, as it were, to guide their perceptions of the minutiae that make up experience, and for whom the satisfaction of egoism and the free play of desire are primary objectives.

Fanny becomes against them the minimal heroine, determined to resist those anarchic forces of desire which threaten to disrupt the stability and quiet of Mansfield Park. She is thus pitted against both those residents of Mansfield whose undisciplined desires threaten to destroy it and, in particular, those outsiders whose charm and energy exploit all of the internal weaknesses. As realism resists the threats of an undisciplined imagination to make a world with no basis in empirical nature and society, so Fanny is the heroine of resistance to the undisciplined fantasies of the Crawfords, who try to transform Mansfield Park into a theater, and Sotherton into an "improved" estate.[23]

Fanny's observations are safeguarded by ideal ordering principles, and thus her resistance allows her novel to affirm an intelligible and ordered reality. The almost perfect "realistic" embodiment of realism's "fear of desire," Fanny reflects realism's basis in a metaphysic that, in the years following *Mansfield Park*, was increasingly exposed and rejected because of developments in positive science. Her world is designed and sustained by a natural and spiritual order, and her success in merely waiting depends on the "real" existence of that world. It is not made by the observer, but preexistent and discoverable. If there is an "epistemology" to be educed from Jane Austen's work, it is this.

But Fanny as heroine is radically different from any of the major Victorian heroines of "realistic" novels. The pattern for such heroines is, in its French incarnation, Emma Bovary, who is seduced both by her false literary dreams and by her own sexual urgency. Or in England we have Maggie Tulliver, similarly misguided by the dreams of romance, similarly passionate. Both die because the world is not like their dreams and cannot respond to their desires. Their narratives are the restraining wisdom; they are exemplars of the disparity between desire and reality. Fanny, however, already contains the restraints that later writers must build into their narratives. She does not risk her feelings.

This difference is particularly important for what it implies about the "reality" to which the heroines do or do not conform. Fanny's restraint allows the narrative to imply a world rational and designed, though—as I have been arguing—deeply threatened. It conforms to Fanny's moral superiority, and it makes sense. But in both *Madame Bovary* and *The Mill on the Floss*, the heroines' failure to know what the narrators know leads to disastrous violence. The heroines' undisciplined passion—however much, as Bersani says, it is feared by the forms of the fiction in which it is embodied—manifests itself in the violence of an irrational world. Certainly, the world *after* the catastrophes makes no more sense than the world before. Flaubert's repugnant Homais thrives. "Nature repairs her ravages" but does not make Maggie's fate any more meaningful. It could all happen again. That is to say, Austen's realism (consonant with the world described in natural theology), resists the consequences of the later realism (consonant with Darwin's antiteleogical vision) in which the moral and material are severed.

The passivity of Austen's realistic ideal is antithetical to the view that nature is half perceived, half created. But this is not to suggest that the novel endorses a merely passive relation to "things as they are." Austen does not admire passivity—the most effective commentary on it may be the wonderfully vegetable Lady Bertram, whose "good luck" initiates the novel. Real strength in Austen's novels resides in the power of choice, so that the dramatic center of the narrative is in Fanny's ability to resist conventional wisdom, in refusing to perform in the play, for example, and more important yet, in her refusal of Henry Crawford.[24]

Fanny's triumphs make sense, and are not inconsistent with the book's (or Austen's) usual directions. She may seem like a hypocritical, passive-aggressive creature as she moves through the last pages profiting from everyone's loss and pain, but it is the passive Fanny, as Tave shows, who saves Mansfield Park and the order it represents.[25] The novel's apparent collapse into her arms and silent needs has, then, been earned by the peculiar powers of passivity and clarity of observation that she represents, and by her acceptance of the reality of an authentic order behind the weaknesses and inadequacies of the material world for which the Bertrams speak and on which narrative must concentrate. As Duckworth puts it, "there is a natural moral order stemming from God," and it is the responsibility of the individual "to maintain faith in 'principles' and 'rules of right' even when these are everywhere ignored and debased."[26] Fanny sustains that faith, locates the "natural moral order," and clarifies the discriminations so erringly made on the first page of *Mansfield Park*.

Within the narrative of her triumph, the details resonate with meaning; chance events turn out to be the consequence of moral choice, and significant. Austen's art provides most of the qualities that would be developed in later realism —the power to explore human consciousness and move flexibly in and out from the individual mind to the narrator's ironies,[27] the easy command of details of everyday living, the affirmation of representation and the possibility of objectivity. But the firm presence of the "natural order" makes irony and description, detail and moral meaning, compatible in ways that later fiction could not duplicate. Austen's novelistic games at the end are not attempts to subvert the ordering ideas that make novels so comfortable and life so painful. Contrast Thackeray's melancholy playfulness in *Vanity Fair* or *The Newcomes.* Austen's playfulness manifests a happy ease that a novel as novel, simplified from life as it may be, reflects life in its ordering power, in its design. The break from realism into self-reflexiveness is a direct exposure of the ordering principles that guided *Mansfield Park* from the start, even where it most effectively demonstrated that they were threatened.

The threats will be more fully and sympathetically explored later in the century. Fanny's is the temporary power of a realism that reads the world as though it can be translated into language, and as though if we could only rid ourselves of personal limitations, it would be seen as meaning intensely, and meaning good. The tenuousness of that position can perhaps best be represented by Fanny's weakness and by the extraordinary appeal generations of critics have found in precisely those forces that would disrupt it. If Mary were the true heroine, *Mansfield Park* would be a post-Darwinian novel.

4

Darwin's Revolution: From Natural Theology to Natural Selection

IF *Mansfield Park* reflects a world rather like Whewell's, it confronts dramatically threats that we might associate with Darwin's. And it is necessary, at last, to look more closely at how Darwin described his world, in what ways it undermined Fanny Price's and made Jane Austen's narrative voice, with its precise ironies, a voice that nineteenth-century fiction could not sustain again. Darwin grew up in Austen's world, but however much he needed and used what he learned from natural theology, his theory became a deliberate inversion of it. Such an inversion challenged the essentialist assumptions of Austen's and Whewell's language and inevitably affected how "reality" could be imagined, how stories could be told. As Beer has argued, even the language of natural history that Darwin inherited was on Paley's side—intentionalist, essentialist, and anthropomorphic;[1] moreover, it implicitly endorsed the analogical thinking that is the foundation of Paley's argument.

I have already in the discussion of Whewell tried to suggest how the shadows of the natural-theological argument—particularly its language and its forms—stretch across the pages of the *Origin*. In one sense the theory of natural selection merely provides a different answer to the natural-theological question of *how* organisms adapt. But in the very process words like "organism" and "adaptation" and "species" get redefined by being plunged into history. Natural theology was an implicit defense of the way things are—a theodicy. Natural selection came to be often used to defend the way things are without invoking God, but the dangers of its procedures and of its secularization were always near the surface. In the *Origin* Darwin tried to avoid extending his biological explanations into social and moral questions, but the extension was unavoidable and he made it himself in *The Descent of Man*. From the start

Darwinism made the human a part of the natural world and subject to scientific analysis.[2] And the tendency, vestigial from natural theology, to believe that the natural is meaningful and value laden was no inevitable part of that treatment of the human. When Darwin subverted natural theology he opened the door to a thoroughly amoral universe that he, with his natural-theological training, could at least partially and temporarily fend off.

The transformation of natural theology into natural selection entailed a complex and imaginative vision and argument, constructed from the very materials it would conclude by subverting. To talk of "contrivances" by which adaptation was accomplished implicitly invoked the argument from design; yet Darwin insisted on the merely metaphorical nature of such language. By the end, he would deny both perfection and ideal classification: there is no closure in the system of nature, for the world is in constant process. Adaptation is but for the moment. Equally important to his whole argument, Darwin believed that teleological explanation, because it led to the end of inquiry by invoking divine intelligence to explain whatever was not apparently lawlike, inhibited scientific investigation.[3]

The consequences for literature were immense: Darwin was in effect changing the way his culture could think. The anti-Aristotelian edge of modern thinking developed from the Darwinian insertion of classification and law into time.[4] And the resistance to sharp and permanent definition seemed to infect the practice of fiction, as well. In the transformation of Austen's kind of domestic fiction into the kind that dominated during the reign of Victoria, there is a parallel to the shift from natural theology to natural selection. The language, the comedies of manners, and the preoccupation with class and inheritance seem generically Austenian. But the form and the directions undergo subtle change.

The change is part of a romantic transformation of Enlightenment ideas and sensibilities. Both Austen and the young Darwin are part of an Enlightenment tradition that believes in the supremacy of rational explanation. But there was a strain of romantic exoticism in the young Darwin, who was inspired partly by his reading of Humboldt into dreams of tropical lushness: "All the while I am writing now my head is running about the Tropics: in the morning I go and gaze at Palm trees in the hot-house and come home and read Humboldt: my enthusiasm is so great that I cannot hardly sit still on my chair . . . I never will be easy till I see the peak of Teneriffe and the great Dragon tree; sandy, dazzling plains, and gloomy silent forest are alternately uppermost in my mind . . . I have written myself into a Tropical glow."[5] The voyage on the *Beagle* was provoked as much by vague youthful dreaming as by scientific aspiration. Temperamentally, while he admired novels like Austen's, which move inexorably toward happy resolution within resolutely rational structures, he seemed to belong to a world of Maggie Tullivers. His theory blurs the sharp edges of natural theological arguments and uses rationalist argument to break loose from rationalist structures. Cautious and romantic at once, he was no mere Enlightenment man; he was becoming very much a Victorian.

To describe the world that heated him to a "tropical glow," he had to invent a way to argue. The *Origin* persuades but in an almost perverse way: it consistently affirms its failure to make its case conclusively or to produce the clinching piece of evidence, abundantly recognizes the most powerful possible objections to the theory, avoids precise definitions of its major terms, sets its most important points in the conditional mode, and concedes, stunningly, "that scarcely a single point is discussed in this volume on which facts cannot be adduced often apparently leading to conclusions directly opposite to those at which I have arrived" (p. 66). The rhetoric of the *Origin* reflects the conditions of the dysteleological world it invokes: various, democratic, multitudinous, constantly transforming, intricately entangled. Read carefully, or perhaps with the eyes of a late twentieth-century critic, the *Origin* seems not only to participate in the celebrated "decentering" of humanity; it develops a quiet, erratic, but forceful attack on what we have learned to call "logocentrism." He found that his argument could not be developed unless he radically undercut the tradition that invested language, definition, and idea, with some ultimate and absolute authority. Before Darwin could even begin to make his case, the word "species" itself had to be disentangled from a timeless reality it was supposed to represent.

Until recently, it had been common to imagine Darwin much as he described himself in his brief autobiography: something of a plodder, just smart enough for what he had to accomplish, rather lucky—in an amateurish sort of way—to have been the first to gain credit for formulating a theory to which the whole of scientific culture was moving anyway. The style of the *Origin* can suggest a rather clumsy or casual relation to language, philosophy, and perhaps even science. But to see it that way is to succumb to its rhetoric, with all its disguises, evasions, and perhaps disingenuous invocations of traditional authority, such as "the creator," or "law."[6] If so apparently straightforward, modest, and clumsy a writer can produce so full an argument, the rhetoric implies, it must be true. But its occasional obvious felicities are no accident; the metaphors operate with anything but casual ambiguity; the consideration of objections to the theory is not merely remarkably honest—though it is that. When Darwin claims at the beginning of the last chapter that the book has been "one long argument" (p. 435), he means that exactly, and the large accumulation of data is part of a precisely argued thesis, held to with extraordinary intellectual agility and tenacity.

Darwin's cautious, charming, self-deprecatory style constructs a large trope of modesty, which is not merely personal. That is, modesty in self-presentation, refusal to advertise self, caution and detail of argument, all are characteristic of a Victorian style that may derive from the proprieties of class, but belongs as well to a theory of knowledge and knowing. Darwin's extension of biological explanation to the human subject is part of the movement of liberal democratic culture, initiated perhaps with Locke, that rejects traditional hierarchy for the natural man, assuming that all humans are potentially equal—anyone can achieve what the great figures of history have achieved given the right conditions. Mill, in his *Autobiography* similarly underplays his personal powers. Trollope, in his *Autobiography*, denigrates art by turning writing into a craft, like cobbling. I will discuss this aspect of Darwin's representativeness in more detail later, here it is sufficient to suggest that the strategies of Darwin's cautious and modest rhetoric are aspects of his theory and help account for its power to take hold, both within and outside the scientific community.[7] It spoke directly to the Victorian turn toward domesticity, and seemed to represent the novelistic conviction that the minute and ordinary could produce the large and heroic.[8] The style is the perfect complement to uniformitarianism.

But the writing of the *Origin* was anything but casual and amateurish. The story has been told frequently: Darwin first developed his theory after reflecting on his experiences on the *Beagle*, in the late 1830s, and the development is traceable in notebooks written at that time. By 1842 he had drafted a relatively brief "abstract," which first finds the language that will be most richly developed in the *Origin*. In 1844 Darwin drafted a much longer formulation, already 230 pages, which he also did not publish.[9] For almost two decades he continued relentlessly to accumulate data and arguments to support the hypothesis he already fully believed. By 1855, under the pressure of friends, particularly Charles Lyell and Joseph Hooker, who were fully informed about his theory, and out of fear that he would be anticipated, he had begun work on the "big" book in which he would at last make his views known to the world; but he had only completed eight and a half long chapters before he really was anticipated by A. R. Wallace in 1858.[10] The "hastily" written *Origin* was thus, in fact, the product of more than twenty years of thinking, writing, planning, revising.

Although the *Origin* can be disconcerting to theorists and philosophers because Darwin did not go any further on general philosophical questions than was required by the exigencies of his "species" arguments, the refusal was essential to his success. He aimed not at establishing theoretical consistency, but at mustering the resources he required, particularly the resource of what all parties could not resist in science, the resource of fact. Darwin's very human and often personal discourse sustains and reenforces that break between scientific and humane discourse that characterizes modern thought. Alert to the pleasures of his facts, he nevertheless sought facts doggedly, the pleasures being irrelevant to his quest. Thus he tried to find out whether seeds eaten by birds and defecated many miles away, or seeds soaked in salt water over long periods, having floated ashore on pieces of driftwood, could germinate. Every "fact" was part of an argument, but not about epistemology or metaphysics or morality. It was about whether natural selection could be shown to be operative and to explain phenomena ostensibly incompatible with it.

Although notoriously cautious, Darwin was determined and confident about his theory. He was out to show that organic nature came under the same rigorous rules that governed the inorganic, as Herschel authoritatively defined those rules. T. H. Huxley concludes his 1860 essay on the *Origin* by noting that Darwin's theory is calculated to exert a large influence "in extending the domination of Science over regions of thought into which she has, as yet, hardly penetrated."[11]

Huxley was only one of the more vociferous spokesmen for the domination of science's way of knowing, and the other "regions" he alludes to are the "human sciences." Darwin seemed to provide the evidence and the theory that might reduce human phenomena to Herschelian order, that might subject the moral and the spiritual to the "law" that science showed reigned everywhere else in nature. Like Herschel he sought only those "facts . . . which happen uniformly and invariably under the same circumstances," only these can "be included in laws" or "achieve universality."[12]

The extension of Herschel's law to humanity, however, creates an "extralegal" quality, and Darwin's self-consciousness about that is implied by the three very safe and conventional epigraphs to the *Origin*. One is by that ultimately respectable protoscientist, Bacon, who is quoted to show that the study of nature is the study of God's works; another is by one of the dominant intellectual influences on the early century, Bishop Butler, quoted to suggest that the "natural" inevitably implies "an intelligent agent to render it so."[13] The other epigraph is the one already noted from Whewell's Bridgewater Treatise.

There is no clearer evidence of the cautious deviousness of Darwin's writing, as he seeks in the heart of natural theology, justification of his secularizing moves. "Law," for these purposes, carries back into science an implicit theological sanction. J. B. Mozley, arguing for "Design," pointed out that "law is indeed a midway position between chance and design, at which many minds find it convenient to stop."[14] Darwin conservatively invoked "law" at every turn and needed the conception to give his theory scientific status. And the *Origin* was most successful in its effort to "elicit faith," as Robert Young has observed, "in the philosophical principle of the uniformity of nature."[15] Darwin's insistence on law—natural selection is, after all, a "law"—was required not only to displace the erratic interpositions of a miracle-making God without leaving the world unintelligible, but also to affirm the scientific validity of the study of organisms (and of the human, too). The Herschel he tried to emulate (and with whom he certainly talked at the Cape of Good Hope)[16] much disappointed him in calling natural selection "the law of higgledy piggledy" (*Life and Letters,* II, 141).

The break from natural theology into natural selection created for Darwin difficulties—linguistic and scientific—that made Herschel's resistance understandable. The central problem was the question of chance, a notion that challenged not only teleology, but law itself. Mayr describes what he calls Darwin's "predicament" in this matter: "Nearly all steps leading to evolutionary change seem to be controlled entirely or largely by accident; yet the final product of evolution is perfection in adaptation."[17] But Darwin could not accept the idea that chance was a "cause." As Edward Manier points out, Darwin attempted to explain the evidence for his theory in ways that might satisfy questions like, "How is it possible?" "Birds," Manier explains, in providing some examples of Darwin's explanatory strategies, "were blown by gales; amphibians or small land animals were carried by sea currents on floating debris; plant seeds could be transported in any of these ways, and even by other organisms being transported in any of these ways." This kind of explanation is in keeping with uniformitarianism: small, familiar facts can explain large phenomena, without recourse to catastrophe theory, or supernatural intrusion. But, Manier emphasizes, chance continues to operate despite the explanations, these "familiar, if somewhat haphazard ways . . . *were haphazard*."[18] Chance for Darwin was a force, despite his own resistance to it.

Darwin's science, continuing self-consciously to expel caprice from the universe by explaining even biological development in naturalistic terms of law and cause and effect, yet affirmed laws that looked to Herschel very much like chance and caprice. "Far from building his doctrine on absolute chance," says Neal Gillespie, "Darwin insisted that all phenomena are governed by laws and so are potentially subject to scientific study and explanation."[19] But he was also arguing for a theory that required the life-giving contribution of apparently random variation. The laws that Darwin affirmed, even if they could be described precisely, would be "laws" operating apart from any intention or meaning. Accepting the attempts by writers like Huxley and Lewes to drain the word of its metaphorical richness, he claimed that "law" means nothing but "the sequence of events as ascertained by us."[20] On these nominalist terms "law," paradoxically, really is the same as "chance" in religious terms, for it implies no intention, no teleology, no meaning, and only order in an exclusively naturalistic sense.

Herschelian resistance, then, might well have been expected. Natural selection was a "law" whose establishment depended on hot and unsettled epistemological issues. Noting one of the crucial differences between inorganic and organic sciences, W. B. Carpenter, defender of physiology as science (against Whewell), argued that "in the mineral or inorganic world . . . *change* is the *exception*, and *permanence* is the *rule*; whilst in the animated kingdoms, *change* is constant and universal, and is indeed essential to our idea of life."[21] The first step in any "science" of life was the creation of a language that made impermanence a condition of meaning. "Law," then, must be of a new kind, encompassing the individual, the vital, the passing. All theorists aspired to the ideal, articulated most fully by John Stuart Mill, of establishing inductive laws (on the basis of experience), which could then serve as axioms from which necessary truths might be deduced. Such an ideal helped lead Darwin to his acceptance of Lyell's uniformitarianism, with its Newtonian and Herschelian biases built in.

Darwin would not worry that no matter how often a phenomenon occurs, one can *never* know whether it has always happened in the same way, or that it will continue to happen. Nor would he allow his thought to be obstructed by the fact that the word "law" was a metaphor with strong metaphysical connotations. "The laws of nature," said Herschel, "are not only permanent, but constant, intelligible, and discoverable."[22] But attempts to drain the word of the metaphorical richness it brought with it from natural theology was at best only partially successful. Huxley was angry at "the tenacity of the wonderful fallacy, that the laws of nature are agents" rather than "a mere record of experience."[23] Law, said G. H. Lewes, "is merely the expression of the relations of coexistence and succession."[24] And Mill, trying bravely, asserted that "the expression, Laws of Nature, *means* nothing but the uniformities which exist among natural phenomena (or, in other words, the results of induction), when reduced to their simplest expression."[25] Darwin used this positivist reading occasionally, but Mozley after all was right (Lewes himself said that "law" tended to suggest "a more impersonal substitute for the Supernatural Power which . . . was believed to superintend all things"). Darwin's metaphorical "selection" needed "law" to redeem it from the chance it was releasing into the world.

Darwin's predicament was aggravated by the irony that scientists' recognition of an ideally organized and thus humanly intelligible world gave support to the view that intelligence itself was exempt from the rules that governed inorganic nature. It was itself the governor. To account for the presence of intelligence in the world one needed a kind of metascience (religion). Not so, said Darwin (with the support of Carpenter and others determined to subject the human to the investigations of science). To argue that "species" and "varieties" and "races" all exist only in time and temporarily, that the human "species" might ultimately be traced back to hermaphroditic, single-celled organisms, is obviously to challenge the structures that exempt human behavior and that imply universality of meaning. Furthermore, to see language as derived from primitive physiological responses is to deprive a word-grounded culture of its source: in the beginning was not the word in its luminous presence, but the twitch, the growl, the whine.[26]

For Darwin, the danger to the practice of science was the intrusion of divine intention disrupting the "laws" which it is the business of science to investigate. For Herschel, a world without divine intention invites mere chance, which also makes science impossible by disrupting the rule of law:

> We can no more accept the principle of arbitrary and casual variation and natural selection as a sufficient account, *per se*, of the past and present organic world, than we can receive the Laputan method of composing books (pushed *à l'outrance*) as a sufficient one of Shakespeare and the Principia. Equally in either case, an intelligence, guided by a purpose, must be continually in action to bias the directions of the steps of change—to regulate their amount—to limit their divergence—and to continue them in a definite course. We do not believe that Mr. Darwin means to deny the necessity of such intelligent direction. But it does not, so far as we can see, enter into the formula of his law; and without it we are unable to conceive how the law can have led to the results.[27]

The absence of intelligent direction, divine or human, has been a continuing concern of anti-Darwinians: of Samuel Butler and Bernard Shaw, of George Mivart and Teilhard de Chardin. It remains a hard pill to swallow, even for evolutionists. So radical was Darwin's argument on this matter that its full impact was not widely felt in literature until the twentieth century, although the possibility itself is intimated in many Victorian narratives. Against the powerful cultural bias that "chance" cannot be operating in the natural world, Darwin was moving toward a notion of the random. His insistence on law imperfectly disguised the fact that he was talking about stochastic processes and had come to recognize at least indirectly that the random and unpredictable—precisely what he could not account for and what seemed to make his argument peculiarly vulnerable—was essential for life. Mill, and other rigorously inclined thinkers, like the Duke of Argyll,[28] tended to believe that the difficulty in predicting the behavior of organisms (and humans) was not that laws did not govern, but that behavioral interactions were too complex to allow predictions in our present state of knowledge.

Darwin only partly subscribed to this view. His beliefs about chance are contradictorily stated throughout his books, his letters, his journals. Manier argues that "his published opinions on the topic fall into two opposed categories"; that "it is wholly incorrect to speak of chance as the cause of anything," and such speaking merely acknowledges our ignorance of causes; and that "certain events may, in the strictest possible sense of the word, be termed 'accidental.' Accidental events are those which result from the haphazard intersection of two or more lines of causality."[29] Nevertheless, the theory allows what Darwin hesitated to affirm. He already understood that regularity and order were essential to the development of laws and yet in a certain sense the greatest threat to his theory. He could not account for those "chance" variations that occur in all organisms, but he knew that they occurred with no Lamarckian connection to particular environmental conditions, to parent stock, to need, or to any recognizable goal. At that time it seemed to Darwin that such variations would likely be absorbed and then lost by large breeding populations, but the evolution and even the survival of species depended on their ability to preserve variations so as to be able to adapt to changing conditions. Chance and the random become great creative forces in Darwin's theory. Natural selection can only be "creative" by extinguishing the inflexible, and hence unadaptable, organisms. The ultimate effect of such extinction is favorable for groups of organisms, but, as Tennyson put it long before the *Origin* was even published, nature seems "So careful of the type . . . / So careless of the single life."

Yet Darwin's stochastic world is oriented to individuals, even if many of them are (apparently arbitrarily) destroyed. "Surely," he wrote to Francis Galton, "Nature does not more carefully regard races than individuals, as . . . evidenced by the multitude of races and species which have become extinct. Would it not be truer to say that Nature cares only for the superior individuals and then makes her new and better races? But we ought both to shudder in using so freely the word 'Nature' " (*More Letters,* II, 44). It was a not inappropriate response by many early critics of Darwin's argument that so negative a force as natural selection seemed to be could not be "creative." What they would not understand is that, while Darwin would have agreed that chance could never be an "agent," chance variations upon which natural selection operated provided the creative energy which they believed had to be attributed to an active intelligence.

Darwin's laws, then, were based on what would have seemed a very strange combination of the random and the orderly. As C. H. Waddington has explained, "The alterations produced in a gene, and the effects which this alteration will have on the phenotype of the individual which develops under its influence, are not causally connected with the natural selective forces which will determine its success or failure in producing offspring in the next generation."[30] But the random had no place in Victorian philosophy or science. And Darwin seemed to risk turning the world into an accident, as he turned his ignorance of the conditions of variation into a condition of the world he observed. It was an imaginative leap that survived Victorian mechanistic explanation, so favored by his own most fervent supporter, Huxley, and it is alive as a hypothesis today: although we now know the mechanisms of variation, they remain inexplicable in the sense of unpredictable and random.

The implications of the Darwinian argument obviously extend far beyond the perhaps parochial contest for scientific authority in which it was set. For to imagine a system in which disorder, dysteleology, and mindlessness are constitutive, and, indeed, the source of all value, is to turn the Western tradition, with its faith that all value inheres in order, design, and intelligence, on its head.

So Darwin's world required a new sort of imagination, even, perhaps, a new sort of politics. The Duke of Argyll, in 1866, can imagine the world as a Reign of Law; in the same year, Arnold reconstitutes order in "Culture," to be posed against the wind-driven forces of "Anarchy"; Carlyle, similarly, watches his society shoot the rapids into anarchy in "Shooting Niagara." But eight years before the second Reform Bill frightened many among the English intelligentsia, Darwin had announced a version of reality that would invert the significance of the images of stability and flux dominating the reaction to Reform. The quick adaptation of Darwin to the reign of law and the confirmation of current social and political structures was paralleled by a subversive understanding of his subversion of "law." Although his theory, as Greta Jones has noted, "was used from the beginning as a defence of 'laissez faire' capitalism," it was also used "in the attack on the remaining areas of special social and political privilege in British society."[31] Value would now be seen to inhere not in permanence, but in change, not in mechanical design but in flexibility and randomness. Natural selection introduced the possibility of incorporating the random into scientific explanation. More than that, as Gregory Bateson put it, stochastic processes of the kind we find in natural selection are characteristic of all biological development, and they make clear that "without the random, there can be no new thing."[32]

Thus, while the uniformitarian, law-bound Darwin transfers metaphorically into a conservative political force, the chance-invoking, change-affirming Darwin poses a major threat to things as they are. The tensions created by the move from natural theology to natural selection, the secularizing of order, the despiritualizing of the human, make an unproblematic reconstruction of Fanny Price's world impossible, even when Darwinian principles of order and development are emphasized. Once the consonance between the natural and the intentional is lost, the space for willed constructions of meaning, like Mary Crawford's, opens up. The principles of secular order that dominate the Victorian novel and seem to parallel Darwin's thought are always accompanied by a threat to that order in Chance, which had hitherto worked in narrative to affirm, not deny, intention.

Here I want to suggest the way Darwin's language helped his ideas subversively enter the culture. The apparently simple language quietly adopts personal anecdote, tentative speculation, hard presentation of data, careful logical argument, broad generalization, metaphorical expansiveness, rigorous scholarship. This mixture of modes, Beer claims, was "no mere stylistic quirk but part of his desire that they should be equivalent to the evidence of the natural world in all its diversity."[33] The title itself of his great book introduced key phrases into the language that have deeply affected the way we think, and did so from the start. Careful consideration of that title offers a convenient approach to the richness, instability, and rhetorical force of his language: *On the Origin of Species by Means of Natural Selection; or, The Preservation of Favoured Races in the Struggle for Life.*

The title tries to tell almost all, dares to address the "mystery of mysteries," and defines natural selection, after the "or," by invoking another metaphor and therefore not *quite* revealing its antidesign intentions. Almost every word raises difficulties. The history of Darwinism and the continuing debate about it have magnified the significance of these words in ways that encourage an almost Talmudic elaboration.

The first problem is with "origin"—a word that spoke directly to an apparently insatiable Victorian urge to determine origins —of language, of the Nile, of the human species, of the cosmos (or, in novels from *Oliver Twist* to *Daniel Deronda,* of parenthood). But Darwin was concerned with origins only to the extent that he wanted to find a mechanism to explain how species developed. In a letter to Charles Lyell, written probably not long before Darwin came to visit him in Africa, Herschel puts the question in an orthodox scientific way, which allows both for religion and a strictly scientific analysis of the problem:

> I allude to the mystery of mysteries, the replacement of extinct species by others. Many will doubtless think your speculations too bold, but it is as well to face the difficulty at once. For my own part, I cannot but think it an inadequate conception of the Creator, to assume it as granted that his combinations are exhausted upon any one of the theatres of their former exercise, though in this, as in all his other works, we are led, by all analogy, to suppose that he operates through

> a series of intermediate causes, and that in consequence the origination of fresh species, could it ever come under our cognizance, would be found to be a natural in contradistinction to a miraculous process.[34]

The question is critical because Lyell and Buckland and Cuvier had made the fact of "extinction" inescapable. How can one account for the disappearance of "divinely created species" in scientific terms. Here was the crux of the battle between catastrophists and uniformitarians: What natural causes, if any, could replace the extinguished species? For Herschel, origins belong to science and "intermediate causes," and Darwin's insistence on naturalistic explanation is consistent with his commitment to Herschelian science. But the question on the table for many readers who were upset by Darwin's unremitting secularity was precisely what was not intermediate. Given Herschel's focus on the intermediate, his objection to Darwin's failure to state the "higher law" of providential arrangement seemed self-contradictory to Darwin, who wrote, "Astronomers do not state that God directs the course of each comet and planet. The view that each variation has been providentially arranged seems to me to make Natural Selection entirely superfluous, and indeed takes the whole case of the appearance of new species out of the range of science" (*More Letters*, I, 191).

The question of beginnings always caused problems for Darwin because any beginning challenges the uniformitarian assumptions on which he built his theory. From the point at which Darwin picks up the narrative, all transmutation and speciation take place according to causes now in operation, and the changes are always very gradual across vast tracts of geological time. But beginnings imply some "catastrophic" transformation—here, from the inorganic to the organic, from nonlife to life. If every creature can only bring forth progeny after its kind, how, short of some divine act, account for the first life? How account for the replacement of extinct species by new ones? One naturalistic theory that might have accounted for such radical origins was spontaneous generation; but this was being finally disproved by Pasteur and John Tyndall.

Nature does not make leaps, Darwin insisted. This denial, thought to be essential not only to his theory but to the integrity of science itself, has been criticized even by strong Darwinists, from T. H. Huxley to Stephen Jay Gould. Gould, in particular, has argued that gradualism—despite Lyell's and Darwin's insistence on empirical evidence—had no empirical ground, was in fact an assumption rather than what was to be proved. With Niles Eldredge, Gould has insisted, "Phyletic gradualism was an a priori assertion from the start—it never was 'seen' in the rocks; it expressed the cultural and political biases of nineteenth-century liberalism."[35] Moreover, in his analysis of the assumptions of uniformitarianism, Gould points out that gradualism was not an essential component. It was not one of the assumptions necessary for the integrity of science but part of Lyell's special argument for a steady-state, not a Darwinian world.

Darwin's strongest faith was the scientifically conservative one that the world can be explained naturalistically, and he was an enthusiast of natural knowledge. The shape of his argument, of his language, of the very titles of his major books, suggests the exuberant range and ambition of his mind. Repeatedly, he clinches his arguments not with some final, decisive piece of evidence, but with the pleased and confident assertion that his explanation simply accounts for more, or opens more questions. When he tells people that certain kinds of questions and answers take a subject "out of the range of science," he pushes his theoretical impulses as far as he wants to go. But is there any reason to assume that an argument is invalid because it takes its subject outside the range of science? Not invalid, perhaps, Darwin would probably have said, but uninteresting. When Hooker was reading the manuscript of the *Origin*, Darwin wrote to him: "You cannot imagine how pleased I am that the notion of Natural Selection has acted as a purgative on your bowels of immutability. Whenever naturalists can look at species changing as certain, what a magnificent field will be open,—on all the laws of variation,—on the genealogy of all living beings,—on their lines of migration, &c. &c." (*More Letters,* II, 128).

Darwin knew that preoccupation with beginnings could only obstruct consideration of the astonishing and awesome phenomena of organic life. His capacity to leave problems alone was a kind of scientific version of negative capability, opening possibilities simply unavailable to the rigorous systematists. The open-ended nature of his world, implicated in time that never ultimately resolves into permanence, is paralleled by the theory itself, which aspires to ask questions as much as to answer them, and which does not claim to know anything of absolute beginnings. Such resistance to closure, such preoccupation with new possibilities as opposed to the comforts of resolution, is one of the distinctive marks of Darwin's break with natural theology.

Perhaps even more interesting (or amusing), as we follow the progress of the title, is the fact that species, on the Darwinian account, have no real existence. Darwin's very effort to use language representatively calls its powers of representation into question; language slips from its Adamic roots toward becoming a set of arbitrary and conventional signs. As I have already pointed out to contrast Darwinian thought with Jane Austen's project, his argument depends on demonstrating that "species" are mere conventions of thought. Here too Darwin finds himself in a major epistemological argument. There is some debate about whether he was a nominalist on the issue,[36] but he was decisive about the thoroughly conventional nature of the term: "I look at the term species, as one arbitrarily given for the sake of convenience to a set of individuals closely resembling each other, and that it does not essentially differ from the term variety, which is given to less distinct and more fluctuating forms. The term variety, again, in comparison with mere individual differences, is also applied arbitrarily, and for mere convenience sake" (*Origin,* p. 108).

The inability to "define" did not impede, and elsewhere, when the rhetorical demands are different, Darwin is comfortable with the term. This kind of apparent inconsistency drew criticism from Max Müller, who insisted that "the word had no right to exist in natural history," and the *Origin* should "have marked the end of all species, at least within the realm of nature." Müller ringingly concludes that the search after what Darwin calls "the undiscovered and undiscoverable essence of the term species is to my mind no more than the search after the hidden essence to Titans and Centaurs."[37] Darwin was working outside the available nonevolutionary definitions toward one that conceived of species in terms of populations rather than "essences." The disturbances of language were, however, all part of Darwin's essential preliminary move—to destabilize the idea of species, to press home to his audience that *they* did not know what species were and that their imagination of systematic classification was empty. Scientists everywhere disagreed about which groups actually constituted species, subspecies, varieties—or even groups. "Certainly," says Darwin, "no clear line of demarcation has as yet been drawn between species and sub-species—that is, the forms which in the opinion of some naturalists come very near to, but do not quite arrive at the rank of species; or again, between sub-species and well-marked varieties, or between lesser varieties and individual differences. These differences blend into each other in an insensible series; and a series impresses the mind with the idea of an actual passage" (*Origin*, p. 107). The theory is built on evidence not available to the senses. Species have no reality; speciation cannot be observed. The Darwinian world is one available to scientist and common sense only through the imaginative and intellectual leaps that make the insensible sensible. The blurring of conventional boundaries is critical here, as Darwin's imagination outruns the evidence. "The change of species cannot be directly proved," he wrote (*Life and Letters,* II, 155). "The idea of an actual passage" is an analogy central to the Darwinian narrative, to which I must shortly return. For the moment, I want to emphasize that Darwin's argument assumes that "species" has not been defined; moreover, he will argue that naturalists' failure to distinguish consistently between varieties and species is a consequence of the basic conditions of organic life. There *is* no demarcation—thus no definition. *The Origin of Species* is not about the origin, and it in effect denies the reality of species.

The subtitles, which should resolve by elaboration whatever ambiguities arise in the main title, simply increase them. The key phrase, "Natural Selection," is an oxymoron. As a metaphor based on the model of artificial selection, in which human breeders select variations for their own purposes, it implies the activity of design and a designer, and yet denies it. The earliest critics asked how there could be selection without a selector. When pressed by such critics for clarification, Darwin took up an almost Comtean positivist position, of the sort Mill, Huxley, and Lewes asserted, claiming that by "nature," he means "only the aggregate action and product of many natural laws, and by laws only the ascertained sequence of events."[38]

Wallace pleaded with Darwin in lengthy correspondence to drop the metaphor. In his own original paper, he had argued the irrelevance of domestic selection to natural selection (and of course Darwin begins the *Origin* with a chapter on domestic selection). "No inferences," said Wallace, "as to varieties in a state of nature can be deduced from observation among domestic animals. The two are so much opposed to each other in every circumstance of their existence, that what applies to the one is almost sure not to apply to the other."[39] Wallace's view on this subject was widely shared, and he was correct to insist that Darwin's use of the metaphor could only cause misunderstanding.

The misunderstanding was intensified by Darwin's tendency to personify not only natural selection but nature itself. Wallace thought that Spencer's term, "survival of the fittest," would be more appropriate, less open to misunderstanding. Darwin, for his part, and much to his credit, wriggled away from the suggestion and, with his quietly perverse doggedness, avoided the progressivist and hierarchical implications of Spencer's phrase. "The term 'Natural Selection,' " he says, "has now been so largely used abroad and at home that I doubt whether it could be given up, and with all its faults I should be sorry to see the attempt made" (*Life and Letters* III, 45–46).

Refusing to give it up meant that Darwin's language would persist in treating natural selection as an active, intelligent life-form, which no disclaimers could negate. The shadow of natural theology lingers. It is remarkable to find that in the 1844 draft of his argument, Darwin blatantly personified "natural selection" as a "being": "Let us now suppose a Being with penetration sufficient to perceive differences in the outer and innermost organization quite imperceptible to man, and with forethought extending over future centuries to watch with unerring care and select for any object the offspring of an organism produced under the foregoing circumstances; I see no conceivable reason why he could not form a new race . . . adapted to new ends."[40] Darwin's modifications of this metaphor reveal his ambivalence about it. In the first edition of the *Origin*, the "he" becomes "Nature" (p. 132), but the personification is still vital and "nature" is juxtaposed to "man," who can "act only on external and visible characters," while "nature" "cares nothing for appearances," and can "act on every internal organ, on every shade of constitutional difference" (p. 132). The extension of this figure sounds like natural theology: "It may be said that natural selection is daily and hourly scrutinising, throughout the world, every variation, even the slightest; rejecting that which is bad, preserving and adding up all that is good; silently and insensibly working, whenever and wherever opportunity offers, at the improvement of each organic being in relation to its organic and inorganic conditions of life" (*Origin*, p. 133).

By the sixth edition, Darwin apologizes for his metaphors, but even then doesn't edit them away. When he introduces "Nature," he asks "if I may be allowed to personify the natural preservation or survival of the fittest" (where Wallace's influence is evident). And he begins the paragraph about natural selection with "It may metaphorically be said."[41] The rest is almost identical with the language of the first edition.

These metaphors survive the criticism and Darwin's caution, even if they lose some of their power, because Darwin knew that his methods were not so pure as he gave them out to be. To one of his young correspondents he once wrote: "I would suggest to you the advantage, at present, of being very sparing in introducing theory in your papers (I formerly erred much in Geology in that way); *let theory guide your observations*, but till your reputation is well established, be sparing of publishing theory. It makes persons doubt your observation" (*More Letters,* II, 323). In another letter Darwin complained about the view that geologists should observe and not theorize. "How odd it is," he remarks, "that anyone should not see that all observation must be for or against some view if it is to be of any service" (*More Letters,* I, 195). Historians of science now on the whole agree that Darwin used what they are calling the hypothetico-deductive method,[42] and that his method corresponds more closely to the modern conception than Bacon's, to which most nineteenth-century historians of science (not, however, Whewell) paid more than lip service. *Non hypothesi fingo*, Newton said, and nineteenth-century science tried hard to emulate. But Darwin could not avoid the fictions of hypothesis.

"Natural Selection" lived for Darwin in the metaphors he never quite abandoned; it was a vital force in a world of organisms entirely plastic. They allowed him to give evidence of design that need not imply a designer. Darwin's revolution inhered precisely in this disanalogy, for it meant that all phenomena were subject to secular, naturalistic study. Nature itself embodies the power formerly attributed to a divine being. "Selection" implies nothing but an analogy whose two terms are irrevocably separated.

The third part of the title, "or, The Preservation of Favoured Races in the Struggle for Life," seems to move from science to melodrama. Here the problem is not only metaphor—favored by whom, what struggle?—but ambiguity and vagueness more or less disguised by implicit narrative. "Races" neatly skirts the problem of species altogether (Darwin indecisively discusses whether human races can be considered species in *The Descent of Man*, pp. 214–250). "Favoured" has different implications from the "fittest" that Darwin was reluctant to adopt. It was potentially redundant, meaning the "preserved"; it could imply intrinsic racial superiority, a concept contrary to the essential thrust of Darwin's theory (even if some of Darwin's explicit commentary in the *Descent* confirms racist interpretations);[43] but it also suggests something of the arbitrary, consistent with Darwin's view that "superiority" consists entirely in the organism's ability to adapt to the peculiar circumstances of its local environment.

"The Struggle for Life" seems to imply that life is an individualistic "war." The theory had, after all, fallen into place after a reading of Malthus, and frequent connections are made between evolutionary biology and laissez-faire economics, noting the complex interweaving of Darwin's scientific thought with the traditions of individualism and divergence of character to be found in Adam Smith and later political economists.[44] But Darwin makes the usual ideological use of his metaphor an oversimplification. "The war of nature" does conclude the *Origin*, but Darwin's theory is also antistruggle, anti-individualist. Darwinian organicism became a biological justification of the moral predominance of altruism, that term borrowed from Comte's positivism, but extended by G. H. Lewes and George Eliot.

Darwin discusses the inadequacy of the "war" model, anticipating the objections that negative forces like natural selection and struggle for life could not be creative. Each of Darwin's "negative" concepts has affirmative implications, and this is particularly true for the figure "struggle for life." Here is how he explains it in the first edition of the *Origin*:

> I should premise that I use the term "Struggle for Existence" in a large and metaphorical sense, including dependence of one being on another, and including (which is more important) not only the life of the individual, but success in leaving progeny. Two canine animals in a time of dearth, may be truly said to struggle with each other which shall get food and live. But a plant on the edge of a desert is said to struggle for life against the drought, though more properly it should be said to be dependent on the moisture. A plant which annually produces a thousand seeds, of which on an average only one comes to maturity, may be more truly said to struggle with the plants of the same and other kinds which already clothe the ground. The misseltoe is dependent on the apple and a few other trees, but can only in a far-fetched sense be said to struggle with the trees, for if too many of these parasites grow on the same tree, it will languish and die. But several seedling misseltoes, growing close together on the same branch, may more truly be said to struggle with each other. As the misseltoe is disseminated by birds, its existence depends on birds; and it may metaphorically be said to struggle with other fruit-bearing plants, in order to tempt birds to devour and thus disseminate its seeds rather than those of other plants. In these several senses, which pass into each other, I use for convenience sake the general term of struggle for existence. (p. 116)

Beginning with the literal "struggle" of the two canines, Darwin moves dizzyingly to a figurative struggle against drought, which is more appropriately understood as *dependence* upon moisture. Although the struggle among the seeds that "clothe the ground" is slightly more literal, the struggle of the parasite against the trees upon which it feeds is not strictly a struggle, but a dependence. Since the parasite cannot afford to destroy its hosts, its "struggle" is a dependence. In the "struggle" between mistletoe and the birds that destroy it, the mistletoe depends on the birds' devouring its seeds; its struggle then is, against other plants, to seduce the birds to destroy it. Struggle and dependence are so closely interwoven that the image of war is entirely inapproriate: struggle in Darwin's sense has as much to do with successful procreation as with any literal battle.

Like organic forms, meanings, as Darwin puts it, "pass into each other." The method and the ways of meaning in Darwin are continuous with the transmuting reality in which he tries to make us believe. In any case, while the meaning of all three aspects of the title is explored in the book which follows them, the ambiguity, metaphorical richness, and fluidity remain.

The antirevolutionary and conservative social and political uses to which the language of Darwinism was often put partly belie the *Origin*, which takes curiosity and life as the true focus of its rhetorical energy. Its subversion of natural theology can be felt in the treatment of stability, perfection, and finality as its true enemies. With all its emphasis on facts and order, the *Origin* is preoccupied with extremes, marginal life, aberrations. Where Paley saw harmony and perfection everywhere, every organism perfectly adapted to its niche in life, and even apparent chance and aberration contrivances for the general good, Darwin sees adaptation as contingent and incomplete, however breathtakingly wonderful it can be. He demonstrates disharmony, maladaptation, imperfection. And where Paley's rationalism points toward an ideal, systematic, and hierarchical structure in both universe and organisms, Darwin disrupts conventional Linnaean taxonomies and replaces the ideal atemporal universal structure that sees evidence of divine system-making in the obvious formal similarities among, say, bat wings, human fingers, tiger paws, fish fins, with temporal, genealogical connection.

Beyond the reach of *logical* movement in its long argument, the *Origin* prepares its ground by suggesting the irreducible profusion, entanglement, and variety of organic life. We have seen that natural theologians, too, used the awesome variety and profusion of species as part of their arguments, but Darwin makes it difficult to assimilate the experience of this fact into a designed benevolence. Paley, focusing on the most visible parts of nature, failed to see the full implications of the adaptations he believed to be divinely ordained. Darwin uses multitudinousness, then, to subvert the idea of perfection. He makes the possibility of separate creation of *all* those species look remote, even unattractive.

More even than Whewell, Darwin wrote like a Victorian sage, knowing that he had to change the way people looked at the world before he could argue them into changing their ideas about it.[45] Although he was "fully convinced of the truth of the views given" in the *Origin*, he did not expect to convince "experienced naturalists," because their "minds are stocked with a multitude of facts all viewed, during a long course of years, from a point of view directly opposite to mine" (p. 453). He needed to change that "point of view," and much of the *Origin* is consequently given over to defamiliarizing experience.

In a famous passage in the third chapter, he establishes the need for new ways to perceive old facts, and in so doing, to disrupt the natural-theological assumption that the evidence of the senses is a valid indication of the "perfect" adaptation of organism and environment. Darwin gives his own untheorized version of the pathetic fallacy, and connects his enterprise with the Victorian literary tradition of realism: "We behold the face of nature bright with gladness, we often see superabundance of food; we do not see, or we forget, that the birds which are idly singing round us mostly live on insects, or seeds, and are thus constantly destroying life; or we forget how largely these songsters, or their eggs, or their nestlings, are destroyed by birds and beasts of prey; we do not always bear in mind, that though food may be now superabundant, it is not so at all seasons of each recurring year" (*Origin,* p. 116). The passage parallels dozens in nineteenth-century writing, from Ruskin's "pathetic fallacy," to Eliot's *Adam Bede*, to Chekhov's "Gooseberries," self-consciously agitating against sentimental egocentricity and against the credibility of single perspectives on experience.

One rhetorical device of defamiliarization Darwin could not use in the *Origin*. Ironically, the scientist most important for making the human a scientific subject dared not do more than mention "man" in his major book, while his bulldog, Huxley, challenged Richard Owen to mortal combat in *Man's Place in Nature* (1863), a full eight years before *The Descent of Man*. The *Origin* as a model of narrative can serve only metaphorically for the human condition: one needs to feel the pressure of the human amidst all that discussion of flora and fauna; the human subject was simply too dangerous. That Darwin's evasive and moderate rhetoric was carefully deliberated for the *Origin* is evident in the differences to be found in his handling of the attempt to defamiliarize ordinary experience in the *Descent*. One of the least gentle of instances comes in discussion of the similarities between the teeth of anthropomorphous apes and those of human beings: "He who rejects with scorn the belief that the shape of his own canines, and their occasional great development in other men, are due to our early progenitors having been provided with these formidable weapons, will probably reveal by sneering the line of his descent" (*Descent,* I, 127). The confident sarcasm here suggests how tightly Darwin had restrained himself in the *Origin*.

When he begins his chapters on difficulties with his theory, he confesses that "to this day I can never reflect on them without being staggered" (*Origin,* p. 205). Quietly as he wanted his revolution to take place, he consistently subverts what would have been taken as common sense. It is not so much that appearances lie, although they do, but that the culture has taken as "natural" the interpretation of facts within a schema of intention. Darwin's minute, relentless observation of phenomena is accompanied by an equally relentless logical investigation; under such scrutiny nature becomes a different thing. It is not, as Bishop Butler would have it, that the spiritual reality is the truer. It is that we have never understood the material reality.

One of the most impressive of Darwin's strategies is to take an argument or fact that seems to invalidate his theory and then to show not that the fact or argument is false, but that it is an indispensable condition of his own position. He confounds possible opponents by showing that if their facts were *not* correct, his theory would be suspect. One of the strongest arguments against Darwin, for example, is that science cannot find either in life or in the fossil record evidence of the innumerable graduated transitional varieties that the gradualist theory of natural selection entails. But Darwin shows that this absence of transitional varieties proclaims not the invalidity of his theory, but its likely truth. For on that theory, species replaced by more adaptable ones could not coexist with their replacements; in addition, they would leave no trace—they would become extinct. Or another example: whereas the presence of complex instincts in animals had been taken to suggest separate creation, Darwin shows that instincts are by no means "perfect and are liable to mistakes," and that, as for organs, we *can* find innumerable gradations of instinctive behavior through the animal kingdom, each one of which is, however, currently valuable. All of these facts do not make instincts evidence for separate creation, but become explicable on none but Darwin's own theory.

Darwin makes us see what has not been there for us before. Absence of fossil records is the *inevitable result* of the way the geological record is created. Vegetable organisms seem incapable of moving, and yet the distribution of flora across great distances is evidence not of creationism but of the necessary effects of mobility of seeds on the surface of the water, or in the digestive systems of animals and birds. Against the criticism that his dependence on extinction of multitudes of species through the ages for explanation of gaps in the living and fossil record was evidence of the theory's weakness, he admits that "no one . . . can have marvelled more at the extinction of species" than he. Yet not so much that the marvelous didn't soon become explicable: "To feel no surprise at the rarity of a species, and yet to marvel greatly when it ceases to exist, is much the same as to admit that sickness in the individual is the forerunner of death —to feel no surprise at sickness, but when the sick man dies, to wonder and to suspect that he died by some unknown deed of violence" (*Origin,* p. 323).

The counterintuitive quality of his world is connected with his obsession with "facts." Common sense may suggest that species are sharply defined, that hybrids are infertile, that such magnificent organs as the eye are unique, that each adaptation is precise. Observation, however, invariably yields facts that blur the margins—infertility in hybrids is not absolute but finely graded; complex organs have their cruder counterparts; not only are there maladaptations, but some adaptive developments are irrelevant to the life of the organism. Multiplicity and difficulty of definition reduce the prima facie case for rationality and simplicity—and hence for intelligent design.

Darwin, as Dwight Culler put it, appealed "not to the formal and final causes of Aristotle, but to the material and efficient causes instead."[46] The effect is to stand the world on its head. The explanation of things in terms of what happens, not of what they come from or where they are going, has all the power of common sense; but it also exposes how deeply common sense had been embedded in the metaphysics of cause, intention, design, mind. Culler argues that Darwin's theory, in its most stringent form, is a kind of irony, a witty inversion of conventional ways of talking, even if Darwin chose not to be witty (he certainly was not as "bland" as Culler takes him to have been). This form is basic to late Victorian literature, the satiric reversal of the commonplace. I would extend the idea: the various strategies of argument and rhetoric in Darwin's attempts at "reversal" are important to the practice of novelists as well, in their attempts to render experience in order to disrupt common assumptions about the nature of things.

To see how these Darwinian strategies work, let us examine one of the less accessible chapters for the lay reader, the detailed and somewhat technical discussion of hybridism. The argument here is critical because the impossibility of cross-breeding was taken as strong evidence of the permanence of species and their separate divine creation. Since Darwin was arguing that "natural selection will never produce in a being anything injurious to itself, for natural selection acts solely by and for the good of each" (*Origin,* p. 229), he needed to account for infertility in some other way. He begins by making distinctions: between two species, infertile when crossed but with their sexual organs in good condition, and hybrids, whose reproductive organs are "functionally impotent" (p. 265). The same effect—sterility—derives from very different causes. To break down the firm distinctions between "species" and "varieties," he proceeds through the gradual steps of a finely wrought argument: "The fertility of varieties, that is of the forms known or believed to have descended from common parents, when intercrossed, and likewise the fertility of their mongrel offspring, is, on my theory, of equal importance with the sterility of species; for it seems to make a broad and clear distinction between varieties and species" (p. 265). Characteristically off center in his questioning, Darwin sees that the large issue is likely to be addressed most successfully by learning how the "marginal" organisms behave. *Some* species are fertile when crossed; *some* hybrids are not sterile. Experimenters have sometimes set up tautological arguments, changing classification from species to variety when crossing worked, and thus using their prior definitions to sustain the arguments that species are defined by their inability to cross-fertilize.

But Darwin finds that "when forms, which must be considered as good and distinct species, are united, their fertility graduates from zero to perfect fertility, or even to fertility under certain conditions in excess" (p. 275). Gradations replace definitions; "strange" connections and "unknown" laws undermine the familiar world of systematic biology. The argument and the rhetoric run together:

> Now do these complex and singular rules indicate that species have been endowed with sterility simply to prevent their becoming confounded in nature? I think not. For why should the sterility be so extremely different in degree, when various species are crossed, all of which we must suppose it would be equally important to keep from

> blending together? Why should the degree of sterility be innately variable in the individuals of the same species? Why should some species cross with facility and yet produce very sterile hybrids; and other species cross with extreme difficulty, and yet produce fairly fertile hybrids? Why should there often be so great a difference in the result of a reciprocal cross between the same two species? Why, it may even be asked, has the production of hybrids been permitted?
>
> To grant to species the special power of producing hybrids, and then to stop their further propagation by different degrees of sterility, not strictly related to the first union between their parents, seems to be a strange arrangement. (p. 276)

The apparent quiet naiveté with which this series of devastating questions closes completes the reversal by which Darwin is demonstrating the total irrationality of the creationist position, the inconsistency of conventional taxonomy, and the potential incoherence of a God who would have deliberately proceeded this way to such "a strange arrangement." The points are made on the basis of very detailed observation, the persistent discrimination of differences, the refusal to allow a general description to survive without empirical testing. It is one thing to say that species do not intercross; it is quite another to look at species with sufficient particularity to note that there are "degrees" of infertility. Like "imperceptible," "gradual," and "insensible," "degree" is a crucial word in Darwin's vocabulary repressing absolute distinctions. And once the element of "degree" is allowed, all the questions challenging the rhetorical questions of natural theology become possible. Assuming a designing mind, why some distinctions and not others? Where is clarity of design, which has such rhetorical force in natural theological argument?

This kind of argument, common throughout the *Origin*, shows how thoroughly Darwin uses his facts as part of a subtly constructed and rhetorically strategic argument. It may, for example, be fascinating that "the common gooseberry . . . cannot be grafted on the currant," but that "the currant will take, though with difficulty, on the gooseberry" (p. 277). But such facts become more than curiosities only insofar as they participate intricately in the one long argument. The questions turn the facts into parts of that argument, and that they need to be asked means that the fact has been turned into a weapon against absolutist positions: sharp definitions blur; laws begins to look like rule of thumb. The rationality of the natural theologian is suddenly slightly mad, creating structures that everywhere deny the richness of life. "For all practical purposes," says

Darwin characteristically, "it is most difficult to say where perfect fertility ends and sterility begins."

While his argument about hybrids is conclusive only in showing that the "facts" are not incompatible with the theory of natural selection, it opens up new possibilities undreamed of in the static, pre-Darwinian world. Darwin has argued that infertility, although it seems such an important quality, apparently "designed" for some important purpose, is in fact only a by-product of other changes effected by natural selection: "There is no more reason to think that species have been specially endowed with various degrees of sterility to prevent them crossing and blending in nature, than to think that trees have been specially endowed with various and somewhat analogous degrees of difficulty in being grafted together in order to prevent them becoming inarched in our forests" (p. 289). The language of purpose leads to dead ends of knowledge, to assumptions about what can be observed that cut off the possibility of understanding what is not immediately visible. While it had been argued that infertility was a condition of the life of species, Darwin's "facts" reveal not only that infertility is a gross simplification of the real conditions of nature, but that "the crossing of forms only slightly different is favourable to the vigour and fertility of their offspring." Crossing makes for life; change makes for increased vigor and fertility, in life and in language (p. 289).

The aberrant fact undermines classification and teleology and the theory of perfect and permanent adaptation of each organism to its particular environment. If we admire the tree-dwelling, hole-drilling woodpecker ("can a more striking instance of adaptation be given?"), we are forced immediately to consider the woodpecker who lives in the desert and "never climbs a tree!" (p. 216). Here again is that charming, understated naiveté: "He who believes that each being has been created as we now see it, must occasionally have felt surprise when he has met with an animal having habits and structure not at all in agreement" (p. 216). There are the webbed feet of the upland goose and the frigate bird, and "we cannot believe that the same bones in the arm of the monkey, in the fore leg of the horse, in the wing of the bat, and in the flipper of the seal, are of special use to these animals" (p. 228). But having demonstrated that the notion of special use is absurd, Darwin goes on to his larger argument so that the details, fascinating as they are, function both logically and rhetorically.

Abstract classification gives way to accumulated material fact. Here is another example of what Beer has called Darwin's tendency "to substantiate metaphor."[47] In an already quoted passage Darwin suggests that the "insensible series" of differences between species and varieties "impresses the mind with the idea of an actual passage." In this metaphor the language of gradualism gets a narrative location. All those minute increments of change, supervised by that nonbeing, nature itself—the gradual, the graded, the movement by degrees, insensible, imperceptible—become manifest. The "actual" passage is the passage of physical inheritance. Subtly graded differences imply no divine, uniform, and timeless system, but steps in a passage through time that would ultimately connect varying organisms to some common parent. And thus the morphologically similar though functionally disparate bones of horse, bat, and seal are indeed connected. But the connection cannot be explained in the language of ideal adaptation, separate creation, and essentialism. Neither traditional natural theology, nor Richard Owen's idealist anatomy, "premissed on a unity of structural type in the vertebrate kingdom, which he happily interpreted in terms of a transcendent archetypal idea,"[48] could account satisfactorily for observed similarities of structure. "We may safely attribute these structures to inheritance," says Darwin quietly (*Origin,* p. 228).

Inheritance becomes the means by which Darwin confirms traditional systematics, and displaces them. Having learned from earlier metaphors, Darwin exposes the metaphorical nature of science, and then, in a necessary shift that modern theorists might think naive, he reinvigorates the metaphors by claiming that they are literal after all. All along we have been Darwinians without knowing it. Once his theory is accepted, he says, "The terms used by naturalists of affinity, relationship, community of type, paternity, morphology, adaptive character, rudimentary and aborted organs, &c., will cease to be metaphorical, and will have a plain signification" (p. 456).

The simple fact of genealogy guides Darwin through a world of tangled particularities. He finds aberrations everywhere: these can only be accounted for in arguments from design, first, if one overlooks the way they do not *appear* to be designed (a crucial fact, since so much of natural theology depends on rhetorical questions about appearances); and, second, if one posits, like Owenite anatomists, a counterintuitive reality that is ultimately more real than the physical. Darwin is fascinated by "rudimentary, atrophied, or aborted organs," with which he must deal if he is to make his case fully. Here are the "mammae of male animals" and the teeth in fetal whales, and even in the language used to describe these useless vestiges is evidence of "descent." Darwin slips comfortably into metaphor at precisely the moment he is making traditional metaphors "plain": "Rudimentary organs may be compared with the letters in a word, still retained in the spelling, but become useless in the pronunciation, but which serve as a clue in seeking for its derivation" (p. 432).

Darwin looks at the same facts available to anatomists and paleontologists. But his eye is off the center. Herschel's requirement that scientific facts be generalizable cannot quite fit Darwin's concern with the aberrant and with individual variations. Such facts ought to be irrelevant to systematists. Instead of focusing on inevitable adaptation of each organism to its niche (if it didn't adapt, it wouldn't be alive), he finds structures irrelevant to adaptation. Eccentricity (as in a Dickensian novel, or in any nineteenth-century realistic narrative that struggles to differentiate character) was for him a means to discovery, as it is a means to life of the organisms he examines. Beer points out that Darwin's son Francis considered Darwin's "recognition of the exception, the anomalous, even minutest instance, to be one of the characterising strengths of his mind."[49] He goes immediately for the exception, not only to the generally accepted rule, but to his own.

The underlying assumption of the procedure is that nothing in life is complete or perfect. Everything on the Darwinian model is conditional (had Darwin been fully consistent, he would not have condoned the association of his theory with ideas of hierarchy or of inevitable progress). Neither he nor the novelists were consistent. Darwin's theory confirms unidirectional change, the condition of most Victorian narrative: Pip cannot go home again. It does not confirm progress.[50] In its almost obsessive engagement with details, as though only accounting for each would substantiate his argument, Darwin's style as well as his argument is democratic, like Thackeray's persistent reminders of the reality of others not touched in the selectivity of his novels.[51] Natural theology could be credible because it chose its facts selectively; natural selection would be plausible, Darwin seems to say, because it embraces all the world's details.

Since natural selection, even as personified, is a metaphor for a mindless temporal process, it cannot be expected to produce timeless and completed forms. It "tends only to make each organic being as perfect as, or slightly more perfect than, the other inhabitants of the same country with which it has to struggle for existence" (*Origin*, p. 229). It "will not produce absolute perfection." The most perfectly adapted organism is perfect only "according to the standard of the country" (p. 232), and native organisms apparently well adapted are specially vulnerable to introduced species from other environments, which may, as the starlings and English sparrows in

America might remind us, almost entirely replace the dominant natives. Species tend to become relatively fixed in nature when they are geographically isolated and comfortably situated in an environment that exerts few pressures.

Darwin's world, like that of the realist novel, is therefore a world of "mixed" conditions. Darwin provides, as it were, the manifesto for the rejection by realism of romance narrative. Imperfection is a condition of time. Paley too believed this, and it would be a mere caricature of the natural-theological position, as Whewell has shown, to deny imperfection in nature. Perpetual change is the universal law of creation. Paley asks, "Why resort to contrivance, where power is omnipotent? Contrivance, by its very definition and nature, is the refuge of imperfection." The answer defers to human *im*perfection, which is one of the dogmas of Christianity. Contrivance, Paley says, is for our sake: "God prescribes limits to his power, that he may let in the exercise, and thereby exhibit demonstrations of his wisdom."[52] One can see why Darwin believed that the faith in a final cause would be so destructive to scientific investigation. As David Bromwich observes in comparing Darwin and Paley, "Paleyan explanations are good for things that exhaust our interest when they reveal the intention of their designer. Darwinian interpretations, on the other hand, are good for things about which we constantly reform our interest as we learn their relations with other things."[53]

Darwinian interpretation is historical. We must learn, Darwin says, "to regard every production of nature as one which has had a history" (p. 456). But every history does not recapitulate an inevitable progress of "lower" to "higher." Rather, it describes particular alterations in form and function, developing from particular variations in relation to particular ecological contexts. "I believe . . . in no law of necessary development," he says, in one of the most ignored sentences in the *Origin* (p. 348). The tale Darwin spins is different from what the teller of it in, say, *The Descent of Man*, tells us about it. The furthest Darwin will go with the idea of progress is to agree that the survival of a species implies that it has "beaten" its "predecessors in the race for life, and is, in so far, higher in the scale of nature; and this may account for that vague yet ill-defined sentiment, felt by many palaeontologists, that organisation on the whole has progressed" (p. 343). "Higher" simply refers to capacity for adaptation, which, for Darwin, usually, but even here not inevitably, means more complex organizations with more elaborate division of labor among the organs. Furthermore, the idea of progress can only mean this *biological* development in adaptation, and has no moral or spiritual significance. And certainly, there is no hint of perfection here, or even of progress, in the way industrial economies adopted the idea.

Darwin's direct combat with the natural-theological idea of perfection and its accompanying ahistorical taxonomic idealism comes in his discussion of the eye. An instance of true perfection—which would mean in effect a condition outside of time, or at least invulnerable to time—would deeply threaten his theory. In chapter 3 of *Natural Theology*, Paley devotes about fifteen pages to the eye, as his first extensive "application" of the argument about the implications of the famous watch, which begins the book. The discussion opens with a comparison of the eye to another mechanical "contrivance," the telescope, to which, of course, despite the wonders it could perform, the eye is infinitely superior. Darwin's treatment of the subject, closing with the same comparison, clearly alludes to Paley.

Once again, Darwin begins by conceding the antagonist's position, even suggesting that once he shared it: "I freely confess" that it seems "absurd in the highest possible degree" to suppose that the eye with all its "inimitable contrivances" "could have been formed by natural selection." Putting common sense on the side of natural theology, Darwin's characteristic counterintuitive argument finally shifts common sense to his side. Even the eye must be assimilated to the undesigning process of natural selection, and Darwin will show that, after all, this instrument of greatest perfection is neither perfect nor unique. Invoking the theme of gradualism, Darwin describes the conditions that will allow him to make the case:

> If numerous gradations from a perfect and complex eye to one very imperfect and simple, each grade being useful to its possessor, can be shown to exist; if further, the eye does vary ever so slightly, and the variations be inherited, which is certainly the case; and if any variation or modification in the organ be ever useful to an animal under changing conditions of life, then the difficulty of believing that a perfect and complex eye could be formed by natural selection, though insuperable by our imagination, can hardly be considered real. (p. 217)

The allegedly unsophisticated mind of Darwin here engages in some very complex conditional arguing, with no reference to observable data. The conditions asserted prove nothing, only lead to a position from which the difficulty of believing would not be "insuperable" to the intelligence. The obstacle, like the condition of belief, is not the rational but the imaginative, for the eye staggers the imagination and common sense resists what rational intelligence can discover. Darwin must then have it both ways: rhetorically he must appeal to the imagination, while claiming to appeal to the intelligence. The object is to make "insensible gradations" replace fixed categories by showing that through such gradations the eye can be understood to be connected with organs not so apparently perfect.

Like Paley's, the argument is a web of analogy, conditional speculation, and metaphor. It begins by suggesting its own weakness: although we ought to look exclusively to the organ's lineal ancestors, we can never know those ancestors, for fossils cannot preserve eyes. Hence Darwin retreats to the lowest form of extant organisms that have optic nerves, the Articulata, and describes for a paragraph various crude organs clearly related to the eye. Since he cannot provide a chronological gradation, he finds a synchronic one. "I can see no great difficulty," Darwin concludes his survey, "in believing that natural selection has converted the simple apparatus of an optic nerve . . . into an optical instrument as perfect as is possessed by any member of the great Articulate class" (p. 218). ("I can see no great difficulty," was the kind of Darwinian formulation that drove rigorous, scientistic thinkers mad. Analogy, derived from natural theology itself, has no empirical authority. It enables a way of thinking, but that is all that Darwin wants at this stage. He can afford the apparent casualness.) Paley, who descends no lower than the eel, has no trouble recognizing other kinds of "eyes" in other creatures. But he rejects the transformation by analogy of synchrony into history. Since, as Darwin himself attempts to show, each organism is adapted to its condition, there is no need to posit development. Darwin sees no difficulty; Paley sees no reason. Thus, although in substance Darwin's reading is actually very close to Paley's, it replaces perfect adaptation with history. Paley looks at the eyes of fish, birds, and eels, assuming that "for different species of animals the faculty we are describing is possessed, in degrees suited to the different range of vision which their modes of life, and of procuring their food, requires."[54]

Darwin's argument depends on Paley's. Ironically, despite his determination to gather empirical evidence, he manages only to affirm the possibility of a different set of assumptions from which to consider the facts. He implies only the possibility that the eye might have developed from some such crude "optic nerve coated with pigment and invested by transparent membrane," and he is forced, therefore, to slip out of the facts overtly into the heuristic mode he has actually been using all along:

> He who will go thus far, if he find on finishing this treatise that large bodies of facts, otherwise inexplicable, can be explained by the theory of descent, ought not to hesitate to go further, and to admit that a structure even as perfect as the eye of an eagle might be formed by natural selection, although in this case he does not know any of the transitional grades. His reason ought to conquer his imagination; though I have felt the difficulty far too keenly to be surprised at any degree of hesitation in extending the principle of natural selection such startling lengths. (pp. 218–219)

The terms "reason" and "imagination" seem reversed. Although one can trace the logical steps of Darwin's argument—if . . . , then . . .—the whole movement is so tenuous and the terminal point so preliminary, that Darwin expresses only a counterfaith. The argument is a construction far more of the imagination than of the reason that is being celebrated. Through this astonishing reversal of language, Darwin suggests that the imagination is a constraining force—of mere subjectivity and prejudice. Reason, by contrast, has the sanction of impersonality and universality and can break through the limits of common sense. Common sense and imagination require a designer for what seems to be designed, but Darwin's reason shows that this move is not really necessary. Darwin will tell another plausible story ("imaginative," also, in our sense of the word).

The story is built on analogy. Although Herschel had argued that analogy was a legitimate device of scientific argument, Darwin's does not meet all the conditions. At issue is what constitutes the ground of argument and of authority itself. Conventionally respectable scientific argument would be inadequate to contend with Paley's position, whose mere logic seemed to Darwin almost invulnerable. And thus Darwin with deliberate reticence, never invoking the name, duplicates the fundamental analogy in Paley's argument, comparison with that other "contrivance," a telescope: "I know no better method of introducing so large a subject, than that of comparing a single thing with a single thing; an eye, for example, with a telescope."[55] Darwin concedes that "it is scarcely possible to avoid comparing the eye to a telescope." Instead of entirely deserting the convention that design entails a designer, he exploits it, as at the the start of the *Origin* he risked the analogy with domestic selection. Using the very language of believers, he challenges the assumption that the analogy is legitimate. The reversals are extraordinary:

> We know that this instrument has been perfected by the long-continued efforts of the highest human intellects; and we naturally infer that the eye has been formed by a somewhat analogous process. But may not this inference be presumptuous? Have we any right to assume that the Creator works by intellectual powers like those of man? If we must compare the eye to an optical instrument, we ought in imagination to take a thick layer of transparent tissue, with a nerve sensitive to light beneath, and then suppose every part of this layer to be continually changing slowly in density, so as to separate into layers of different densities and thicknesses, placed at different distances from each other, and with the surfaces of each layer slowly changing in form. Further we must suppose that there is a power always intently watching each slight accidental alteration in the transparent layers; and carefully selecting each alteration which, under varied circumstances, may in any way, or in any degree, tend to produce a distincter image. (p. 219)

Darwin wrests the high religious ground from the natural theologians. How dare we assume that God thinks as humans do? Analogy becomes a challenge to analogy, subverting the entire project of natural theology, making useless the distinction between the natural and the man-made by proposing yet another analogy. The reputed act of intelligence is made into history: think of the millions of organic forms and the millions of years in which the organ might have been tried out. One sees—Darwin concludes—how superficial the analogy is: "May we not believe that a living optical instrument might thus be formed as superior to one of glass, as the works of the Creator are to those of man?" (p. 219). The case against teleology is thus set in the strictest teleological language: "Natural selection will pick out with unerring skill each improvement."

The metaphors are devious and forceful. A "power" (natural selection) "intently watches," "carefully" selects, challenging natural theology with its own rhetorical weapons. The metaphor implying intelligence embodies the imagination of unintelligent action. Darwin seems to *accept* the natural-theological argument of the superiority of the creator to human intelligence, but suggests a naturalistic third party—the "power" that could create a living optical instrument. In later editions he is more cautious about the metaphorical nature of his antimetaphorical arguments—and less interesting: the "power intently watching" becomes "a power, represented by natural selection or the survival of the fittest."[56] But the original imaginative leap keeps even the diluted version alive, and the "reason" Darwin claims as the condition of his argument exposes the assumptions that made culturally determining arguments out of mere metaphors.

Using metaphors to reverse their implications *re*mythologizes the world he has tried to demystify. The scrutinizing, meticulous, always alert "natural selection" is usefully imagined as a living active force, ranging through a world shaped not by some external, designing intelligence, but by the multitudinous possibilities and chance collocations of local and individual entities and conditions. "Natural selection" answers no prayers, and though Darwin makes it motherly, careful, intent, seizing the best opportunities for its children, it can be relentless. For this reason, the theory itself became available on both sides of the argument about whether nature was meaningful and inherently value laden. On the one side, one could argue that there is no connection between morality and biological fact; on the other, one could argue that biological fact is normative, and all value derives from it.

But nature in Darwin's "mindless" story no longer was obviously normative, and the extension on naturalistic explanation to the distinguishing mark of the human—language—evoked Max Müller's continuing resistance. While accepting some naturalistic explanation, Müller rejected the suggestion that language could have emerged from mere natural noise. As for Whewell, intellect precedes all experience:

> If the words of our language could be derived straight from imitative or interjectional sounds, such as *bow wow* or *pooh pooh*, then I should say that Hume was right against Kant, and that Mr. Darwin was right in representing the change of animal into human language as a mere question of time. If, on the contrary, it is a fact which no scholar would venture to deny, that, after deducting the purely onomatopoeic portion of the dictionary, the real bulk of our language is derived from roots, definite in their form and general in their meaning, then that period in the history of language which gave rise to these roots, and which I call the *Radical Period*, forms the frontier—be it broad or narrow—between man and beast.[57]

Again, opponents of Darwin (although the Darwin here is Charles's son, George) require a line of demarcation—between language and noise, between human and nonhuman. The "roots," or origins, reconnect with a divine fiat, affirm a presence that makes language and nature meaningful. Appropriately, Müller's story is degenerative, telling of a fall from the initial luminosity of root meaning into metaphor and corruptions of language. Darwin's, if not philosophically tied to progressivism, is a story of gradualist progress, of minute, insensible gradations out of primitive single-celled organisms to the larger organism of civilized community, with no points of absolute demarcation and no intimation of an end. His greatest imaginative achievement was the construction of a world whose apparent intentionality and teleology disguise (even in Darwin's own intentional and teleological language) randomness, mindlessness, and irrevocably mixed conditions.

This rhetorical and imaginatively reasoned inversion of the natural-theological scheme in Darwin parallels developments in the novel. Darwinism is secularized design, widening the chasm between the "is" and the "ought," making nature less than trustworthy, making impossible the sorts of Austenian discriminations that could be taken as the norm of order, morality, and meaning. The natural is no longer obviously value laden or meaningful; the ordering of morals and of art, while they may seem to be rooted in nature, are rooted only in the sense of historically derived. Intelligence is opposed to "nature"; and human structures, like language itself, are conventional, arbitrary, vulnerably unstable. The consequences of this are felt in the Mary Crawfords of Victorian fiction. Deprived of the authority of the divine, natural theology becomes natural selection, romance structures veer toward an increasingly destabilizing realism. If Jane Austen can be taken as a representative of the kind of narrative that follows from the ideals of natural theology (while intimating Darwinian possibilities to be resisted), Anthony Trollope might be taken as representative of the kind that follows from the destabilizing world of natural selection, in which the intimations of chance and discontinuity are repressed. Dickens, with his individualist insistence constantly compromised by the moral demands of community, his efforts toward ideal narrative resolution undermined by his proliferating imagination of likely alternatives, plays out cultural ambivalences about the coming of the Darwinian vision.

5

Dickens and Darwin

DICKENS is the great novelist of entanglement, finding in the mysteries of the urban landscape those very connections of interdependence and genealogy that characterize Darwin's tangled bank. Certainly, Dickens is not self-evidently a Darwinian novelist—much of his catastrophist and apocalyptic imagination is incompatible with Darwin's gradualist world. Yet in many respects, particularly in his energetic tendencies to multitudinousness and the mysteries of imperceptible connection, he is close indeed to Darwin's "nature," far from the ordered world of natural theology. Even his "catastrophism," with its implicit recognition of progressive change rather than Lyellian stasis, belongs to Darwin's world, for, as I have suggested, Darwin's achievement was in part the absorption into uniformitarianism of catastrophist progression.

From the start Dickens's preoccupation with irrepressible multiplicity contends against an aspiration to order and meaning. When Mr. Pickwick slams the door on the suffering outside his prison room, Dickens dramatizes the loss of an unambiguous sense that the world makes sense and is ultimately ordered and just. He yearns for a "nature" that is indeed God's second book, as in the tradition of natural theology. But, like Darwin, he describes a world that resists such ordering. Unlike Darwin, he is often driven to arbitrary manipulation of plot to reinstate what his imagination has expelled.

The refrain "What connexion can there be?"[1] which echoes implicitly through all of *Bleak House* is answered by genealogy, just as Darwin's question about the meaning of the "natural system" is answered: "All true classification is genealogical; . . . community of descent is the hidden bond which naturalists have been unconsciously seeking, and not some unknown plan of creation" (*Origin,* p. 404). The juxtaposition of the separate worlds of Chesney Wold and Tom-All-Alone's in sequential chapters implies just such a "hidden bond," which is laden with moral implications.

Esther is the natural daughter who links the apparently unrelated city and rural life, poverty and wealth, lower class and aristocracy, and she is a figure for the moral bond that society ignores. Many in Dickens's society thought that Darwin's establishment of such natural connections of descent implied the destruction of the very moral bonds Dickens used genealogy to affirm. Both *Bleak House* and the *Origin* bespeak, in their different ways, the culture's preoccupations with "connexion" where physical juxtaposition, as in the cities, seemed to reveal startling spiritual, even biological discontinuity. What has Jo the crossing sweep got to do with Tulkinghorn the rich and powerful lawyer? Much of the battle about evolutionary theory implied the culture's deep discomfort with its new social juxtapositions, its attempt to deny the implicit religious context of the "hidden bond" as it appears in Dickens, its unwillingness to know that we are all literally one family.[2] Dickens's preoccupation with discovering connections links him in one way with a tradition of narrative that goes back to Oedipus, in another, with the Judaeo-Christian insistence that we are our brothers' keepers, and in yet another, to Darwinian styles of investigation and explanation.

Dickens certainly admired Darwin's theory, as Darwin took pleasure in Dickens's novels. There is no evidence that Dickens, like the more austere and dogmatic Carlyle, found Darwinism anathema. And it has been suggested that "the organisation of *The Origin of Species* seems to owe a good deal to the example of one of Darwin's most frequently read authors, Charles Dickens."[3] No literate person living between 1836 and 1870 could have escaped knowing about Dickens. After 1859, the same would have been true about Darwin. While Darwin rewrote for nineteenth-century culture the myth of human origins, secularizing it yet giving it a comic grandeur and a tragic potential, Dickens was the great mythmaker of the new urban middle class, finding in the minutiae of the lives of the shabby genteel, the civil servants, the "ignobly decent," as Gissing's novelist Biffin called them, great comic patterns of love and community, and great tragic possibilities of dehumanization and impersonal loss.

Given the pervasiveness of their fame, Dickens and Darwin had to have known each other's myths. In the crucial period of the late 1830s, in the notebooks that show him developing his theory, Darwin recorded that reading a review of Comte "made me endeavour to remember and think deeply," an activity that gave him an "intense headache." In contrast, he noted "the immediate manner in which my head got well when reading article by Boz."[4] The pleasures of Dickens remained with Darwin permanently. Although Darwin claimed that later in life he lost the power to enjoy poetry, he was a constant reader of literature from his youth,[5] and to the end, as he indicates in his *Autobiography*, he read novels steadily—or had them read to him.[6] In a well-known passage in his letters, he returns casually to *Pickwick Papers*—one of his favorite books—for a little philosophy: "As a turnkey remarked in one of Dickens's novels, 'Life is a rum thing' " (*Life and Letters,* II, 446). He even uses a Dickens description of a snarling mob in *Oliver Twist* to support his argument that human expressions are ultimately derived from rudimentary animal behavior.

What matters far more is that Dickens's development implies a confrontation with the very kinds of problems that Darwin, in his much different way, was also addressing. Dickens would turn the preoccupation with connections into moral parables, but his major narrative and moral difficulty had to do with the problem of change, about which he was much more ambivalent than Darwin. Although he supported the developments of the new science, that greatest instrument of change, and he despised the ignorance and prejudices of the past, there remains a strain of essentialism in his writing that led to trouble when he tried to imagine change of character; and though he brilliantly satirizes those who deny change, his style itself often denies it.

Dickens greeted with eagerness the radical developments in knowledge and communications that marked the nineteenth century. It would have been impossible for anyone, no less someone as imaginatively alive as Dickens, to have written without absorbing into his language something of the way science had been changing it. But he always regarded science as means to a human end, and he characteristically used scientific fact and method for moral purposes. According to Jonathan Arac, Dickens absorbs and transmutes the development in late eighteenth-century discourse by which scientific language was transferred to social theory. Arac points out, for example, how in the description of Tom-All-Alone's, Dickens "conveys less a specific physical description of the slum . . . than an attitude of scientific precision about it . . . Dickens's insistence on 'truth' in his preface to *Bleak House* . . . leads him to draw wherever possible on scientific authorities, for he was convinced that there was no conflict between science, rightly understood, and the imagination."[7] The megalosaurus waddling up Holborn Hill, to take an obvious but minor example, was a discovery of nineteenth-century geology and paleontology, and was named by William Buckland, apparently no earlier than 1824. Dickens's friend, the famous anatomist Richard Owen, made megalosaurs an important element in his own theorizing and regarded them as the highest forms of reptile, with real affinities to mammals. Moreover, there are signs on the very first page of *Bleak House* that Dickens was aware of and could use for his own purposes the early-century debate among geologists over the question of whether the mineral world and the fossil record are to be accounted for by flood or fire. At first Dickens seems a Neptunist, as Ann Wilkinson points out,[8] but his Neptunism is opposed by Vulcanism, as, for example, in the fires of Rouncewell's mill, Mr. Krook's spontaneous combustion, and the "transferred" spontaneous combustion of the whole Jarndyce and Jarndyce case. These more or less plausible and respectable geological positions were scientifically argued and subserved traditional religious ends. Part of Dickens's materials for imagining the world, they are evidence that he used science as much for metaphor as for the latest news about the cosmos. But he did turn to it, he would not be reckless about what science had already revealed. His Neptunism and Vulcanism are a literary convenience that required no belief, but "spontaneous combustion" did. On that, too, Dickens thought he had science on his side.[9]

He was in fact extremely alert to modern scientific and technological developments. As Alexander Welsh has noted, it would be unwise "to underestimate the degree to which Dickens was aware of the intellectual ferment of his time."[10] Harvey Sucksmith points out that Dickens was "receptive to biological ideas throughout his life."[11] Unlike Carlyle and Ruskin, with whom he is often associated, Dickens does not look back nostalgically to a golden past. There is a strain in him that does praise "merrie olde England" and revere the old-fashioned. But the old, old fashion, Paul Dombey discovers, is death. Dickens was very much a man of his time, "a pure modernist," Ruskin notoriously complained,—"a leader of the steam-whistle party *par excellence*—and he had no understanding of any power of antiquity except a sort of jackdaw sentiment for cathedral towers."[12] Despite the wonderful extravagance, Ruskin was right. The savage satire at the start of book II of *The Tale of Two Cities* is only one example of Dickens's attitude toward the past; the more complex celebration of the railroad in *Dombey and Son* is another.

The bias of Dickens's world is toward the new. His attack on modern bureaucracy is more often than not an attack on a system that madly repeats the worst of ancient practices and traditions: the circumlocution office, chancery, charity schools, and new poor law, almost invariably reenforce the values and methods of the old. Even Dickens's vendetta against utilitarianism and laissez-faire economics is directed not at the new and industrious middle class, but, rather, at the heartlessness of bureaucratic and institutional England. And these Dickens shows to be reflected and abetted by obstinate support of obsolete procedures and structures that confirm old class divisions and generate new ones. Society sets up against competent and innovative minds like Daniel Doyce and Mr. Rouncewell obstructive relics like Tite Barnacle or (with some vestiges of dignity) Sir Leicester Dedlock. Worse, it produces a new breed of villain, ostensibly "modern" but by gestures at respectability merely exploiting ancient injustices in pursuit of success for Number One: Fagin, Mr. Carker, Uriah Heep, Mr. Vholes, Mr. Bounderby, Bradley Headstone.

Science, for Dickens, was a means to help dispel superstition and ancient prejudice and habit. Ignorance is the enemy of morality. In a speech as late as 1869, at a point in his career when, if he had been as disillusioned with contemporary materialism as he is sometimes purported to have been, that disillusion would have emerged, he objects strenuously to the characterization of the age as "materialistic." Instead, he celebrates the scientific and technological discoveries that had improved the quality of life. The speech was an implicit attack on a recent speech by Francis Close, Dean of Carlisle, who had complained about the secularization of knowledge. "There were those," the dean had complained, "who would prefer any dream, however foolish or vain, to the testimony of God respecting the origin of our species."[13] Dickens argues energetically for the continuing expansion of scientific knowledge, always seeing it as a means to important human ends. "I confess," he says,

> that I do not understand this much used phrase, a "material age," I cannot comprehend—if anybody can: which I very much doubt—its logical signification. For instance: has electricity become more material in the mind of any sane, or moderately insane man, woman, or child, because of the discovery that in the good providence of God it was made available for the service and use of man to an immeasurably greater extent than for his destruction? Do I make a more material journey to the bedside of my dying parents or my dying child, when I travel at the rate of sixty miles an hour, than when I travel thither at the rate of six?[14]

Here, if anywhere, is the credo of the "steam-whistle party." But Dickens was not unambivalent, and the treatment of Dombey's ride in the train that seems hurtling toward death, if not its personification, can give some sense of why the problem of change was never a simple one for him. Nevertheless, the speech is unequivocal in embracing the new. And it is not merely an endorsement of technological application of scientific ideas. Practical as Dickens's orientation was, the speech shows that he believed that the practical grew from a willingness to entertain and seek new ideas, whatever their apparent application. Darwinism and the secular interpretation of nature are not the problem; the problem is dogmatic traditionalism. "Do not let us be discouraged or deceived by vapid empty words," he urges. "The true material age is the stupid Chinese age, in which no new grand revelation of nature is granted, because such revelations are ignorantly and insolently repelled, instead of being humbly and diligently sought."[15] Dickens, too, believed that science is compatible with religion: the true irreligion is conventional dogmatic religiosity.

That he was not an intellectual, in our usual sense of the word, is obvious enough. Although some of his more mature comments on science (particularly on "spontaneous combustion") may seem both ignorant and prejudiced, Dickens maintained a warm relation with science and scientists. He enlisted important scientists for help with his weekly journals—not only Owen, who wrote several pieces for *Household Words*, but also, for example, Michael Faraday, who sent him the notes for his famous lectures on the candle, which eventually became *The Chemical History of a Candle*. Dickens published a kind of summary of it in *Household Words*, a summary that Wilkinson has found useful in understanding the structure and significance of *Bleak House*.

The details of Dickens's novels often reveal that he had absorbed, like an intelligent layman, some of the key ideas issuing from contemporary developments in geology, astronomy, and physics. The evidence is most obvious in *Household Words* and *All the Year Round*, where scientific matters are taken as significant despite the homely and domestic emphases. Often surprisingly sophisticated despite their popularizing strategies, the scientific essays stressed the relation of science to ordinary life and made his journals important popularizers of scientific ideas.

This is not to deny the complicating antiscientific strain in his writing. Mr. Pickwick begins as a butt of satire, and one of his persistently satirized characteristics is scientific ambition. He is introduced, in a gently Swiftian way, as the author of "Speculations on the Source of the Hampstead Ponds, with some Observations on the Theory of Tittlebats." On his first adventure, a cab ride, he solemnly accepts and notes the cabman's sardonic exaggeration that his horse is forty-two years old, or that the horse stays out two or three weeks at a time and only can keep standing because the cab supports it. A bit later, he becomes deeply excited about an "archaeological" discovery, which turns out to read BILL STUMPS HIS MARK. Some of the animosity to trivial science is diverted later in the book to "the scientific gentleman" who manages to mistake Mr. Pickwick's lantern in the garden for "some extraordinary and wonderful phenomenon of nature."[16]

In 1846, Dickens published his last Christmas book, *The Haunted Man*. It gave him the opportunity to carry out further that attack on the scientific character comically announced in *Pickwick Papers*, for the central character is the chemist, Redlaw, who has bargained away his power of memory. The connection between the scientific pursuit and dehumanization that follows is implicit, but Dickens makes very little of it. Here was a subject designed to explore the anaesthetizing consequences of exclusively analytic mental activity such as we find in the actual autobiographies of Darwin and Mill. But Dickens does not seem very interested in pursuing it. The hard look at life is painful, so the moral goes; implicitly, the scientist sees the pain and, sensitive enough, feels it. For the most part in this very thin tale Redlaw is a sympathetic figure, whose decision to give up memory is treated with understanding, although, of course, implicitly criticized. A willingness to take a good look at the worst is as essential as a celebration of the virtuous. Thus, in appealing for a restoration of his memory, Redlaw cries: "In the material world, as I have long taught, nothing can be spared; no step or atom in the wondrous structure could be lost, without a blank being made in the great universe. I know, now, that it is the same with good and evil, happiness and sorrow, in the memories of men."[17] With the conventional moral application of excessive reliance on science, the speech is nevertheless couched in the terms of science itself; and it reveals Dickens's awareness of one of the fundamental principles of contemporary science, the conservation of matter, demonstrated by Lavoisier late in the eighteenth century. Lavoisier had shown that the actual amount of material in a chemical transformation remains the same before and after: "We must always suppose an exact equality between the elements of the body examined and those of the product of its analysis," Lavoisier said.[18] In Dickens, the indestructibility of the physical universe, like all other scientifically affirmed ideas, becomes moral metaphor. For Redlaw and Dickens the physical world signifies, as it did for the natural theologians. Here, at least, Dickens's ambivalence about "God" does not inhibit him from using the physical as a sanction and even a model for the moral; rather, as for the natural theologians, faith that what turned up would be meaningful encourages further scientific pursuit of knowledge.

Similarly, in the very year of the *Origin*, Dickens published in *All the Year Round* an essay called "Gamekeeper's Natural History," which mocks in a traditional way the abstractness of most scientific thought. "No one can paint a thing which is not before him as he paints," says the author, and "natural history is not to be written by professors in spectacles—timid, twittering, unsophisticated men—from stuffed animals and bleached skeletons."[19] But it *is* to be written from life, by naturalists, so the author says, like "Audubon, White of Selbourne, Gould." Sharing his culture's Baconian commitment to "experience" as the source of knowledge, Dickens implicitly sees the writer's and the scientist's task of representing the real as deriving from the same powers, leading to the same places.

The coverage of science in the journals does not suggest that scientific thought and experiment were dehumanizing. Taken together the essays show that Dickens was familiar with and sympathetic to the large ideas which, though not strictly anticipations of Darwin's theory, were conditions for it. For example, Darwin needed, above all, the large infusion of time that Lyell's *Principles of Geology* gave him; and Dickens was not retrograde in accepting it, as is manifest in the comfortable allusions in *Bleak House* to geological time. The essay "The World of Water" talks about the "thousand, thousand years ago" in which fossil creatures lived. It casually refers to man as a latecomer into the world (although this is true even in Genesis), and it accepts the position of Cuvier and Lyell that there has been large-scale extinction of species, even forecasting the ultimate extinction of man himself.[20]

In the essays Dickens seems particularly fascinated by the minutiae revealed under the microscope—the dramatic disparity between what is visible to the naked eye and what is really there. In essay drawing on Philip Gosse's *Evening with the Microscope* Dickens describes the similarity in all vertebrate blood. And he gives a series of dramatic and pleasantly horrific pictures of the natural world, as, for example: "We venture to say that the poet who spoke of butterflies kissing the sweet lips of the flower &c. never looked through a microscope at that flat coiled tongue bristling with hairs and armed with hooks, rifling and spoiling like a thing of worse fame, but of no worse life."[21] Dickensian gothic is merely an entertaining way to emphasize that the natural wonders revealed by science were evidence of its value, and, indeed, of its value as entertainment; it further expresses Dickens's instinctive view that matter of fact is really mysterious and wonderful and not fully visible to any but an intense and imaginative moral vision.

Thus, while Dickens was willing to consider the pursuit of knowledge for its own sake, there was a touch of the Gradgrind in him; he always wanted to know to what use the knowledge would be put. His dislike of Gradgrindian "science" is dislike for the privileging of the intellect, which turns human complexity into abstraction and allows brutality under the sanction of "Truth." Dickens wants science in his fictions as metaphor, and this is true even for such burning theological questions as whether man is a child, cousin, or sibling of the apes. The opening chapter of *Martin Chuzzlewit*, as Sucksmith reminds us, alludes comically to the theory that man is descended from the apes (probably drawing on Dr. Johnson's description of Lord Monboddo).[22] Concluding his genealogical chapter about the Chuzzlewits, Dickens writes: "It may be safely asserted, and yet without implying any direct participation in the Mondboddo doctrine touching the probability of the human race having once been monkeys, that men do play very strange and extraordinary tricks."[23]

A much later essay in *All the Year Round* called "Our Nearest Relation" comfortably accepts the biological closeness of gorilla to man. Even the essayist's misapprehension that the gorilla is a ferocious and aggressive animal is what most naturalists' accounts at the time would have asserted. Dickens and his journal take the facts where they find them, but convert them quickly into moral metaphor. The gorilla essay concludes with this passage:

> Again and again it strikes the fancy—strikes deeper than the fancy—that the honey-making architectural bee, low down in the scale of life with its insignificant head, its little boneless body, and gauzy wing, is our type of industry and skill: while this apex in the pyramid of the brute creation, this near approach to the human form, what can it do? The great hands have no skill but to clutch and strangle; the complex brain is kindled by no divine spark; there, amid the unwholesome luxuriance of a tropical forest, the creature can do nothing but pass its life in fierce sullen isolation—eat, drink, and die?[24]

The essay takes up the idealist anatomy or transcendental biology of Richard Owen, the view, as Peter Bowler describes it, that similarity in structure among living creatures expresses an " 'archetype' or ground plan on which all forms of life . . . are modeled."[25] Several essays in *Household Words* expound and argue for transcendental biology. One such concludes exuberantly in this way: "Thus, beyond and above the law of design in creation, stands the law of unity of type, and unity of structure. No function so various, no labours so rude, so elaborate, so dissimilar, but this cell can build up the instrument, and this model prescribes the limits of its shape. Through all creation the microscope detects the handwriting of power and of ordnance. It has become the instrument of a new revelation in science, and speaks clearly to the soul as to the mind of man."[26] The similarity among organisms, like the similarity between gorilla and human, does not imply consanguinity, and certainly not descent, for the essential pre-Darwinian tenet of almost all thinkers aware of the similarities was that there is, nevertheless, an absolute gap between humanity and anthropoid, a gap to be filled only by the "divine spark" so manifestly missing in the gorilla's brain (a position which, remarkably, A. R. Wallace also took up later in his career, to Darwin's deep disappointment).[27] Even the most partisan Darwinians would concede what Huxley, for example, called "the vast intellectual chasm" between Man and the ape; but Huxley was to argue that the similarity in physical structure of the brains was evidence of consanguinity. There was no need to assume that an intellectual difference would entail "an equally immense difference between their brains."[28]

Nevertheless, Dickens's enthusiasm for new "grand revelations" of nature seems to have led him to publish in *All the Year Round* a remarkably fair-minded review of the *The Origin of Species* only a few months after the book first appeared. The review congratulates Darwin for living not "in the sixteenth century" and not in "Austria, Naples, or Rome," but in "more tolerant times." It proceeds to a reasonably skeptical but very careful presentation of the theory (using, without quotation marks, much of Darwin's own language), and concludes in a splendidly Victorian way, with sentiments worthy of Dickens:

> Timid persons, who purposely cultivate a certain inertia of mind, and who love to cling to their preconceived ideas fearing to look at such a mighty subject from an unauthorized and unwonted point of view, may be reassured by the reflection that, for theories, as for organised beings, there is also a Natural Selection and a Struggle for Life. The world has seen all sorts of theories rise, have their day, and fall into neglect. Those theories only survive which are based on truth, as far as our intellectual faculties can at present ascertain; such as the

> Newtonian theory of universal gravitation. If Mr. Darwin's theory be true, nothing can prevent its ultimate and general reception, however much it may pain and shock those to whom it is propounded for the first time.[29]

Although Dickens tried to avoid controversy in any of the essays he published in the journals, one can only infer that he was willing to risk controversy on this issue. A month earlier he had published another essay, called "Species," which, without reference to Darwin, quotes him at length as though in the essay writer's voice. The prose is judiciously impartial, but it employs Darwin's own words as its own: "It may be just as noble a conception of the Deity to believe that he created a few original forms capable of self-development into other and needful forms, as to believe that He required a fresh act of creation to supply the voids caused by the action of His laws."[30] Two essays so generously indulgent of the development theory in a journal as tightly controlled as *All the Year Round* seem very unlikely unless Dickens were ready to endorse the idea himself. The strategy of the review, carefully considering objections, but proceeding with a long and unquestioned set of quotations from Darwin in the voice of the reviewer, and concluding with an open evocation of a Darwinian metaphor, suggests a far greater commitment to the idea of evolution by natural selection than is explicitly affirmed. Even as it questions Darwin's theory, it uses his dominant metaphor to predict its future.

The attitudes implicit in the language and structure of Dickens's books are, like the attitudes essayed in his journals and afloat in scientific thought in the 1830s and 1840s, premonitory of the argument Darwin was constructing; they are also often in tension with it.

Dickens's openness to science is reflected in the qualities that characterize his fiction. His novels, in their way, work with the materials that Darwin transformed in another. What Dickens could not have accepted—and *Hard Times*, for example, is in part a tract against it—is the "scientific" treatment of the human subject, although in Bucket, an ultimately sympathetic character, Dickens prepares the way even for this; and the satirical strain of the third person narrator in *Bleak House*, like the sardonic voice at the opening of *Oliver Twist*, provides rhetorical form for such detached treatment. The human in Dickens largely escapes the reduction Darwin's theory implies, but the bleaker his vision the more ready Dickens is to regard the human as scientific (that is, merely material) subject. To avoid such a fate, he leaves open rationally inexplicable avenues of plotting and characterization. Nevertheless, many of the major characteristics of Dickens's way of seeing and writing about the world are reflected in major elements of Darwin's theory. The cultural theme of connection, with its implication in genealogy, is a major concern of both writers, for example, and suggests again that the possibilities of imagination in science and literature are mutually bounded, mutually derived. Science and literature help create the conditions necessary for each other's development.

The differences between Dickens and Austen are not merely the differences of individual genius. Both may have used contrivances necessary to resolve narrative problems; but Austen's self-consciousness allows her to affirm the intelligible design of the world. Parody is possible for her because she is easy with what is parodied. Dickens, by contrast, thinks less about the contrivance of the coincidences that drive so much of his plotting because they are essential to him if he is to find any shape for a world of profusion, multiplicity, and apparent disorder, a world in which, despite his celebrations of order, he is at home. The landscapes and the architecture of these worlds are far from those ordered eighteenth-century houses and gardens that define and place the characters in *Mansfield Park*. They are the view from Todgers, the chaos of Barnaby's London, Tom-All-Alone's, the dust heaps of *Our Mutual Friend*. Dickens is closer to Darwin than Austen, and not merely chronologically.

Some of the elements of Darwin's vision that I isolated in the first chapter have their counterparts in Dickensian narrative, and for the rest of this chapter I want to consider the parallels and the points of divergence. What is true for Dickens, a writer brilliantly outside the main stream of Victorian realism, is true more emphatically for the realists. Discussion of the Darwinian elements in Dickens, even of the ways he averts Darwinian treatment, should throw light on the other novelists, as well.

First, Dickens the "catastrophist" has much of the Darwinian uniformitarian in his vision. Like the great domestic novelists of the century, Dickens is fascinated by the most trivial domestic and social details—food, furnishings, manners, and all the particularities of ordinary life. The whole movement of narrative toward these details is very much part of the movement that led to evolutionary theory, and it is evident in the rhetoric of Darwin's own argument. Second, the emphasis on the ordinary is often accompanied by a preoccupation with mystery. Somehow, the familiar resonates through all of Dickens with tones of the unfamiliar; things are and are not what they seem. As Dickens himself says (in the Preface of *Bleak House*), he is concerned with "the romantic side of familiar things." To this I juxtapose Darwin's program of defamiliarization, discussed in detail in the preceding chapter, the attempt to discover new principles of order in the midst of what we have long taken for granted. Third, the mystery of the familiar seems to generate complicated plots, full of coincidence, as amidst the multitudinous populations of each novel jostling against each other, new relationships are perceived. The whole seems to move toward catastrophe and a reversal when everything is explained; yet, on the whole, everything *is* explained and what has seemed like chance at the level of story acquires a meaning in an overall plot or design. Dickens does not quite accept the Darwinian rejection of teleology and the need for chance as explanation; chance is there, to be sure, but Dickens makes it work for teleology, even if under strain. Fourth, Dickens struggles with the cultural and Darwinian tendency to blur boundaries. The familiar Dickensian "character" has a sharply defined nature, a singular essence normally conveyed in a few tricks of manner: Pecksniff is invariably a hypocrite, although his hypocritical invention is wonderfully various; Mr. Dombey is invariably proud, Amy Dorrit invariably angelic. The reading of the essential nature of characters seems related by contrast to Darwin's nominalism, and here the question of "change," raised at the start of this chapter becomes prominent. Fifth, the question of connection is critical in both writers: things hang together in Dickens's world, stories converge, unlikely connections are made, entanglements and dependencies are inevitable. In modern jargon, Dickens has an ecological vision; and so, of course, has Darwin. Finally, sixth, all of the elements I have been noting become part of a world overwhelmingly vital because abundant, multitudinous, diverse, full of aberration, distortions, irrationality, which may or may not be ultimately reducible to the large patterns.

The importance of uniformitarianism to the Darwinian argument should by now be clear; it is worth emphasizing, however, that preoccupation with the ordinary is the very heart of romanticism, Wordsworth's responsibility in the division of labor in the *Lyrical Ballads*. Wordsworth's songs and ballads began to emerge only a few years after James Hutton's paper, in 1785, giving the gist of his position in *The Theory of the Earth* (1795). Dickens begins his career as a reporter whose skills are based on his powers of observation, with an uncanny eye for the ordinary. In his eyes the ordinary is transformed, not by miraculous or catastrophic intrusions, but by intense and minute perception. So in his sketches he examines door knobs and reports on the behavior of cabbies, shopkeepers, marginal gentlemen. Wherever he looks, even in the Vauxhall Gardens by daylight, when the ordinariness leads to pervasive disenchantment, the ordinary carries its own enchantment. Describing "early coaches," for example, he notes that "the passengers change as often in the course of one journey as the figures in a kaleidoscope, and though not so glittering, are far more amusing."[31]

The extraordinary popularity of such trivia presupposes a shift to an audience concerned with middle-class domesticity and to the recognition from that perspective of how completely the largest events of our lives evolve from the accumulation of precisely such minutiae. The essay from *Household Words* already alluded to, "Nature's Greatness in Small Things," explores the similarity between the minutest microscopic organism and the largest. "Not unfrequently," says the author, "it is seen that forms the most minute are most essential," capable of working "immeasurable changes."[32] The popular fascination with books about what the microscope revealed is also related to the preoccupation with the domestic and the ordinary. All of these phenomena are part of the same movement that made concern for the domestic the dominant motif of the self-consciously "realist" fiction of the high Victorian period.

Aesthetically, the fulfillment of the uniformitarian vision was articulated in the Victorian novel's constant reversion to the ordinary, and to its treatment of it as normative. We find it most completely formulated in George Eliot's celebration of the art of the Dutch realist school of painting as a kind of model for her fiction. The antirevolutionary implications of this aesthetic are worked out in Eliot (see her handling of politics in *Felix Holt* as the most obvious example). Later, when Razumov of Conrad's *Under Western Eyes* scrawls "Evolution not Revolution," Conrad is affirming both a political and an aesthetic tradition that, by late in the century, was breaking down. I will be taking Razumov's attempt to affirm evolution against the revolutionary substance and style of the novel he occupies as a convenient marking point for a shift from Darwinian thinking in fiction. Evolutionary theory, Victorian realism, and antirevolutionary ideology go together very tightly through the century. Ironically, the materialist and secularizing implications of the revolutionary views that Jane Austen was resisting, when embodied in evolutionary theory, become conservative.

While it is common to see realist and Dickensian art in opposition,

Dickens, with what Arac describes as a "scientific" attitude, seems even more concerned to insist on the literal truth of his writings than the more conventionally realistic writers.[33] For however much Dickens is to be regarded as a great entertainer or as metaphysical novelist, he *claimed* that he was a realist. Perhaps the earliest claim is in the preface to *Oliver Twist*, in which he attacks those who cannot stand the unhappy truths he has revealed. "There are people of so refined and delicate a nature, that they cannot bear the contemplations of such horrors," he says contemptuously. But he would not for those readers "abate one hole in the Dodger's coat, or one scrap of curl-paper in the girl's dishevelled hair." And as for the character of Nancy, "it is useless to discuss whether the conduct and character of the girl seems natural or unnatural, probable or improbable, right or wrong. IT IS TRUE." He bases this claim on his own experience of watching "these melancholy shades of life." Notice that here, in the defense of the reality of his fiction, Dickens rejects romance literature, which ignores surface details, and that this rejection entails mimetic particularity, attention to the minutiae of ordinary life. Have these sordid facts he has revealed "no lesson," Dickens asks, "do they not whisper something beyond the little-regarded warning of an abstract moral precept?"[34] The ordinary—the hole in the Dodger's coat, Nancy's disheveled hair—is given in Dickens some of the quality of allegory.

In the preface to *Martin Chuzzlewit* Dickens makes a similar point, emphaizing how perspective determines what is to be considered "realistic." "What is exaggeration to one class of minds and perceptions, is plain truth to another . . . I sometimes ask myself whether it is *always* the writer who colours highly, or whether it is now and then the reader whose eye for colour is a little dull." This eagerness to assert the literal truth of his fictions continued to the end of Dickens's career. In the postscript to *Our Mutual Friend* he talks of the "odd disposition in this country to dispute as improbable in fiction, what are the commonest experiences in fact," and he proceeds to defend as realistic old Harmon's will and his treatment of the Poor Law with evidence from *The Lancet*.[35]

Perhaps the most famous and egregious instance of Dickens's defense of the literal reality of his stories comes in the preface to *Bleak House*, where he defends the scientific validity of spontaneous combustion. "Before I wrote that description," he says, "I took pains to investigate the subject" (p. 4). It is particularly strange that an episode that has such coherent symbolic significances should seem to require from Dickens a defense of its literal truth. The spontaneous combustion of Krook is formally like the shooting of the albatross, and in the novel it is self-evidently the physical equivalent of the consumption in "costs" of the case of Jarndyce and Jarndyce, and the externalization of the moral nature of "justice" in Chancery. But again it suggests that while a "Coleridgean" novelist, showing himself most advantageously in extreme and quasi-supernatural situations, Dickens always saw himself as a realist, committed to the truthful representation of commonly experienced particulars. Thematically, his enterprise *was* very similar to that of the realists.

But Dickens had the confidence of natural theology, in which material reality corresponds meaningfully to a moral reality. The great analogy of natural theology, between physical and spiritual nature, is embedded in his imagination; the Darwinian disanalogy is the threat. If it is not quite the designed world of the natural theologians, it is nevertheless a world in which the fall into secularity is not inevitably a fall from grace. Allegory is not so much an invention as a representation, a mirror as much as a lamp.

In realist fiction of the kind Eliot wrote, "Nature has her language, and she is not unveracious; but we don't know all the intricacies of her syntax just yet."[36] In Darwin's writing nature is not illegible, but its syntax is difficult, and its meaning does not imply a moral reality inherent in the material, only its own nature. Eliot, through her conception of nemesis, often tries to infuse nature with moral meaning. But equally often she can sound like Darwin reminding us of the difference between "the face of nature bright with gladness"—the face that the natural theologian tends to see—and the destruction and devouring that accompany that "gladness." So she tells also of "what a glad world this looks like," but how "hidden behind the apple blossoms, or among the golden corn, or under the shrouding boughs of the wood, there might be a human heart beating with anguish."[37] In the wooden roadside cross Eliot finds a fit "image of agony" for the representation of what lies beneath the visible loveliness. It is a human symbol for a nonhuman nature. That is, the realist can find symbolic representations of the moral implication, but the symbol and the moral reality are human inventions. Nature is Darwinian. For Eliot, as for Dickens, the novelist was to make the ordinary resonant with myth, to show that the dream of romance is an absurd distortion and inferior to the romance of the ordinary, which contains within it forms of myth. But the romance of the ordinary is never inherent in nature. Nature's language is neutral.

The romantic-uniformitarian leaning of Dickens is partly undercut in the longer novels in which the traditions of stage melodrama are used to allow for quite literal catastrophe. Yet the distance between the uniformitarian and catastrophist position is much less absolute than it may at first seem. Both are romantic positions—the Wordsworth and Coleridge of science, as it were. In Dickens, the "catastrophe" of the murder of Tulkinghorn, for example, or the literal collapse of the Clennam house, can be seen as a metaphor for the consequences of the tedious daily accumulation of depressing facts—the slow grinding of Chancery, the moral bankruptcy of the deadening, static, circumlocutory world of bureaucracy and business. Dickens saw that the ordinary world was full of the extraordinary; he saw, too, that the extraordinary was the inevitable consequence of what seemed merely trivial, as an earthquake is caused by minute, almost undetectable movements over long periods of time. The argument between uniformitarians and catastrophists was, thus, double-edged, and we can feel analogous ambivalence in Dickens. If all extremes are merely accumulations of the ordinary, all the ordinary is potentially extreme.

The ordinary, then, is latent with possibilities of the extraordinary. It is a trick of contemporary horror movies, whose fundamental strategy is to focus on recognizable people in recognizable situations and then intrude something monstrous upon them. In Dickens, it is not only such gothic strategies (the talking chair in the *Pickwick Papers,* for example). But it is also Mrs. Copperfield bringing home a second husband who becomes, in his Puritanical austerity, a monster to the child. It is Boffin's dust heap, the dreary refuse of a recognizably ugly city, which becomes a mysterious treasure to Wegg; it is the clock greeting young Paul Dombey with "How-is-my-little-friend?"; it is Boz's superb account of the clothing in the window of the "emporium for second hand wearing apparel," which suddenly enact their melodramatic and yet commonplace histories. Like Darwin, who said, we must "no longer look at an organic being as a savage looks at a ship, as at something wholly beyond his comprehension," and must learn "to regard every production of nature as one which has had a history" (*Origin,* p. 456), Dickens makes us see the history—and the melodrama—in the commonplace object. He often affirms a world beyond the secular, but his works for the most part lose their touch with that world beyond, and with any authority except time, chance, and personal avarice. The world he creates—even Amy Dorrit's—is, like Darwin's, time-bound. Truth is not on the surface, after all, except as the surface offers to the keen observer clues to its history. All things imply histories but hide their pasts. By the time of *Bleak House*, only the professionally trained—police inspectors, like Bucket—can pierce through appearances with any confidence.

As he puts it at the start of his sketch "A Visit to Newgate," "force of habit" exercises great power "over the minds of men" and prohibits them from "reflection" on "subjects with which every day's experience has rendered them familiar." The essay "Character" is almost archaeological in that it infers whole lives from mere surfaces: "There was something in the man's manner and appearance which told us, we fancied, his whole life, or rather his whole day."[38] These attitudes and strategies are characteristic of Dickens's method throughout his career.

Such strategies parallel the views of the most advanced thinkers about science at the time. Herschel does sound occasionally like a romantic poet, or, perhaps more precisely, he formulates in the language of science ideas that were powerful in both poetry and fiction. If undisciplined experience leaves us open to our prejudices of opinion and of sense, a close look at nature—the glitter of a soap bubble, the fall of an apple—under the restraint of rational discipline, transforms it into a wonderland. "To the natural philosopher there is no natural object unimportant or trifling. From the least of nature's work he may learn the greatest lessons."[39] But the mysteries of the ordinary are only there for those who, like the readers of Dickens's novels, have been taught to look for clues.

As I pointed out in the last chapter, Darwin learned from Herschel and tried to emulate him. And although Herschel was not entirely pleased with Darwin's theory, he would have found in Darwin's work the same fascination with details, the same recognition of the miraculous nature of the ordinary, that he had tried to imbue in his readers. Darwin not only investigated the most ordinary phenomena—seeds in his garden, worm castings, bird excrement, bees' nests, pigeon breeding—but as a consequence discovered and persistently revealed that the details are not what they seem: plants travel; and organisms are frequently maladapted to their environment, or have organs irrelevant to adaptation. The strategy of defamiliarization so central to the *Origin* in its reeducating of natural philosophers and weaning them from creationism and natural theology, is akin to the strategies of domestic novels—as in Eliot's reminder in "Amos Barton" to learn to see some of "the poetry and the pathos . . . lying in the experience of a human soul that looks out through dull grey eyes."[40] The poetry and the pathos of natural history are evident in this piece of domesticated science, in which Darwin builds his argument about the way seeds can be transported: "I took in February three tablespoonfuls of mud from three different points, beneath water, on the edge of a little pond; this mud when dry weighed 6 ¾ ounces; I kept it covered up in my study for six months, pulling up and counting each plant as it grew; the plants were of many kinds, and were altogether 537 in number; and yet the viscid mud was all contained in a breakfast cup!" (*Origin*, p. 377). Such defamiliarization, characteristic of Darwin as well as of Dickens, makes it impossible to tuck nature into the neat formulas of natural theology. It is not only that the distribution of vegetation all over the world can be accounted for by natural means, but that nature is extravagant, wasteful, busy in activities not perceptible to the casual observer. Domestic detail changes under rigorous scrutiny. And such a passage is part of an overall strategy that suggests once more that there is nothing stable in the world around us. Species are not fixed but endlessly varying. All the stable elements of our gardens, our domestic animals, our own bodies are mysteriously active, aberrant, plastic.

On the issue of teleology Dickens tried not to be Darwinian. In novels so chance-ridden as his, one would expect to find real compatibility with Darwin, whose theory posited a world without design, generated out of chance variations. But since Darwinian variation occurs without reference to need, environment, or end, Darwin's chance is antiteleological. Contrarily, in traditional narrative of the sort Dickens wrote, chance serves the purpose not of disorder, but of meaning—from Oedipus slaying his father to the catastrophic flood at the end of *The Mill on the Floss*. The order "inside" the fiction might be disrupted—Oedipus's reign, or the life of Maggie—but the larger order of the narrative depends on such disruptions.

The difference might best be indicated by the fact that while both Dickens and Darwin describe worlds in which chance encounters among the myriad beings who populate them are characteristic, for Dickens chance is a dramatic expression of the value and ultimate order in nature, and it belongs recognizably to a tradition that goes back to Oedipus. Each coincidence leads characters appropriately to catastrophe or triumph and suggests a designing hand that sets things right in the course of nature. The "contrivances" in Darwin, however, though they tend to move the species toward its current state of adaptation or extinction, appear to be undesigned. Chance in nature drains it of meaning and value. The variations even in domestic animals, carefully bred, are inexplicable. Only close attention of a breeder, who discards variations he doesn't want, leads to the appearance of design. But the breeder is entirely dependent on the accidents of variation. Darwin and Dickens in a way tell the same story, yet the implications are reversed.

Working in a theatrical and literary tradition, Dickens must use apparent chance to create a story with a beginning, a middle, and an end. And it is a story much like that told by natural theologians, which makes "chance" part of a larger moral design, thus effectively denying its chanciness by making it rationally explicable in terms of a larger structure. The feeling of coincidence is merely local. Such manipulation is a condition of storytelling, where "chance" must always contradict the implications of the medium itself. Even in narratives that seem to emphasize the power of chance over human design, narrative makes chance impossible. Design is intrinsic to the language of storytelling, with its use of a narrative past tense. "Once upon a time" already implies design. Moreover, the focus of narrative attention on particular characters makes everything that happens in the narrative relate to them. It may be that the relation is a negative one: the character, like Micawber, waits for something to turn up, and it never does; but in the end, of course, Micawber has been in the right place at the right time, and while the narrator might applaud Micawber's sudden energy, what happened is not because he chose it. At the same time, the narrative certainly did choose it, both because it in fact helps Micawber achieve the condition to which he has always aspired, and because it allows the exposure of Uriah Heep and the righting of all the wrongs with which that part of the story has been concerned. Ultimately, it is all for the sake of David Copperfield as the happy resolutions of *Mansfield Park* are for Fanny.

Chance in narrative has at least two contradictory aspects. When Eliot's narrator condemns "Favourable Chance" in *Silas Marner,* she is among other things suggesting (what Darwin would have agreed to) that the world is not designed for any individual's interest. What Godfrey or Dunstan wants has no more to do with the way the world operates than what the giraffe wants. The giraffe's long neck does not develop because he wants it, but because longer-necked giraffes had on the whole survived better than shorter-necked ones. Nothing is going to shorten the trees for any given giraffe; nothing, presumably, will put gold in Dunstan's hands or rid Godfrey of his wife. Dickens, I believe, would subscribe to this way of seeing, although his attacks on chance are less obvious and direct. Yet in *Silas Marner* all the major events are the result of "chance."[41] The narrative does not make credible a necessary connection between the events and the behavior of the characters, nor does it try. The fabular structure of the story is outside the realistic mode that the expressed sentiments of the narrator affirm. In being much more self-consciously a "tale," and less a "realistic" representation of the world in all its complexity, *Silas Marner* exposes boldly what is usually more disguised in realistic fiction, where the necessary "coincidences" are normally made to appear natural and causally related.

The "chance" events in *Silas Marner* self-consciously work out a parable (complex as it becomes), in which they all reflect moral conditions and shadow forth a world in which the principle of nemesis, works, in which we bear moral responsibility for what we do; and that moral responsibility is worked out in nature and society. The effect of the narrative is to convey the sense that the "chance" events were determined by a designing power, intrinsic to nature itself, that used to be called God.

Narrative, it is assumed, is different from life, however, and presumably "real" coincidences would not imply the design of some "author." But any language used to describe events will turn into narrative and import design once more. Sudden catastrophes invariably evoke the question, "Why?" "Why did he have to die?" The question implies that there are "reasons" beyond the physiological and that the explanation, he was hit by a car, or his heart stopped, is not satisfactory or complete. What moral end was served? Where is the justice in the death? Or, if catastrophe is avoided, the language is full of "luck," the remarkable luck that we canceled off the plane that crashed, and the accompanying sense that we weren't "meant" to die yet. Often, others' catastrophes inspire guilt, as though the survivors are responsible, or could have managed to swap positions had they the courage. Even in trivial affairs, this tendency of ordinary language is powerful. We talk about bad weather as though it were designed to ruin our one day off, or we carry umbrellas and half believe that this will trick the rain away.

Such anthropocentric language is characteristic of natural theology, and Dickens does not resist it. But Dickens still uses chance to project a world governed by a great designer, even if he often has difficulty doing so. Putting aside the random abundance of the earlier works, we find that the self-consciously less episodic and more thematically coherent later novels use mysterious and apparently inexplicable details for the sake of human significance. Inevitably, Dickens does produce a Darwinian excess, which he needs to ignore or compress into order to achieve the comfort of significance; but his plotting is determined by the illumination and intelligible explanation of *apparently* random detail. The collapse of Mrs. Clennam's house is both literal and figurative, of course. Oliver's innocence is preserved and triumphant. Carker is crushed by the new railway, which we earlier learned opens for all to see the ugliness and misery of London. Dickensian narrative derives much of its energy from the gradual revelation of the design that incorporates all accidents, just as Herschelian science derives its energy from the attempt to explain all of the minutest natural phenomena in terms of general law. Characters struggle to discover its existence, and to work out its particular meaning, while the reader is always several steps ahead of the characters and several behind the author. We know that there is meant to be nothing chancy about Dickensian chance.[42]

Whereas Dickens, then, could exploit the metaphorical implications of language with confidence in its power to reveal design, Darwin had to resist language's intentionality and implications of design in order to describe a world merely there—without design or meaning. We have seen how in the very act of developing his theory and rejecting arguments from design he fell into the metaphorical and storytelling structures of the language to talk about "Natural Selection" as a "being." But the development of genetic variations, as Waddington points out, is not causally connected with the selective process that will determine whether the variations survive. In narrative terms this suggests that there can be no moral explanation, no superphysical "justification" of the development. The gene and its phenotype develop regardless of their narrative context. Such a separation drains nature of its moral significance and links Darwinism with the realist project that Dickens resisted even as he more than half participated in it. The matter of chance and teleology constituted the core of Dickens's defense, his attempt to keep nature from being merely neutral.

But Dickens did not reject science in order to resist that cold neutrality. It was Darwin, most effectively, who split scientific from theological discourse on this issue: science would not allow any "explanation" that depended on unknown principles that might be invoked, erratically, whenever empirical investigation failed. Scientific faith in law need not extend to scientific belief in the good intentions of the natural world. Dickens, like Darwin, would exclude mere caprice from the universe, but Darwinian "law" might well be regarded as capricious from the human point of view.

Yet another essay in *All the Year Round* provides a typical Victorian affirmation of the value of science, which grows from "Patience," while "Magic," its ancient forebear in the quest for meaning and control, is based on "Credulity." For science to emerge, the essay argues, "the phenomena of Nature, at least all the most ordinary phenomena, must have been disengaged from this conception of an arbitrary and *capricious* power, similar to human will, and must have been recognized as *constant*, always succeeding each other with fatal regularity."[43] These are certainly principles to which Darwin, like Lyell and Herschel before him, would have subscribed. The writer here, in eagerness to dispel "caprice," is not considering the full human implications of this apparently unimpeachable, modern, scientific position. In narrative, ironically, the ultimate effect of the "scientific" view of order and regularity is that the world begins to feel humanly erratic. That is, it becomes a fatalistic or deterministic world, like Hardy's, in which events do indeed develop inexorably from the slow accumulation of causes; yet they are, from the human perspective, entirely a matter of chance, because they are not subject to the control either of will or consciousness. Such a world is not, strictly, disordered, but it is, as Mayr has argued, probabilistic: "No one will ever understand natural selection until he realizes that it is a statistical phenomenon."[44] Not only does it work regardless of the interests of individual members of the population, but its working can only be described statistically, without explanation of why in any case or in the majority of cases, things develop as they do. Natural selection is humanly meaningless. Narrative forces what abstract discourse can avoid, a recognition of the difficulty and potential self-contradictoriness of the very ideas of chance and order.

The radical difference between Darwin and Dickens, despite Dickens's predisposition both to science and to the overall Darwinian vision, is simply in that Darwin's "laws" have no moral significance. Although they can be adapted for moral purposes (and were, immediately and continuingly), they do not answer questions like "Why?" except in physical or probabilistic terms. Birds can carry seeds in their talons, or deposit them thousands of miles away in their excrement. But what design is there in these particular seeds, these particular species making the trip? Why did the bird eat this plant rather than that, travel to this island rather than that? Survival in Darwin's nature is not *morally* significant. Adaptiveness is not designed, being the mere adjustment of the organism to its particular environment, and it has no direction. There is no perfection in Darwin's world, no intelligent design, no purpose. Fact may not be converted to meaning.[45]

This is a very tough sort of "chance," and its toughness evoked resistance from scientists as well as writers and theologians. In Dickens, while Darwinian chance threatens almost instinctively to overwhelm order, chance largely derives from another tradition, the one, in fact, that Darwin was self-consciously combatting. His novels tend to act out the arbitrariness of the connections they want to suggest are natural (and in that unintended sense, even here they are Darwinian). Dickens tries to tie event to meaning in a way that removes from chance its edge of inhumanity. This is the very tradition that Darwin identifies when, in *The Descent of Man*, he explains why he had perhaps overestimated the power of natural selection in the early editions of the *Origin:* "I was not able to annul the influence of my former belief, then widely prevalent, that each species had been purposely created; and this led to my tacitly assuming that every detail of structure, excepting rudiments, was of some special, though unrecognised, service" (I, 15). Every detail, on that earlier model, means something. Insofar as Dickens's later novels begin to suggest a chasm between event and meaning (delicately intimated in the "usual uproar" that concludes *Little Dorrit*), Dickens moves, like the later Darwin, away from the natural-theological tradition that had dominated his imaginative vision.

It is perhaps a measure of how far Dickens has traveled from Austen's way of seeing, however, that even where he persists in the contrivances of coincidence, their discontinuity with the worlds he is creating is disturbing. Such discontinuity is particularly striking in *Little Dorrit* and *Our Mutual Friend*. In most cases, while there are no naturalistic laws by which to account for the "chances" in Dickens's novels, coincidence feels too often like a matter of the conventions of narrative. Of course, Lady Dedlock *must* die at the gate of the wretched source of all plagues, where Jo had given off one ray of light in his gratitude to the now dead Nemo. The characters cannot perceive the design, but it is really there. Still, though there are scientific laws that make the development of organisms intelligible, the comfort of intelligibility does not lead to the comfort of meaning and purpose: in Darwin's world, it is random.

Darwin could get nowhere with his theory as long as language was taken to imply an essential reality it merely named. As Gillian Beer points out, in this respect, as with the question of chance, Darwin was forced to use a language that resisted the implications of his argument. Language, she says, "always includes agency, and agency and intention are frequently impossible to distinguish in language."[46] Yet more generally, to borrow a page from Derrida, language implies "presence." It assumes an originary reality ultimately accessible. Here as elsewhere, Darwin avoided epistemology to stick to his biological business, and here again he was forced to resist the implications of the language with which he made his arguments.

I have shown how at the center of his theory is a redefinition of the word "species," by which he almost undefined it. Species can have no Platonic essence, and Darwin was content to use the word as others used it, while demonstrating that species could be nothing but time-bound and perpetually transforming aggregations of organisms, all of which are individually different. For the most part, Darwin tried to do without a definition of species at all, for a definition would have got him into the kinds of serious difficulties already discussed, leading to the view that his book, as Louis Agassiz claimed, was about nothing: if "species" is merely an arbitrary term not corresponding to anything in nature, then *The Origin of Species* is about nothing.

Definition would have implied an essentialist view of the world, one entirely compatible with natural theology, and incompatible with evolution by natural selection. Essentialism, as Mayr has noted, implies a "belief in discontinuous, immutable essences,"[47] and this belief is reenforced by the reifying nouns characteristic of our language. John Beatty, in a revision of his argument that Darwin in fact was denying the existence of species, points out that Darwin could "use the term 'species' in a way that agreed with the use of the term by his contemporaries, but not in a way that agreed with his contemporaries' *definitions* of the term."[48] Darwin was not a poststructuralist and would not have argued that there is nothing out there to correspond to his language. But he knew he would have been paralyzed by accepting the definitions of "species" current among fellow naturalists. Recall that when Darwin talks of the "something more" naturalists think is implied by the natural system, he is working with their general nonevolutionary understanding of "natural system" and classifications within it; what he does, to follow Beatty's point, is accept their usage but not their definition so that he can replace the essentialist "something more" with the "hidden bond" of "propinquity of descent."

Essentialism was the enemy of evolutionary thinking, creating the greatest obstacle to conceptions of change. Mayr singles out Platonic essentialism, "the belief in constant *eide*, fixed ideas, separate from and independent of the phenomena of appearance," as having had "a particularly deleterious impact on biology through the ensuing two thousand years." Essentialism made it almost impossible to name a *kind* of animal—say, horse—without implying both its permanence and the "real" nature of its identity in all important qualities with all other horses, regardless of its merely accidental, that is, its particular physical and living characteristics. "Genuine change, according to essentialism" notes Mayr, "is possible only through the saltational origin of new essences,"[49] and clearly Darwin, for whom nature made no leaps, found essentialism a large obstruction.[50]

In plotting and characterization, change (as I have earlier said) was Dickens's greatest difficulty. His narratives and his characters seem to belong to a saltational world. For the narratives do make leaps, and when characters change they often do so (particularly in the earlier work) through abrupt conversion, as, for example, Scrooge. As opposed to a realist like Eliot, who writes from within a tradition much more clearly related to Darwinian thought and to the advanced science and psychology of the time, and whose narrator claims that "character is not cut in marble," Dickens writes out of an essentialist tradition. Barbara Hardy has pointed out that his novels rarely escape some tinge of the tradition of the *Bildungsroman*, but Dickens's use of that tradition of character development and change rarely explores the slow processes by which characters in that tradition learn and grow.[51]

Typically, Dickensian characters behave as though they had single, discoverable selves that constitute their essence. Mr. Jarndyce is a good and generous man, all of whose strategies in the world are designed to reaffirm that goodness. To be sure, this also entails a certain deviousness, for if he is to accept congratulations on his goodness, he can no longer regard himself as disinterested. But this ambivalence is built into his essence, and one of his most characteristic self-expressions is his complaint about the wind being in the East, which signals either bad news or the self-division that comes when he is about to receive praise or gratitude. So it is, in other ways, with most of Dickens's characters, who have been criticized through the years by critics seeking more fluid, complex, unstable, and I would say Darwinian, "selves."

The essentialist nature of Dickens's imagination is perhaps most evident in the clarity with which he usually distinguishes goodness and badness. Dickens's tendency is to read character into these categories, even when by virtue of his extraordinary sympathetic imagination he creates sequences like that of Sikes on the run or Fagin in prison, which shift our perspective on the melodramatic narrative. But the moral borders are firmly drawn. As Leo Bersani has observed, "In Dickens, the mental faculties dramatized in allegory are concealed behind behavior which *represents* those faculties. And the critical method appropriate to this literary strategy is one which treats the words and acts of literary characters as signs of the allegorical entities which make up these characters."[52] In this respect Dickens is most distant from Darwin and realism. Eliot's emphasis on mixed conditions, mixed natures, and her virtual incapacity (until Grandcourt in *Daniel Deronda*) to create a figure of unequivocal evil, fairly represents the difference. Dickens, writing within the "metaphysical mode" as Edward Eigner defines it, is heavily dependent on plot and emphasizes the external rather than the internal, but only because he counts on the adequacy of the natural to express meaning. In keeping with the natural-theological tradition, the emphasis on the external itself depends on strong confidence in the legibility of the material world, its expression of spiritual and moral realities comprehensible to those who choose to see. Ironically, when the world is secularized, as in the Darwinian scheme, narrative must turn inward because the material world becomes increasingly unintelligible. In Dickens, whom I have been characterizing as essentialist, there is visible a growing inability to be satisfied with the essentialist imagination.

The process of making narrative more literal by turning from allegorical representation to psychological mimesis under the pressures of secularization parallels the strategies Darwin uses in breaking with essentialism. One of the great Christian metaphors, and one of the central concerns of Victorian writers, becomes in Darwin a literal fact: we are all one family. Not the idea, but physical inheritance connects all living organisms. The move severed event from meaning (in a way contrary to Dickens's largely allegorical use of event) and destabilized all apparently permanent values by thrusting them into nature and time. Essentialism and nominalism were, therefore, no merely abstract metaphysical problems. On the whole, common sense and tradition required a world in which the ultimate realities remained outside time, and in which an ideal essence (as opposed to biological inheritance) defined the self. The concept of "character" itself implies such an essence.

The implications of this distinction extend into every aspect of narrative art. The essentialist mode is, for the most part, metaphorical. On the one hand, it depends on the likeness between physical and moral states, and the Dickensian emphasis on the physical peculiarities belongs to such a metaphorical tradition. On the other hand, the nominalist position, like Darwin's, severs the physical from the moral, and Darwin begins to make the connection metonymically. That is, in *The Expression of the Emotions in Man and Animals*, Darwin tries to read feeling from expression. But his reading is predominantly physiological. For example:

> Although . . . we must look at weeping as an incidental result, as purposely as the secretion of tears from a blow outside the eye, or as a sneeze from the retina being affected by a bright light, yet this does not present any difficulty in our understanding how the secretion of tears serves as a relief of suffering. And by as much as the weeping is more violent or hysterical, by so much will the relief be greater—on the same principle that the writhing of the whole body, the grinding of the teeth, and the uttering of piercing shrieks, all give relief under an agony of pain.[53]

Just as, for Darwin, organisms are connected by physical inheritance, so moral and emotional states are expressed by physiological activity directed at physical defense and relief. All those aspects of human identity and experience that are traditionally regarded as uniquely human, connected with spiritual states unavailable to lower organisms, are in fact physical conditions shared by other organisms. Darwin had observed monkeys in zoos, for example, to discover whether "the contraction of the orbicular muscles" was similarly connected with "violent expiration and the secretion of tears." He notes that elephants sometimes weep and contract their orbicular muscles! Of course, novelists in the realist tradition did not need to accept Darwin's extension of the "uniquely" human to the rest of the natural world, but their emphasis on close analysis of character, increasingly from the inside, corresponded to a decreasing (but never extinguished) reliance on the conformity of physical and moral states, of the sort so characteristic of Dickens.

Nevertheless, Dickens's fiction does participate in the move toward a Darwinian imagination of the world, a growing uncertainty about the notion of an "essential" self or about the possibility of detecting the moral through the physical (a quest that is increasingly professionalized, requiring a Bucket to do the work); and he is thematically urgent about the need for change. Abruptness remains characteristic, yet his preoccupation with the slow but inevitable movements in nature toward change, and with the consequences of refusing it is something more than a throwback to old comic literary traditions. It is as though he accepts uniformitarianism but rejects the gradualism that Lyell imposed upon it; like Darwin he seems to be reconciling the progressivism of catastrophism with the naturalism of uniformitarianism. Several of his most wonderful narratives focus on this problem: Mrs. Skewton in *Dombey and Son*, Miss Havisham in *Great Expectations*, and Mrs. Clennam all succumb to the forces of change their whole lives would have denied. And yet the very figures Dickens uses to thematize change are static (essentialist in conception) and require extravagances of plot to force them into time. They dwell in worlds not where change evolves slowly through time, but where it comes catastrophically, through melodrama, revelation, conversion. In a world of Bagstock and Barnacle, Captain Cuttle and Flora Finching, Mr. Toots and Mr. Merdle, it is hard to imagine that each of us does not have some essential, inescapable selfhood. But Clennam, Sidney Carton, and John Harmon, not to speak of Pip and Eugene Wrayburn, all flirt with doubts about the self so profound that they verge on self-annihilation. Each of them either literally or metaphorically dies, almost as though it were suicide. The question of change, even of the reality of the self, moves from the periphery to the center of Dickens's art and brings him to the edge of the Darwinian world, which feels like a threat, but can also, as for Clennam, be a liberation.

In other respects Dickens's worlds often seem to be narrative enactments of Darwin's theory. Beer makes the connection by pointing to the "apparently unruly superfluity of material" in Dickens's novels, "gradually and retrospectively revealing itself as order, its superfecundity of instances serving an argument which can reveal itself only through instance and relations."[54] *Bleak House* is only Dickens's most elaborate working out of the way all things are connected, and connected by virtue of mutual dependence and relationship. The answer to the question, "What connexion can there be?" we have noted, is a genealogical one, that Esther is Lady Dedlock's daughter. Mr. Guppy and Mr. Tulkinghorn detect it immediately. But all characters eventually connect, in other literal ways, from the brickmakers, who deceive Bucket in his pursuit, to Jo, the literal bearer of the plague, to the lawyers who drain the life from Richard, to Mr. Jarndyce, Skimpole, Boythorn, and Sir Leicester himself.

Bleak House embodies in every aspect of its plots and themes the preoccupation with connections, across place, class, and institutions. But almost all of Dickens's novels from *Dombey and Son* to *Our Mutual Friend* are novels of crossing and interconnection and responsibility. And throughout these works Dickens employs devices beyond the plot to enforce an overwhelming sense of hidden bonds, dependencies that have about them the quality of mystery appropriate to both religious intensity and gothic narratives.

The great dramatic moments in Dickens are often framed by apparently irrelevant natural scenes. Among the most vivid is the passage in *Bleak House* that precedes the ominous image of Allegory pointing above Mr. Tulkinghorn's heads on the evening of the murder. It wanders, by way of the moon, far from Mr. Tulkinghorn: He looks up casually, thinking "what a fine night, what a bright large moon, what multitudes of stars! A quiet night, too." And the passage moves across the whole of the English landscape, hill summits, water meadows, gardens, woods, island, weirs, shore, the steeples of London. The universal silence is violated by the shot, and when the dogs stop barking, "the fine night, the bright large moon, and multitudes of stars, are left at peace again" (pp. 584–585).

The passage effectively intensifies the event, implying merely by description some larger significance, but it does not express in its physical nature the moral condition at the center of the narrative. The vast silent panorama, disrupted only by the barking of the dogs, is antithetical to the murder going on. And surely it is not mere ornamentation. Passages such as this (there are equivalent ones in *Little Dorrit*) are the scenic counterpart of Dickens's commitment to multiple plots. The natural world, for Dickens, contains and limits human action. Often, as in the passage just quoted, it comments ironically on the action. But the effect is always larger than irony. The world is larger than anyone's imagination of it; connections extend out endlessly. In its vast and serene movement, it seems indifferent to human ambition. Regardless of the arbitrariness and violence of human action, nature continues its regular movement, has its own plot, as it were, which inevitably crosses with and absorbs the human plot.

The novels from Dickens's great middle period forward, with one or two possible exceptions, build thematically on the conception of society as integrally unified, with the revelation of that unification a central element of plot as well as theme. It is clear in the relation between the Toodles and the Dombeys, as in that between Esther and the Dedlocks. Even in *Hard Times* the circus and Mr. Bounderby's past confirm connections among classes and types denied by social convention. *Little Dorrit* explores in William Dorrit's genteel ambitions and in the fate of the Marshalsea and Bleeding Heart Yard the inevitability of connections. *The Tale of Two Cities* acts out melodramatically and in the context of revolutionary action the undeniable mutual dependence of class on class. And *Our Mutual Friend*, in its river and dust heaps, in its tale of the crossing of classes, makes the theme of "connections" both symbolic and literal. The mutual dependencies on which organic life depends in Darwin are dramatized socially in Dickens through his elaborate and multiple plotting and through his gradual revelations, often through the structure of a mystery plot, of the intricacy of relations disguised by sharp demarcations and definitions of classes.

Dickens's world, then, is as much a tangled bank as that evoked by Darwin at the end of the *Origin*. Of course, Dickens takes the metaphorical, Christian view, that Darwin was to make literal, that we are all one and deny our brotherhood at our peril. And he strains his plotlines to do it. Darwin tells us that seeing all organisms as "lineal descendants of some few beings which lived long before the first bed of Cambrian system was deposited," makes them seem to him "ennobled" (*Origin,* p. 458). The world is a tangled bank on which "elaborately constructed forms, so different from each other, and dependent upon each other in so complex a manner," struggle and evolve yet further into the "most beautiful and wonderful of forms" (p. 460). Had Darwin not written this passage in almost the same form ten years before *Bleak House*, one might have thought he was trying to sum up that great novel.

It is not, however, simply in the fact of complex interrelationship and interdependence that Dickens's and Darwin's worlds seem akin. In both there is an almost uncontrollable energy for life. Dickens's novels are densely populated, full of eccentrics, variations from the norm, and, as in the case of Jo or the retarded Maggie of *Little Dorrit*, or Barnaby Rudge, or Smike of *Nicholas Nickleby*, marginal figures who test the validity of the whole society. Darwin, for his part, needs to locate the unaccountable variation, the deviant figure or organ that will not be accommodated in an ideal and essentialist taxonomy, like woodpeckers who do not peck wood, vestigial organs that serve no adaptive function. Relentless in the pursuit of detail, he qualifies almost all generalization, even his own charting of the descent of species, with the muted and powerful "nature is never that simple."

Victorian clutter and Dickensian grotesque are akin to Darwin's encyclopedic urge to move beyond the typical. In this quality Darwin transcends again the kind of rationalist law-bound science that he inherited from Herschel. Darwin is not very interested in types. I would argue that although Dickens's characters seem to be "types," they are atypical in their excesses. Old Joey B, Josh Bagstock, is not so much a type as a grotesque—an aberration from the norm whose fictional strength lies in the way the excesses echo recognizable and more apparently normal human behavior. That sort of grotesque, in its multiple manifestations, is very much in the Darwinian mode. As Michael Ghiselin has suggested, the Darwinian revolution, in its overthrow of essentialism, lays "great emphasis on the 'atypical' variants which the older taxonomy ignored."[55] Distortion, excess, and clutter are the marks of Victorian design, of Dickens's novels, and of Darwin's world.

For both Dickens and Darwin, knowledge (and humanity) are not attainable unless one learns to see the multiplicity of variants that lie beyond the merely typical. Biology revolutionized nineteenth-century science because it displaced mathematical models (however briefly) as an ultimate resting place for belief. Darwin's book insisted that one could not understand the development of species unless one recognized that individual variations were always occurring, and that the world is filled with developments from variations which might once have been considered aberrant.

Thus, for Darwin—and the pressure for this is evident in the sheer abundance of Dickens's novels as well—variety is not aberration but the condition for life. Dombey attempts to limit the Dombey blood, and in rejecting the fresh blood and milk of Mrs. Toodles causes his son's death. *Our Mutual Friend* concludes with one of the most remarkable moments in Dickens: Wrayburn is redeemed and brought back to life by crossing the class boundary to marry Lizzie Hexam. In the final scene Twemlow is forced to redefine language in a Darwinian way, moving the idea of "gentleman" from a fixed and permanent class definition to a vital (and indeed sexual) one that becomes so wide-ranging as to lose its exclusivity. "Sir," returns Twemlow to the permanently Podsnapian Podsnap, "if this gentleman's feelings of gratitude, of respect, of admiration, and affection, induced him (as I presume they did) to marry this lady . . . I think he is the greater gentleman for the action, and makes the greater lady. I beg to say, that when I use the word, gentleman, I use it in the sense in which the degree may be attained by any man" (p. 891).

Mixing and denial of absolute boundaries become the conditions of life in Dickens's novels as they will be in Darwin's biology. In Darwin, the unaccountable variation, the crossing—of sexes and varieties, if not of species—increases vitality. On the old model life was determined by separate creation and eternal separation into ideal and timeless orders. On the Darwinian model life is enhanced by slight disturbances of equilibrium, by change.

Learning to confront change and to make it a principle of life was Dickens's great trial as a novelist. It is a long way from Pickwick to the timid Twemlow, who affirms the value of mixing in a society constructed to deny it. Twemlow is, moreover, one of the very few minor characters in Dickens who actually change, and whose change is not abrupt reversal, but the quiet consequence of a long accumulation of frustrations. While Pickwick's novel happens episodically and manages to return to the moral condition of innocence by slamming the door, the later Dickens moves increasingly to the structure of Paradise Lost—you can't go home again; change is irrevocable. The Darwinian revolution entailed the rejection of cyclical history, or ideal history: time moved in one direction only (and in this respect, it made a powerful companion to the otherwise initially hostile science of thermodynamics, which introduced the irreversible "arrow of time" into physics—a development I will discuss in the next chapter, on *Little Dorrit*). And yet Dickens goes on, as in the Christian dispensation, to make the change he can no longer avoid facing, the loss of stasis and ideal innocence, into the condition of a greater redemption.

The resistance of Victorian intellectuals to Dickens as a truly serious novelist might be atttributed, partially, to two different emphases that can be traced conveniently in the reading of Darwin. The one, the more strictly intellectual, exemplified by G. H. Lewes and George Eliot, absorbed Darwin into the uniformitarian-scientific mode and, as it were, domesticated him and the chancy dysteleological world he offered. The other reads Darwin within a fully comic vision, lights instinctively on the aberrant and extreme cases of the sort that Darwin had to emphasize to disrupt contemporary religious and scientific thought and to reveal a world prolific and dynamic in the production of endless and sometimes grotesque varieties of life. Yet Dickens always struggled back toward the possibilities of essentialist thought and morality. The aberrant comes round to the ideal, at last.

While Dickens often strains toward the comfort of design, he has an astonishing eye for the aberrant and energy for abundance and for life. Perhaps equally remarkable, like Darwin, he managed to use inherited idealist and design-permeated conventions to build almost mythic structures of crossing and recrossing appropriate to the sense of modern urban life (Ruskin deplores this in his "Fiction, Fair and Foul," where he totes up the number of deaths and diseases in *Bleak House*). Dickens's capacity to imagine himself beyond the conventions of order that dominated Victorian social and political life allowed him to write in a way that helped open the culture for the Darwinian vision toward which, in his increasingly courageous confrontation with change, he himself was moving. Podsnap's relation to Twemlow parallels society's first response to Darwin. We should not be surprised that Dickens did not play Podsnap himself.

6

Little Dorrit and Three Kinds of Science

BLEAK HOUSE works through many of the crises that Darwin's thought made explicit. It confronts a world where law and regularity are both valuable, as the figure of the scientific Bucket implies; and the voice of the third-person narrator regards his human subjects from a cool distance that one might regard as "scientific" (disguising a deep anger from which the coolness protects). Law and regularity are, however, inadequate to human need, and the complex world of the novel implies a nonhuman vastness and complexity that seems like disorder. In a pattern almost the reverse of *Mansfield Park*'s, life and redemption become possible through the violation of established institutions. *Bleak House* imagines a world threatened by moral meaninglessness and the disanalogies implict in Darwinian rejection of the relation between the physical and the spiritual world.

There is a strong tension between the narrator's and Esther's and Bucket's persistent quest for knowledge, and the apparently approved sense that protection depends on a refusal to know, as Jarndyce withdraws into his Growlery, and Esther speaks with the wisdom of innocence. The price of knowledge in a world subject entirely to naturalistic analysis is high. And the novel as a form, particularly the realistic novel of the nineteenth century, typically dramatizes the cost, treating knowledge as dangerous. *Bleak House*'s anti-intellectual strain is evident in more consistently realistic novels, like George Eliot's, and I shall be returning to this issue in the last chapter. The point here is that *Bleak House* is deeply divided about knowledge. It empowers Tulkinghorn to blackmail; it almost empowers Bucket to save Esther's mother. Krook and Chancery collect and bury knowledge. The issue for Dickens is not so much knowledge as its uses. The amorality of Darwinian knowledge is a danger;

the importance of knowledge—as in Woodcourt's medical skills—is manifest. The pervasive science of *Bleak House* is assimilated to a religious ideal, and the narrative itself approaches allegory to affirm a reality more real than the material world.

Dickens, then, hovers half way between the conventions of reality implicit in natural theology (analogical, teleological, catastrophist, and beneficent) and those developing in modern science toward Darwinism (metonymic, organic, dsyteleological, uniformitarian). It is a cliché that *Bleak House* is organicist, that it works out in its very structure the almost physiological interdependence of all society and, implicitly, of all nature. Organicism, however, becomes an injunction (as in Eliot) to mutual responsibility and recognition. *Bleak House* fights off the threat of a purely materialist and secular reading of nature, but the threat is a consequence of knowledge—of Tulkinghorn, who is demonic, and of Bucket, who is not quite. Ironically, it fights off that vision, which inheres also in the rhetoric of the third-person narrator, with a representative of the Victorian conventions of realism.

Bleak House, that is to say, is as divided as its two narratives, requiring new knowledge and distrusting it, valuing the domestic but setting it in a world of catastrophic leaps, leary of a uniformitarian dullness that can grind the spirit down or domesticate sublimity. Dickens experiments with the new knowledge and assimilates it to old ideals. But his imagination entails along the way a prodigality of creation and connection that creates what appears to be a very un-Austenian shapelessness (or a new kind of entangled shape) and sanctions the resistance to conventions that defeated Mary Crawford.

Little Dorrit, while including all those Darwinian elements I discussed in the last chapter, takes the experiment further. It manifests a significant shift in Dickens's imagination: the catastrophist possibilities of an escape from system into moral freedom are further tested and appear even less likely. It describes, and through its main characters seems to enact, a diminution of energy. Most important, the potential harmony between the natural and the human, already threatened by the Darwinian vision, is subjected to yet greater skepticism. Willed ignorance of the facts of nature provides no protection and becomes, in the Circumlocution Office, a central disintegrative force. The harmony between the natural and the human constructed laboriously at the end of *Bleak House* is only symbolically affirmed at the end of *Little Dorrit*, and not allowed the last word, after all.

Little Dorrit also struggles movingly against the bleakness of a world in which change is the rule and knowledge and moral meaning are separated. The Darwinian model can still in many respects apply, but perhaps an even better model for the surrender to the secularization of nature and its inhuman energies is thermodynamics. It is as extravagant to think of Dickens in relation to thermodynamics as in relation to Darwin; yet, not only was Dickens well attuned to the attitudes and tendencies of science, but society as a whole was creating the conditions that made both Darwinism and thermodynamics conceptually possible. Different, even antagonistic as they occasionally were, the two systems were equally part of the progressive secularization of knowledge and society; equally enforced the denial that nature was God's second book; equally cut off the possibility of locating an originating source that would ground knowledge and value; equally thrust all phenomena, including the human, into time; and equally decentered the human. Ironically, the great romantic displacement of the human with a sublime and impersonal nature found its scientific counterpart in such theories. In this chapter I will be looking at *Little Dorrit* as implying a world coherent with that asserted by thermodynamics—seeing it, then, in the context of three sciences: Darwinism, natural theology, and thermodynamics. These three often overlap, but at least as frequently are opposed, and their juxtaposition suggests something of the tensions within the novel itself.

The strangeness of *Little Dorrit* can be registered in its self-contradictions: the most religious of Dickens's novels, it most relentlessly explores the unredeemed secularity of human society. With much of the abundance and extravagance characteristic of the Darwinian novel, it implies a world impersonal, austere, and restrictive. Consistent with the thrust of Darwinian themata, it manifests a diminishing faith in the authenticity and value of selfhood and initiates in the Dickens canon a series of studies of the artificiality and ultimate breakdown of the self, like those in *Great Expectations*, *A Tale of Two Cities*, and *Our Mutual Friend.* Like other Dickens novels, it is filled with aberrant eccentrics, yet most of these sustain their identities through willed retreat from nature and time. While it seems to condemn the lack of will of William Dorrit and the society he represents, it attempts to dramatize the redemptive capacity of the absence of will. This breakdown of selfhood and the contradictions that impel it are related to the conflicting ground plans of reality that contemporary scientific thought was drawing up.

Little Dorrit's prose is energetically preoccupied with disorder and loss of energy. Although it would be absurd to claim that Dickens had in mind the developments in thermodynamics of the 1840s and 1850s, when it was being formulated so as to change radically the way scientists and lay people thought about the nature of the physical world,[1] it comes quickly to mind as an appropriate metaphor for what happens and the way it is expressed in the novel. The source of the narrative is weakness—Clennam's and William Dorrit's. Through them from the start, it thematizes failure of energy. To be sure, much of the book's moral energy, particularly as it is embodied in the plot, is directed at countering that failure: Dickens appears to want to deny the absence of energy that he uses and exposes. At its most literal, *Little Dorrit* attempts, like most of Dickens's earlier work, to affirm a world of abundance, growth, and multiplicity, a world far closer to Darwin's than to Helmholtz's. But Clennam is a curiously weak protagonist, unable to make claims, passive except in the attempt to alleviate his guilt for nothing that he has done or knows, unwilling because he is "nobody" to express his love to either of the two women who attract him. In certain ways a study for Pip, he has none of Pip's early ambition or passion, certainly none of his vitality. And Amy as heroine constructs her heroism from an utter refusal to express desire, a total selflessness. Energy in the novel belongs to a world furiously moving into disorder and meaninglessness. The saving figures must be saved by energies not their own, relying on a disintegrating nature (and society). The strangeness and self-contradictions of the novel enact a conflict between two mythic structures, the progressive vision of Darwinism and the degenerative vision of thermodynamics; it is a conflict for control of a world Dickens was still trying to save—as he did in *Bleak House*—for order, stability, and God.

The question of direct influence by, say, the work of Lyell, Robert Chambers, Nicolas Carnot, Robert Mayer, James Prescott Joule, or William Thomson, does not enter here, though thermodynamics was, like evolution, in the air. The implications of Helmholtz's critical formulation of thermodynamic theory of 1847 probably did not reach the level of popular exposition in England until the late 1850s or early 1860s,[2] when John Tyndall, in particular, disseminated the idea broadly. But the theory of the conservation of force, as Helmholtz called it, that energy could not be destroyed, and that energy within a closed system never increased without the introduction of new force into that system, was becoming part of scientific consciousness and was unsystematically making its way in the world.

Household Words is once again a useful place to look for what Dickens might have known and thought about science. "Nature's Greatness in Small Things" (1857) tells of the unity of type that underlies all variation among living organisms, echoing Matthias Jakob Schleiden's views on cell theory and Richard Owen's on comparative anatomy. To the point of thermodynamics, "The Mysteries of a Tea-Kettle" (1850) tells many of the same things that Joule—another key figure, with Helmholtz, in the development of thermodynamic theory—told his audiences in his important lecture "On Matter, Living Force, and Heat" in 1847, or that Tyndall, in the major English popularization of thermodynamics and the law of the conservation of energy, was to tell in his *Heat: A Mode of Motion* (1863).[3]

The first of the "three sciences" I want to impose on *Little Dorrit* is our old friend, natural theology, which was the framework for all popular natural history, and for most science in England.[4] Although the Bridgewater Treatises were almost the last deep breath of natural theology, they helped perpetuate the endangered tradition of reading matter as evidence of spirit. One could be a physicist or even a Darwinian (as the example of Asa Gray makes evident) and retain natural-theological belief in the evidences of the creator, although that often entailed strategic maneuvering of the sort that brought epicycles into Ptolemaic cosmology. However rationalistic, natural theology allowed for a continued narrative tradition of affirming a world spiritually alive behind the appearances of matter, of quasi-allegorical and teleological realism.

The second of the sciences is thermodynamics. While evolution seemed immediately threatening, thermodynamics was not taken as incompatible with religious faith, witness the very religious Joule, or William Thomson, later Lord Kelvin. Yet thermodynamics would ultimately prove equally damaging to an anthropocentric view of the world, and the second law seemed from the start to run counter to the optimistic "progressivist" directions of most contemporary science, particularly evolution. Not long before *Little Dorrit*, Thomson was foreseeing the death of the sun.

The third of the sciences is evolutionary biology. What needs emphasis here is its tendency toward progressivism (although I have shown that a strict reading of the *Origin* does not justify progressivism). The critical question for those concerned with the history of organisms on the earth—the mechanism by which new species appeared in successive geological strata—was answered in one way by the catastrophists and in another by Darwinism, which also rejected Lyell's steady-state uniformitarian view to recognize development. Chambers needed spontaneous generation as

Darwin was to need Malthusian overpopulation, but the direction implied by the fossil record—with some losses at least for extinction—was onward and upward. Dickens was probably an Owenite (a potentially progressivist position), yet certain elements of *Little Dorrit* suggest a break with pre-Darwinian attitudes. While Darwin's theory deprives life of its spiritual ancestry and subjects everything to natural law (just as Chambers's did), the progressivist direction of evolutionary thought was easily adaptable by some scientists and clerics to teleology and design;[5] but the degenerationist directions of thermodynamics, which also applied theories of energy source and expenditure to the human body, led to the bleakness of, for example, Huxley's *Evolution and Ethics* forty years later.

These three sciences suggest three different versions of the world: one, a world rational, just, divinely meaningful; one a world fallen from a golden age; one, a world moving toward that golden age. Each myth posed a different sort of threat for Dickens and his audience. Evolutionary theory and thermodynamics came into conflict directly when Thomson deprived Darwin of the time he needed for natural selection to create the organic world we know.[6] Kelvin, and physicists who followed him, waged a continuing war against Darwinism through most of the last half of the nineteenth century, primarily by challenging the uniformitarian grounds of the argument. There simply could not have been anything like the three hundred million years Darwin inferred for his evolutionary processes. Not only was there inadequate time at the beginning, but the end was coming sooner than we would like to imagine. Thermodynamics—at least its second law—came to suggest a tragic narrative, evolution a comic one. Even Herbert Spencer, who managed to reason his way out of the thermodynamic tragedy, wrote in 1852 of the impending catastrophe juxtaposing evolutionary to thermodynamic thought:

> If evolution of every kind is an increase in complexity of structure and function that is incidental to the universal process of equilibration, and if equilibration must end in complete rest, what is the fate toward which all things tend? If the solar system is slowly dissipating its forces—if the sun is losing his heat at a rate which will tell in millions of years—if with diminution of the sun's radiations there must go on a diminution in the activity of geologic and meteorologic processes as well as in the quantity of vegetal and animal existence—if man and society are similarly dependent on this supply of force that is gradually coming to an end, are we not manifestly progressing toward omnipresent death?[7]

In Spencer's argument, as in science as a whole, physics took priority over geology. Stephen Brush points out that the more theoretical and general science, physics, intimidated the more empirical and observational science of natural history, geology.[8] No geological evidence could withstand Kelvin's abstract theoretical argument about the universe's dissipation of energy. This dominance of the general over the particular, which we have noticed elsewhere, importantly affected narrative, with its emphasis on observation and empiricism. In Dickens, the emphasis on the particular is life-giving, on the general (and the institutional, which seems to dominate in *Little Dorrit*) is deadly. Nevertheless, the tension between thermodynamic and evolutionary theory seemed for a brief period almost absolute.

The theories agree, however, in their emphasis on time—not on its extent but on its unidirectionality. While Darwin characterized all things by their continuing and evanescent movement through time, thermodynamics was almost equally important in thrusting time into "natural law." The formulas of Newtonian mechanics worked with equal effectiveness backward or forward. But the second law of thermodynamics introduced the irreversible arrow of time, the inevitable movement from warmer to colder, from order to disorder, from concentrated to dissipated energy.

It is important to keep clear a distinction between the idea in thermodynamics that there is never an increase or a decrease in the total amount of energy in a system, only a redistribution, and the idea of entropy. "Entropy" is the process that brought thermodynamics into conflict with Darwinism because while sustaining the idea that there is neither loss nor gain in total energy within a system, it projects a world in which the transformation of heat into energy always entails a loss, an inefficiency that makes the *perpetuum mobile* chimerical. While the total energy remains the same, the total *available* energy diminishes so that within any closed system (like the world) the movement is always toward increased cooling and increased disorder and decreased energy available to do work. The irreversible arrow of time is always pointing downward. Joule could reconcile the law of conservation of energy to his religious belief because, without taking into account the inevitable decline in usable energy, he could affirm a world in perfect balance, actually not unlike the uniformitarian, cyclical world of Lyell, with which (barring entropy) thermodynamics is actually compatible in many ways.[9] The thermodynamics that Ann Wilkinson discovers in *Bleak House* might be called Jouleian, in this respect, for the novel posits a world in which the lost energy of the chancery suit is recompensed by the new energy of "negentropic" Esther (how would she or Dickens feel about such an epithet!) and Alan Woodcourt, who reenergize the system, fully compensate for its losses.

But the slow descent into cold and disorder, the dissipation of energy projected by the second law of thermodynamics at least *seems* incompatible with Darwinism. Evolution, of course, is subject to the same physical constraints that apply to the inorganic world. But the movement of evolution is, as Darwin put it, from "the war of nature, famine and death" to produce "the most exalted object which we are capable of conceiving" (*Origin,* p. 459). Within a universe conceived as slowly dying, biological evolution seems to make a negentropic island.

But even here there is a difference. Although Darwin's time may entail loss, it tends to transform loss into gain and implies movement from lower to higher forms of order. The Spencerian formulation is that all organisms (and societies) develop through increased integration to increased heterogeneity, complexity and interdependence of parts. The energy dissipated is transformed into a greater integration of matter. For Spencer, then, development is progressive and implicitly teleological.[10] Whereas Darwin's theory implies abundance, excess, and multitudinousness, thermodynamics is a rather stingy theory. If energy does not perish, it does not increase and is only redistributed. Biology and physics were for a while at odds since biology studied objects (akin only in physics to the sun, which posed a problem because it was not consuming itself at a rate compatible with the known laws of chemistry), which seemed negentropic. That is, they did not run down the way inorganic matter ran down; hence, physicists like Tyndall were eager to show that the human body was subject to the same laws of energy expenditure and loss as all other matter. In thermodynamics for every gain there must be a loss, and inefficiency is universal; the quantity of heat never translates into work without some loss. "Nature," says Tyndall in his splendid popular treatise on heat, "is a constant quantity, and the utmost man can do in the pursuit of physical truth, or in the application of physical knowledge, is to shift the constituents of the never-varying total, sacrificing one if he would produce another. The law of conservation rigidly excludes both creation and annihilation."[11] (Nevertheless, Tyndall has very little to say about the bleak possibilities of the heat death that must ensue in the universal entropic movement of nature; his use of thermodynamic theory has little of the evidence of *dissipation* of energy that I find in *Little Dorrit.*) Evolution, by contrast, is profligate and regenerative. Thermodynamics has something of the severity of the austere Calvinist world that almost crushes Clennam, that imprisons his mother, that punishes inexorably every spontaneous outburst of feeling, every excess.

But the ostensible incompatibility between aspects of thermodynamics and evolution is probably of less significance than the incompatibility between natural theology and both thermodynamics and evolution. Both sciences participate in the positivist extension of natural law to human behavior. Darwin's uniformitarianism, we have seen, was partly a commitment to an entirely secular means of explanation of natural phenomena. At all costs, he had to avoid explanations that depended on singular intrusions into the "laws" of nature, and was therefore resistant to any explanation dependent on causes not *now* in operation.[12] Although Joule believed that his argument about the indestructibility of living force could have been derived *a priori* on the strength of our knowledge of the creator, and Kelvin himself allowed his science to support the view of the compatibility between religious and material explanations, the success of material explanation gradually removed the necessity to invoke God or design or intention. The loopholes for spirit were closing rapidly.

The second law not only seems to exclude design, but also reduces, perhaps eliminates, the possibility of human intervention in its relentless processes. Thomson managed to remain a progressivist and a believer only by positing the possibility of action *outside* the system of nature as we now understand it. Such a move, Darwin would have argued, is a betrayal of science and would have been, as he liked to say, "fatal" to his own theory. Although, Thomson says, "mechanical energy is *indestructible*, there is a universal tendency to its dissipation . . . The result would inevitably be a state of universal rest and death, if the universe were finite and left to obey existing laws."[13] For this reason Thomson led an attack on uniformitarianism and continued the strong early century tradition that regarded science and religion as complementary. Yet there is a difference. To account for the way the world is, it will be necessary to discover laws not now in operation. Unlike Paleyan thinkers, Thomson was not arguing for divine intelligence *in* the laws of nature. He affirmed the divine presence by giving the game away to positivism: religion and science take parallel paths in which their mutual claims do not conflict or converge.

> It is impossible to conceive a limit to the extent of matter in the universe . . . Science points rather to an endless progress, through an endless space, of action involving the transformation of potential energy into palpable motion and thence into heat . . . It is also impossible to conceive either the beginning or the continuance of life

> without an overruling creative power; and therefore no conclusions of dynamical science regarding the future condition of the earth can be held to give dispiriting views as to the destiny of the race of intelligent beings by which it is at present inhabited.[14]

This intellectual dead end has its narrative counterpart. What separates both thermodynamics and evolutionary theory from natural theology is their mutual readiness to read all experience in inhuman but nondivine terms. Human life is determined by its intake and expenditure of energy. The laws that govern the expiration of a candle govern the expiration of human life.[15] Together thermodynamics and evolution offer themselves as modes of explanation that omit what Dickens would think of as the distinctively human; and they pose two threats: that a moral, voluntarist reading of experience is no longer possible, and that the meanings constructed by narrative resolution will be arbitrary human impositions on forces that owe allegiance only to the laws of matter, not of spirit.

We take for granted George Eliot's concern with the large impersonal forces, social and psychological, by which the individual will is bent to the service of its primary animal nature and constrained by the weblike, irrational, and powerful community. The bleakness of *Middlemarch* reflects the power of what she once called "undeviating law," and the apparent powerlessness of individuals not only in increasingly complex social structures but against the unconscious strategies of their own psyches. Given the representative status of *Middlemarch*, it is easy to assume that the Victorian novel, as a form, consistently struggles with such impersonal constraints. Yet even with Scott, Thackeray, and a remarkable series of his own novels behind him, Dickens did not have available a fully developed tradition of this kind. Through most of his earlier work, he had celebrated exuberantly the powers of innocence, of good intentions, of change of heart, although he had always been uneasy about strong will. Even in *Bleak House* although the individual is powerless against institutional constraints, the narrative ultimately affirms the possibility of keeping free. In *Little Dorrit*, there is far more uneasiness, and no celebration. It manifests again Dickens's deep distrust of institutions but extends the power of institutions into the individual wills and consciousnesses of the best kinds of people. The institution in *Little Dorrit* is no longer a merely external force; its pervasiveness is almost total because it has become internalized. Strong will becomes ineffectual. Energy dissipates and landscapes fragment into disorder.

In *Little Dorrit* Dickens seems to make discoveries that subvert his intention to affirm the power of love over system, the possibility of harmony amidst disintegration: faith in energy and free choice is partly denied at the moment it is affirmed. Choosing to see nature from within the context of natural theology, he reveals a disorder that is not a disguise of ultimate design—as, say, even the chaos of Tom-All-Alone's is; the disorder is a natural condition of the loss of energy everywhere observable. The web of connections determining modern life is reckless of individual will and entails a reimagination of "character"; the pressures of the impersonal assert themselves in the self and threaten identity; the mystery required formally to sustain Dickensian narratives and thematically to construct a new urban world no longer legible even to its most familiar citizens is not so obviously penetrable. Choice begins to lose its significance, and the natural world threatens to mean nothing but itself. We find in *Little Dorrit* ambivalences about modern, particularly urban life, with its enormous potential for creative change and its apparently inevitable move to decay. These shadow forth a crisis of selfhood and personality, later celebrated by D. H. Lawrence, but here uneasily wavering between religious and secular forces, and echoing the contentions among the scientific theories I've been laying out.

The development of these ambivalences and contradictions can be traced from Dickens's initial conception of the novel. As Harvey Sucksmith notes in his introduction to *Little Dorrit,* the original title was to be "Nobody's Fault," and its central character was to be a man "who should bring about all the mischief, lay it all on Providence, and say at every fresh calamity, '. . . nobody was to blame.' "[16] Its central idea as proposed in the notebooks was "the people who lay all their sins negligences and ignorances on Providence." The title, of course, was to be ironic, and in Gowan's cynical stance we have the sort of thing Dickens must originally have intended. But Dickens had great difficulties with the conception. The problem, Sucksmith argues, was that "the central idea he had chosen for the novel was incapable of organizing the material into an integrated and meaningful structure and vision."[17] My emphasis would be different. Dickens was finding that the initiating idea was becoming literal rather than ironic. The world he was constructing was making individual choice, or action, almost impossible. Action in the novel often becomes little more than exaggerated passivity, particularly with Arthur Clennam, possibly, too, with Little Dorrit herself. Dickens, that is to say, had moved into a world where "nobody's fault" had become an almost inevitable reading of experience, where the power of the will was indeed in question, where the voluntaristic model of behavior was put to the test.[18]

In what follows I examine how some of these pervasive difficulties, intimated by reiterated images and motifs, can be seen as reflections of the new thermodynamic model of nature, and how the efforts to reject this model by reaffirming the design and meaning of natural theology lead to creative incoherences in the way Dickens imagines "character," self, and the action of the will. A narrative, of course, is not a cluster of independent images and metaphors, and in focusing on these and, for the most part, ignoring Dickens's narrative organization of them, there are obvious dangers. A literal reading of the novel would need to take into account the slow "natural" growth in the love of Amy Dorrit for Clennam and the triumph of that love over the indifferent flow of life all around them.[19] But the patterns with which I am concerned constitute, as it were, a second "plot," crossing and contesting with the one (or ones) Dickens obviously labored to achieve. The very laboriousness of the construction of plot in *Little Dorrit* suggests to me difficulties of the sort I have already intimated: what Dickens wanted to demonstrate was being confuted by the materials with which, in the spirit of quasi-scientific disinterest, he worked.

The science plot implicit in the images of *Little Dorrit* shows nature resisting and indeed overwhelming the human intention. Taken together, the images imply that the primary force in the world is a mysterious nature that creates its own impersonal plot of entropic decline and cuts like fate itself across the conventional narrative through which characters and narrator alike aspire to meaning. More repressive than society itself, because more embracing, nature manifests itself particularly through ineluctable time. As Thomson invoked the infinite to deny the spiritual authority of the thermodynamic laws he formulates, Dickens attempts to make the natural and the divine compatible. In a certain sense *Little Dorrit* struggles to reestablish the analogical connection between the material and the spiritual that had been one of the assumptions of natural theology. But the option of timelessness, intimated through a natural-theological view of laws of nature and through the prison bars not of society but of nature itself, is unavailable. Uniformitarian geology or modern thermodynamics entail a recognition of the dominance of time in all areas of experience. The risk and the power of *Little Dorrit* is in Dickens's dramatization of the primacy of the secular: the protagonists must reengage in time. Despite the ultimate convergence in love between Amy and Clennam, only on the margins of the narrative, perhaps in Doyce, certainly in Physician, is there the suggestion of a world less austere, where the irreversible arrow does not inevitably point downward.

The primary physical feature of this world is entropic decline: the novel is full of fragmented land- and city-scapes, from the "Babel" of foul-smelling Marseilles, to the dissonance, soot, and death carts of Clennam's plague-ridden London, to the ruins of Rome, through the Alps, where there was loveliness on the surface but "dirt and poverty within," to Venice, with its mouldering rooms and fading glories. Matter, objects, ominously symbolic in the texture of the prose, resist the meanings to which they are assigned, do not seem evidence of the power, goodness, and wisdom of the creator. The sun that beat down on Marseilles that August day is raw, oppressive energy, exhausting all it touches, signifying nothing but itself—it "was no greater rarity in Southern France then, than at any other time, before or since."[20]

Each book begins with extreme images of disorder. The powerful Marseilles sun beats down on dust, disharmony, fatigue, and the prison, which itself holds a cluster of heterogeneous and waste objects, vermin, human and rodent. The second book opens at the Great St. Bernard, in a chaos of cold, of mules biting each other, of men racing about; and "in the midst of this, the great stable of the convent, occupying the basement story and entered by the basement door, outside where all the disorder was, poured forth its contribution of cloud, as if the whole rugged edifice were filled with nothing else, and would collapse as soon as it had emptied itself, leaving the snow to fall upon the bare mountain summit" (p. 484). The book's obsessive preoccupation with such scenes carries into minor episodes and almost gratuitously unstable scenes. Visiting Fanny's theater, Amy came "into a maze of dust, where a quantity of people were tumbling over one another, and where there was such a confusion of unaccountable shapes of beams, bulkheads, brick walls, ropes, and rollers, and such a mixing of gaslight and daylight, that they seemed to have got on the wrong side of the pattern of the universe" (p. 279). The Gowans' house in Venice seems to have "floated by chance into its present anchorage," and is surrounded by decaying buildings, washed linen, houses "at odds with one another," and a "feverish bewilderment of windows" (p. 543). The packet dock at Calais is extensively described in its bleak and weather-worn decay; Miss Wade's house, with its cracked door knocker, dead shrubs, dry fountain, statueless grotto (p. 716) is only another and lesser version of Mrs. Clennam's house, held up tentatively by a great timber and ominously creaking with decay. And the metaphorical disorder of the Merdle house parallels the literal degeneration of Mrs. Clennam's.

Passages like this go on at great length and acquire great emphasis. Some, of course, have important work to do, like the metaphor of the disease that inspires the city's speculation, or Mr. Dorrit's travels, which suggest that "the whole business of the human race, between London and Dover, was spoliation" (p. 695). Amy's experience of the ruins of Rome puts the narrative and physical decay together in a way representative, I believe, of the whole story. And while Amy is contrasted with such ruins, she has no way to combat their quite literal vastness in space and time (p. 671). Here Dickens's characteristic distrust of the past extends the metaphor of entropic decay across the span of Western history.

The images of disorder, while suggesting a moral condition, are also couched in the language of mystery. We expect of Dickens, especially after *Bleak House*, that he will build his narrative around mysteries, whose secrets will be revealed either gradually or abruptly. Unlike *Bleak House*, *Little Dorrit* goes out of its way to keep its secrets obscure. It does not offer interpretable clues, such as the likeness of Esther to the portrait of her mother, which sets Guppy off on his proposals. The characters are swept up, unable to retreat or to understand. The world of *Little Dorrit* is all mystery, and its secrets oppress and overwhelm. The Circumlocution Office is the obvious comic-satiric expression of the institutional source and power of this disorder and mystery. But perhaps the richest figure for it comes in a passage that is sharply particular, dramatically located in Clennam's consciousness at a moment when he is preoccupied with secrets, particularly secrets about money. But as the passage registers Clennam's perceptions in his walk through the city toward his mother's house, the images speak with remarkable appropriateness as representative both of the plot(s) of *Little Dorrit* and of the texture of its world:

> As he went along, upon a dreary night, the dim streets by which he went seemed all depositories of oppressive secrets. The deserted counting-houses, with their secrets of books and papers locked up in chests and safes; the banking-houses, with their secrets of strong rooms and wells, the keys of which were in a very few secret pockets and a very few secret breasts; the secrets of all the dispersed grinders in the vast mill, among whom there were doubtless plunderers, forgers, and trust-betrayers of many sorts, whom the light of any day that dawned might reveal; he could have fancied that these things, in hiding, imparted a heaviness to the air. The shadow thickening and

> thickening as he approached its source, he thought of the secrets of the lonely church-vaults, where the people who had hoarded and secreted in iron coffers were in their turn similarly hoarded, not yet at rest from doing harm; and then of the secrets of the river, as it rolled its turbid tide between two frowning wildernesses of secrets, extending, thick and dense, for many miles, and warding off the free air and the free country swept by winds and wings of birds. (pp. 596–597)

But this cluster of images, with its deliberately apocalyptic intimations of a time when the graves will open to offer up their secrets, provides no religious alternative. The quasi-religious rhetoric is part of the irony and bleakness of the passage. It further intensifies that tension in the novel between its actual achievement and its overt direction, which is in part an opening of Clennam to love and to a less bleak, less Calvinist vision of reality. The London of the novel is dramatically shown to be precisely the kind of city Clennam here is imagining it to be. And amid such secrecy, it might be fair to say of Clennam that whatever it is, it is not his fault. Precisely, it is the "nobody's fault" that Dickens was attempting to deny.

The natural world that seems an alternative in this passage is elsewhere no real alternative. The natural world is often invoked, by narrator and characters alike, as a possible escape from the oppression of the particular scene. Yet in almost every instance, when nature is appealed to as that kind of harmonious and meaningful world we are familiar with in early Dickens and in natural theology, it is as a dream rather than an enactment. When the novel is located in "nature," nature participates as furiously in secrets and oppression. So Amy, watching a sunrise, sees "the spikes on the wall . . . tipped with red, then made a sullen purple pattern on the sun as it came flaming up into the heavens. The spikes had never looked so sharp and cruel, nor the bars so heavy, nor the prison space so gloomy and contracted. She thought of the sunrise on rolling rivers, of the sunrise on wide seas, of the sunrise on rich landscapes, of the sunrise on great forests where the birds were waking and the trees were rustling; and she looked down into the living grave on which the sun had risen, with her father in it three-and-twenty years" (p. 276). The idyllic imagination of nature has no correspondence in the nature of the novel, and Amy of course has never seen such things as she imagines. It is imagination itself that turns nature into value, juxtaposing it to the apparently singular brutality of human institutional repression. Indeed, in the second half of the book, real nature appears and remains as forcefully imprisoning as the Marshalsea. The imagination, in *Little Dorrit*, is the creator, not the discoverer of moral alternatives. Similarly, there is a passage in which Clennam is given what seems indeed to be a peaceful and a beautiful nature. "Everything within his view was lovely and placid," and the very lyricism of the prose seems to affirm the loveliness: "the rich foliage of the trees, the luxuriant grass diversified with wild flowers, the little green islands in the river, the beds of rushes, the water-lilies floating on the surface of the stream," and so on. Clennam finds "no division" between "the real landscape and its shadow in the water." Both, he feels, are untroubled and clear "while so fraught with solemn mystery of life and death, so hopefully reassuring to the gazer's soothed heart, because so tenderly and mercifully beautiful" (p. 382). There is no gainsaying here the felt beauty of the scene, a beauty that makes "mystery," otherwise so painful, somehow "reassuring." But into that landscape of peace and reassurance—almost like the image of the cross, or Hetty Sorrel's anguish in the pastoral scene of *Adam Bede*—comes Minnie. It is at this very peaceful moment that she lets Clennam know she will marry Gowan and not only shatter Clennam's unstated hope but her own life. Whatever Dickens's atttitude toward the beauties of nature, they emerge as a mysterious irony on human dreams and desires. Nature is available to the imagination, but it writes its own plot against the plotting of the characters.

Secrecy, death, absence of self-definition are linked as conditions of life. Clennam literally turns himself into a nobody by the end of the scene with Minnie. As the conventional self disintegrates under the pressure of the impersonal forces that diminish it to a spot in the landscape, or to a Clennam-like "nobody," so the conventional narrative unfolding toward a clarifying revelation is put under extreme pressure. The plot of nature resists Dickens's plot.

The takeover, as it were, by natural law moving toward a merely material and determined universe such as that implied by thermodynamic theory, is further developed in the narrative's fatalistic language. We might expect Dickens to dismiss such language as belonging precisely to those figures "who lay all their sins negligences and ignorances on Providence," that is, to the ironic objects of the original title. Yet that is not how it works. True, Miss Wade, the most willful of characters, announces the motif first: "In our course of life we shall meet the people who are coming to meet *us*, from many strange places and by many strange roads . . . and what it is set to us to do to them, and what it is set to them to do to us, will all be done" (p. 63). But several chapters later, when Clennam first sees Amy, it is the narrator who talks about "the destined interweaving of their stories" (p. 140). Rigaud talks of "destiny's dice-box" (p. 175). Yet, in discussing Mrs. Clennam's time-locked room and the candle burning in it, the narrator echoes him:

> Strange, if the little sick-room fire were in effect a beacon summoning some one, and that the most unlikely some one in the world, to the spot that *must* be come to. Strange, if the little sick room light were in effect a watch-light, burning in that place every night until an appointed event should be watched out! Which of the vast multitude of travellers, under the sun and the stars, climbing the dusty hills and toiling along the weary plains, journeying by land and journeying by sea, coming and going so strangely, to meet and to act and react on one another; which of the host may, with no suspicion of the journey's end, be travelling surely hither?
>
> Time shall show us . . . (p. 221)

Such ominously fatalistic language emphasizes the powerlessness of the will to effect change. So Clennam, translating himself into nobody, looks down the road as he walks with Gowan and thinks: "Where are we driving, he and I, I wonder, on the darker road of life? How will it be with us, and with her, in the obscure distance" (p. 367).

This language, however reverberant with mystery, invokes again the irreversible arrow of time and is fully consonant with that self-conscious elimination of the divine toward which science was moving. So, for example, at a meeting of the Geological Society of London in 1852, William Hopkins, noting the irreversible loss of heat through geological time, put it flatly: "I am unable in any manner to recognize the seal and impress of eternity stamped on the physical universe."[21] "The seal and impress of eternity" is missing; the language of *Little Dorrit* emphasizes deterministic restriction and the likelihood that after all what is to happen is not foreknown or planned: it is merely inevitable.

Jerome Beaty notes that the first image of the book places the narrative within the context of nature, as, we have seen, so often happens in Dickens.[22] It is not the prison, but the sun. And nature swirls in the breezes that touch even the Marshalsea, floods under the iron bridge, rises with the sun. It manifests itself in the prisons in Marseilles and London, in the creaking timbers of Mrs. Clennam's house, over the streets of London at Little Dorrit's party. If you granted the sun "but a chink or keyhole," "it shot in like a white-hot arrow" (p. 40). But nature is not unremittingly oppressive. It often becomes the only source of joy outside human love that the novel allows. Although nature images emphasize the smallness of the human and the social, they also imply an unwonted expansiveness. In a shift characteristically self-contradictory, Dickens seems to turn to the very nature that constrains, that limits meaning and the possibility of action, to provide the way back beyond the merely natural to a more traditionally natural-theological nature.

The dream of the world outside the Marshalsea, "the free air" and "the free country," is an aspect of the prison metaphor. Little Dorrit spends her free hours on the iron bridge, watching the river, the only occasions of solitude, freedom, and peace in her Marshalsea life. (The river, though viscid and mysterious for Clennam, on the whole serves to suggest the possibility both of movement with change, and regularity and order, and Little Dorrit's instinctive attraction to it implies the book's commitment to or quest for these qualities through the figures of nature.) The Marshalsea is not at all separate from the natural world, although it is touchingly juxtaposed with the fields of Surrey that lie within the imagination of some of the characters. Of course, it is enclosed and restricting, and limits the moral as well as the physical vision, but it participates as all life must in the natural world. The static image of the Marshalsea is in part created by its constant relation to the motions of nature itself. So, "the equinoctial gales were blowing out at sea, and the impartial south-west wind, in its flight, would not neglect even the narrow Marshalsea. While it roared through the steeple of St George's Church, and twirled all the cowls in the neighbourhood, it makes a swoop to beat the Southwark smoke into the jail; and plunging down the chimneys of the few early collegians who were yet lighting their fires, half suffocated them" (p. 130). The solidity and unchanging quality of the prison and the city are regularly countered by bursts of wind and rushing clouds, as at Little Dorrit's "party." Despite the human constriction and avoidance, Dickens's city is part of nature, and equinoctial gales at sea blow smoke into the faces of early rising collegians. In counterpointing the narrow lives of the characters, who are shut up in dark rooms or in prisons, or at dinner parties, blind in Venice to the constant movement of the waters, huddled against the chill of the Alps, nature affirms the reality they try to deny. Nature is normative here not so much as a moral model but as a constantly moving and active force which by its very presence denies the social constructions and deceptions that constitute the substance and plot. Nature is, simply, a fact that most of the characters ignore at their peril. It crosses their lives, betrays their ambitions, ages and kills them. Or, like the stars over the rubble of the

Clennam ruins, it speaks of a larger world. In its vastness and in the sureness of its movement, it seems indifferent to human ambition. Regardless of the arbitrariness and violence of human action, nature continues its regular movement, has its own plot, which inevitably crosses with and absorbs the human plot.

Dickens's ambivalence about nature, and an ultimate commitment to the secular, is evident even when the language reaches for spiritual meaning. The last double number returns to the sun, but with a very different image of it as it rises in London: "Far aslant across the city, over its jumbled roofs, and through the open tracery of its church towers, struck the long bright rays, bars of the prison of this lower world." Here, the language seems to shift from the aggressively secular metaphors of the opening passage to an emblem traditionally sacred. But the "jumble" of roofs reasserts the pervasive secular disorder; the juxtaposed tracery of the church towers, ostensibly different, implies disorder in two ways, first in the incongruity of its relation to the jumble, second in its containment by the "bars of the prison of this lower world," the bars of nature itself. Tentativeness about the religious implications of such passages is appropriate. "Lower world" implies a higher one, but we are given only the lower. The laws of nature are determining for both the secular and the religious.

Through the clustered images, nature asserts itself primarily in time and movement with their figuring in the road. It is a commonplace that although William Dorrit takes to the road, he is always in prison: he denies the past and lives in a stasis similar to that of other prominent figures. The most obvious is Mrs. Clennam, who remains locked in her room and in the unforgiving self she has invented, and enacts the consequences of the refusal of nature; certainly, her house does. The ignored creaking sounds that do so much work of mystery are the material signs of time's arrow: stasis itself is an artificial construction of identity, but only a temporary stay against, or more accurately a disguise of, the movement of time toward the collapse of order in the dissipation of energy. "I am not subject to change," says Mrs. Clennam (p. 389). In less melodramatic versions there is the absurdity of Flora, pretending to be a young lover, and of the patriarch, Mr. Casby, glowing contentedly, unwithered by time. More painfully, we have the stopped time of poor Maggie, who believes she is ten years old. Against the artificiality of stasis and fixed identity are the images of road and street, into which, appropriately, Clennam and Little Dorrit descend in the last page of the novel. Time is, simply, natural, and the characters who live in it must accede to its power. Change, which is movement in time, thus becomes essential to whatever possibilities the austere world of *Little Dorrit* will allow. But acquiescence in time undermines selfhood. Time breaks down the artificially constructed character (like Mrs. Clennam) whose rigid self denies the inevitabilities of thermodynamic or evolutionary movement.

But there are negentropic forces in the novel. Dickens's ambivalence toward nature is most particularly evident in his handling of the representatives of "true" science. These figures have the quality of saviors, even though they are confined to dramatically marginal roles. I mean, of course, Daniel Doyce and the figure called "Physician."

With the "scientists" the languages of religion and of nature cross. Doyce, to be sure, belongs in the thermodynamic schema as the book's one principle of both energy and order. Pitted in a losing battle against the overwhelming entropic force of the parasitical Barnacles and the Circumlocution Office, he has something oddly otherworldly about him. Dickens has put his faith in science, but consistent with the contradictions I have been locating, he thinks of science as entirely compatible—in the mode of natural theology—with the world of spirit. He deliberately gives to the word "practical" some of the same mystique Caleb Garth gives to the word "business" in *Middlemarch*. So the negentropic Doyce is, in his reversal of the irreversible arrow of time, also a spiritual figure. "He never said, I discovered this adaptation or invented that combination; but showed the whole thing as if the Divine artificer had made it, and he had happened to find it; so modest he was about it, such a pleasant touch of respect was mingled with his quiet admiration of it, and so calmly convinced he was that it was established on irrefragable laws" (p. 570). There is, nevertheless, a tricky ambivalence here: the "Divine Artificer" may be read as an invention of Doyce himself. Tricky or not, the language assimilates science and religion: like Paley finding a watch that evidences design, Doyce *finds* the "irrefragable laws"—principles of order and stability—that Dickens seeks in both science and religion. The tradition of natural theology, like science itself, valorizes what the narrator calls the "regularity and order" (p. 736) that the rest of the novel cannot find. (One should note here in passing that even the "good man" refuses to make claims for the self or to claim responsibility.)

If there is some slight uncontrolled ambiguity about the religious implications of what Doyce represents, there is none in regard to

Physician. It is he who counters disorder and decay, not through denial but through frank recognition of the realities of nature. As scientist, he is given the burden of peace, stability, and spirit. A figure who would seem to belong to the cluster of impersonal caricatures who are aspects of the systemic decline the book traces, Physician emerges as almost divine. The extravagance of the praise is astonishing and suggests how thoroughly Dickens relied on science as he worried the problem of the relation of the physical to the moral and religious world: "Many wonderful things did he see and hear, and much irreconcilable moral contradiction did he pass his life among; yet his equality of compassion was no more disturbed than the Divine Master's of all healing was. He went, like the rain, among the just and the unjust, doing all the good he could, and neither proclaiming it in the synagogues nor at the corner of streets" (p. 768). The indifference of nature and of society itself becomes a spiritual condition, akin to that of Amy Dorrit. Dickens returns to a quasi-allegorical mode to express this pre-Darwinian conjunction of science and religion. Here the confusion of the novel that I have been intimating, between a selflessness which reflects merely secular failure of will and the ideal of Christian selflessness, is most clearly figured. Here, too, there is a kind of Darwinian extension of the laws of nature to the laws of human nature, and to the physical possibilities of human life. Like nature, and yet also like Christ, Physician is the most explicit expression of the saving possibility of the power to accept the natural without disguising the facts of one's physical being—even, as it might turn out, one's apelike ancestry.

Dickens accedes to the irreversible arrow of time. Before Darwin and against the grain of the great sages who refused the full implications of science's descent into time, Dickens saw the ideal of permanence as vicious, change as the condition for life. Better to move with time at the sacrifice of clear and willed identity, than to deny it, as Mrs. Clennam and the whole of aristocratic society do.

Yet if life is determined not by religious energy from outside the system but by the system itself, reliance on time as redeemer is belied by the second law of thermodynamics. There was, however, a secular alternative: evolution. Only a culture released from an Aristotelian position that identifies permanence with the divine, and change with corruption, could have been ready for the Darwinian argument. And although *Little Dorrit* shows us a Dickens divided on the question, his desire for peace, stability, "the hearth," is countered by his instinct for change, and life.

The moral language remains in *Little Dorrit* in spite of the evidence that time runs all down and nobody will be at fault. At the end Dickens attempts to transform the meaningless sun into an intimation of the divine, while the imprisoned characters seem all, as Edward Eigner suggests, to burst into rebellion against imprisonment—Affery, Pancks, Frederick Dorrit, Mrs. Clennam herself: "If, as virtually every reader has noted, imprisonment is the controlling metaphor of the first eighteen numbers of the book . . . , then the final double number is characterized by a succession of stunning jailbreaks."[23] These splendidly and traditionally conceived "characters" all renounce the prisons of their imposed identities, although each rebellion is not realistically accounted for. Similarly, as the novel struggles to a conventional ending, it is thoroughly unconvincing in explaining the powerfully imagined mysteries. The tradition of natural theology reasserts itself against the new imprisoning systems that reduce all experience to the terms of natural law. We have seen Kelvin's evasion of the problem; through Mrs. Clennam's confession, Dickens resorts to forced imposition of meaning on an experience too complicated to be resolved through these narrative means.

Finally, then, the contradictions in *Little Dorrit* are significant of its courage. The narrative seeks signs of the creator in the forced conventions of an earlier conception of character and plot, but follows the direction of its images elsewhere. These images reflect a world irredeemably secular, one in which both Darwinian theory and thermodynamics would find a home, in which the self has lost its clear outlines as well as its powers. The text is at the intersection of positivist and what I have been calling natural-theological ways of knowing and imagining. But following the logic of its own vividly registered sense of the material world, it almost chooses a positivist imagination, and does so by using the way of seeing that was to inform Darwin's *Origin*. Dickens clearly could not have sustained unmodified the thermodynamic structure so pervasively at work in the book's images. Neither could he deny the force of its vision. Redemption entailed looking *within* time, not beyond it, but within a time not dominated by the entropic system, the irreversible arrow moving toward disorder and the dissipation of energy. Clennam and Little Dorrit may be inseparable and blessed as they move into the streets, into movement, and into the stream of time, and the sun may finally shine on them "through the painted figure of Our Saviour on the window" (p. 894), but they go *down*. *Little Dorrit* itself is a descent into time, and in the course of the descent Dickens loses firm control over at least three central aspects of his art: first, over the centrality of character, with the adhering notions of selfhood and moral responsibility; second, over the narrative connection between action and responsibility; third, over the *meaningful* movement of narrative time (that is, from the image of time as water, obeying universal laws, and thus universally significant even in its temporality, to time as the image of the movement on the streets, among the arrogant, the forward, and the vain, time which is thus continuous and meaningless).

The secular-scientific way out of the crisis was evolutionary science, whose arrow seemed to point upward (until later in the century it was assimilated to entropy and "degeneration"). Evolution, a force that Jane Austen would have found disruptive and which certainly was associated with forces hostile to established philosophical and social institutions, was also perceived to be what we might call a negentropic force. Life seemed not to obey (or only very tardily) the iron rules of cooling and disorganization, and even the inevitable individual death produced, in Darwin's view, "the most exalted object which we are capable of conceiving." Whereas both evolution and thermodynamics presented a radical difficulty for the traditional conception of character, evolution could at least suggest not decline but growth and improvement. This arrow, too, moves through nature, despite the prisons and the urban and claustrophobic texture of so much of the novel. Nature counterpoints the stasis of will that characterizes the society of *Little Dorrit*, and it determines our understanding of it. Unsentimentally, and erratically, nature introduces into the novel the possibility of change in a negentropic direction. Like society itself, impersonal and indifferent to the individual needs of the protagonists, nature is nevertheless the possibility of life: it becomes at the end not a road to inevitable decline, but a very secular street chancily, vitally leading to the possibility of joy and love.

Insofar as he could return to the religious interpretation of nature, as is partly implied by the late images of the sun (seeming rather fatigued, however, without the secular energy of the opening pages), Dickens could avoid temporarily, at least, many of the bleak scientific implications of what his imagination was revealing to him. The courage of *Little Dorrit* is in its confrontation of the possibility that the religious account could not stand against the pressure of those irrefragable laws, denying both God and self, that he wanted to celebrate, and his willingness to risk that the arrow of "time" would point at last beyond his own imagination of character and narrative to life and regeneration.

7

The Darwinian World of Anthony Trollope

TROLLOPE, whose associations included some of the great scientists and avant-garde thinkers of his time, who edited *St. Paul's Magazine* and helped found the *Fortnightly Review*,[1] probably knew less and certainly cared less about science than Dickens. Politics and institutions were not, for Trollope, objects of contempt and satire—no Boodles and Coodles, no Circumlocution Offices fill his pages, although he recognized and described much toadyism and bureaucracy. Very much the writer of compromise, Trollope expected and demanded less from the world than Dickens. Far from Dickens's engaged and passionate vision, he indulged Dickensian satire primarily in his late work *The Way We Live Now* (1875). For the most part, that hateful institution, Law, was for him, with all its abuses, a condition of civilization and a mark of its humanity. Nevertheless, I take the very unscientific Trollope as my central example of Darwinian novelist.[2]

Trollope had published at least eight novels by 1859, including the first three of the Barchester series, which made his fame. But almost from the start, his fiction fit Darwinian patterns. Certainly, to use crudely the dualism invoked frequently here, his social world is uniformitarian rather than catastrophist: "Nothing surely is so potent as a law that may not be disobeyed. It has the force of the water-drop that hollows the stone. A small daily task, if it be really daily, will beat the labours of a spasmodic Hercules. It is the tortoise which always catches the hare. The hare has no chance" (*Autobiography*, pp. 103–104). Politically, as opposed to *Bleak House*, his novels imply so strong a faith in the English traditions of precedent and convention that they leave no space for abrupt violations of the social or natural order. Ironically, the new Darwinian vision, while seeming to disrupt, in the manner of Mary Crawford, the fixed structures of traditional society, becomes conservative itself, not through any faith in the divine and fixed nature of the way things are, but because its gradualism implies that abrupt change can only make things worse. Darwinian fiction is good, then, at demonstrating the irrationality of social structures but finds that irrationality no disqualification. Although critics like Bernard Shaw or Jack Lindsay overdo it by claiming that Dickens was a protosocialist, his novels, as we have seen in *Little Dorrit*, tend to encourage disruption and value Affery-like bursts of liberation. Fictions of discontinuity or reversal are more likely to incorporate ideas of radical change thematically than are fictions of continuity. While Trollope is very good at recognizing the *temptation* to break loose from convention, law, and the accumulated bonds of culture, his narratives normally thwart all such moves as incompatible with civilization, either merely silly or wicked.[3] Trollope's greatest achievements are dramatizations of the inescapability of the ordinary, as when Plantaganet Palliser, in *Can You Forgive Her?* reconciles himself to domesticity, offers to take Glencora on a trip, and refuses to accept the political position he has been aspiring to all his life. Being inconsistent, irrational, or self-interested is not a serious mark against person, institution, or idea.

Trollope's social conservatism parallels the conservatism that quickly assimilated Darwinian science and made it so easily adaptable to the antirevolutionary character of English politics and society. Despite his deep interest in the life of the clergy, Trollope imagines a world entirely secular in texture. Secularity, for Trollope, did not entail a denial of church or an Eliot-like humanist alternative. When he defines his own political position in his *Autobiography,* he lays out the assumptions of the Darwinian world view as they tended to be applied analogically to social and political matters. The liberal, says Trollope, like the conservative, is

> averse to any sudden disruption of society in quest of some Utopian blessedness;—but he is alive to the fact that these distances are day by day becoming less, and he regards this continual diminution as a series of steps towards that human millennium of which he dreams. He is even willing to help the many to ascend the ladder a little, though he knows, as they come up towards him, he must go down to meet them. What is really in his mind is,—I will not say equality, for the word is offensive, and presents to the imaginations of men ideas of communism, or ruin, and insane democracy,—but a tendency towards equality. In following that, however, he knows that he must be hemmed in by safeguards, lest he be tempted to travel too quickly;

> and therefore he is glad to be accompanied on his way by the repressive action of a Conservative opponent. Holding such views, I think I am guilty of no absurdity in calling myself an advanced conservative liberal. (p. 253)

There are to be no narrative equivalents to divine intervention, and there are to be no disruptions of the current order of things, only slight, imperceptible gradations of change worked out in constant, quiet struggle.

The politics find expression in what might be called the ideal realist form. In his novels Trollope self-consciously abjures surprises, coincidences, melodramatic disruptions of continuous narrative surface. Moreover, despite his continuing use of well established narrative conventions of organization and closure, Trollope is much more comfortable with loose ends than Dickens. I have argued that Dickens's irrational world signifies and affirms a rational design, but Trollope is casual about meaning, less symbolic, not at all allegorical, and with the worldly-wise tone of the rational onlooker, describes irrational worlds and affirms systems of order that defy rationality—like politics, the church, or any established institution. For Trollope, extant systems of law and order emerge from history and are sustained by multifarious relations in constant flux, not rationally designed but ultimately producing imperfect yet satisfactory adaptations of individual and society to each other. With all his casual insight into the madnesses and ironies of contemporary order, Trollope is perhaps the best Victorian example of the way post-Darwinian assumptions fed into a political and social conservatism that could, in narrative, only be disrupted by an anti-Darwinian narrative form—or by a recognition that there is no moral or social imperative embedded in the evolutionary scheme, or anywhere in nature.

The peculiarly "unscientific" nature of Trollope's "Darwinism" needs to be confronted first. Trollope's novels self-consciously resist "ideas." Not, of course, that his characters fail to have ideas and discuss them, or that his novels fail to explore particular moral and political positions; but they make absolutely no pretense to the kind of philosophical status claimed, say, by George Eliot. In this respect, perhaps, he is no different than Henry James. Nevertheless, his nonfiction, where ideas are the substance and get themselves expressed, is uniformly uninteresting (except as it reveals something other than the ideas expressed). Trollope's description of himself in his *Autobiography* as an altogether practical man was, after all, only partly disingenuous.[4]

His novels aspire to render not ideas but the texture of social intercourse, characters thoroughly observed, love affairs, fox hunting, wills, politics. Although they are more schematic than Trollope would admit, they are rather loose in form. The related subplots often make their connections tenuously, through problems of romantic, class, social, and political relationships.[5] When ideas matter—that is, when they matter to any of the characters—they tend to be at odds with the dramatized shape of Trollope's world. There is a certain mad charm, for example, to Palliser's long commitment to the idea of decimal coinage. But its ostensible silliness is part of the point. The commitment matters not because of its substance, but because of what it reveals about Palliser's character. (Has Palliser been vindicated by the United Kingdom's turn to decimal coinage in our generation?)

Trollope's resistance to ideas was remarkably self-conscious for a writer who put himself in the middle of the Victorian intellectual scene. As editor of *St. Paul's Magazine* he wrote to a potential contributor who had asked about publishing an essay on Darwin, "I am afraid of the subject of Darwin. I am myself so ignorant on it, that I should fear to be in the position of editing a paper on the subject."[6] His friendship with G. H. Lewes and George Eliot seems to be the closest he ever came to scientific thought, but when Lewes died and Trollope was to write a memoir, he asked that the scientific side of Lewes's work be confided "to some one less wholly ignorant of philosophic research than myself."[7] The distance between him and George Eliot on such matters is suggested by the way he portrays his most famous provincial doctor, Doctor Thorne. The good doctor is, like Lydgate, from the best and oldest of families; he too feuds with less competent and more old-fashioned local doctors. Yet, unlike Lydgate, Thorne has no large scientific ambitions. "He might certainly be seen," Trollope says, "compounding medicines in the shop at the left hand of his front door; not making experiments philosophically in materia medica for the benefit of coming ages—which, if he did, he should have done in the seclusion of his study far from profane eyes—but positively putting together common powders for rural bowels, or spreading vulgar ointments for agricultural ailments."[8] No Rosamond Vincy for him! Lydgate, it might be added, ran into trouble precisely because he did not think it right for doctors to compound their own medicines.[9]

Given the enormous differences between them, it is striking how much Darwin and Trollope's ways of talking about themselves and about the complex and abundant worlds they described converge.[10] One of the most interesting places to look for that convergence, before proceeding to Trollope's fiction, is in their autobiographies. These slight volumes suggest conveniently the connections between Darwin's theory and the fundamental cultural assumptions that seem to drive Trollope's fictions as well as his conception of self.

In a uniformitarian world large effects are produced by small causes; no single event matters apart from thousands of others. To carry this view over into social matters (as realist novelists tended to do), is to minimize the importance of individual politicians, individual genius.[11] Of course, Darwin—leisured, retiring, noncombative, scientific, and even "anes-thetic"—was an utterly different sort of person from Trollope, whose noisy bluffness and deliberately vulgar social manners were notorious.[12] But both adopted a rhetoric of self-deprecation in their autobiographies, both wrote simply, lucidly, directly. Darwin the scientist seeks knowledge in empirical particulars rather than in abstract theorizing; so too does Trollope the writer. Both writers have been severely underestimated in part because of the myths created by their strangely unassertive autobiographical narratives. Gertrude Himmelfarb notes that "Darwin must take his place alongside Anthony Trollope and the other great Victorians whose creativity has been impugned by their methodicalness."[13] Both describe great achievements as resulting from the drudgery of ordinary people with ordinary talents. Trollope is confident that "in making my boast to quantity" it will not be thought that "I have endeavoured to lay claim to any literary excellence." He does, however, "lay claim to whatever merit should be accorded to me for persevering diligence in my profession . . . A constancy in labour will conquer all difficulties" (*Autobiography*, p. 313). Darwin finds a kindred spirit in the persevering Trollope. In a letter of 1877, he turns to Trollope to make his point: "Trollope in one of his novels [*The Last Chronicle of Barset*] gives as a maxim of constant use by a brickmaker—'It is dogged as does it'—and I have often and often thought that this is the motto for every scientific worker" (*More Letters,* I, 370–371).

Odd, at first, to find Trollope providing a maxim to guide all scientists. But it is a typical strategy of Victorian genius to refuse both genius and the self-dramatizing stance of genius. This strategy is most notorious in Mill's *Autobiography*, which argues that anyone trained as Mill was could have achieved what he did. Darwin certainly indulges it, calling himself a poor critic with a weak memory and "no great quickness of apprehension or wit."[14] He concedes, however, that he is a good observer and protests that he has a greater power of reasoning than many critics have credited him with. Even here the affirmation is couched in self-abnegating rhetoric. Against the suggestion that he has "no power of reasoning," he says meekly but with a steely confidence, "I do not think that this can be true, for *The Origin of Species* is one long argument from beginning to end and it has convinced not a few able men." But the boast is soon modified: "I have a fair share of invention, and of common sense of judgment, such as every successful lawyer or doctor must have, but not, I believe, in any higher degree" (*Autobiography,* p. 140). To Francis Galton, whose book *Hereditary Genius* was to argue on the *inheritance* of the qualities necessary for success, Darwin wrote, "I have always maintained that, except fools, men did not differ much in intellect, only in zeal and hard work" (*More Letters* II, 41). The uniformitarian world of the autobiographies is egalitarian—all men (and women) are created equal.

In refusing the "extraordinary" as a name for their obviously extraordinary achievements, and in indicating that these achievements resulted from the mere accumulation of ordinary qualities, Trollope and Darwin apply their gradualism and uniformitarianism to their own identities. Personal modesty in these cases implies the positivist episteme. The apparently extraordinary must be understood as the product of everyday experience, as the great geological phenomena of the world are the product of causes now in operation. This kind of explanation deromanticizes the self and the natural world. Not miracle nor catastrophe nor genius needs to be invoked to explain the way things and people are. At the same time, deromanticizing the world requires that we reimagine, revalue, and observe freshly with heightened intensity the particulars of every day. The world then becomes explicable to anyone who tries hard enough, and is not the domain of the mystic hierarchies of prescientific culture. But the plain and modest style as a cultural norm and value depends on a faith in the universal availability through observation and hard work of the facts and of their meanings. Trollope's realism is very much a product of a new scientific culture.

Trollope and Darwin agree too on what constitute the most important activities of scientist and novelist. For Trollope "the portion of a novelist's work which is of all the most essential to success" is "observation." Without the work of "observation and reception," his "power cannot be continued—which work should be going on not only when he is at his desk, but in all his walks abroad, in all his movement through the world, in all his intercourse with his fellow creatures" (*Autobiography,* p. 198).

This emphasis on observation is an essential theme of science, and Darwin makes his one claim of superiority here: "I am superior to the common run of men in noticing things which easily escape attention, and in observing them carefully. My industry has been nearly as great as it could have been in the observation and collection of facts" (*Autobiography*, pp. 140–141). The two writers converge in the shared Victorian view that the common basis of authority was to be empirical evidence—the record of what is available to the senses. Here is a critical point of intersection between science and literature, and its obviousness does not diminish its significance. The apparent coincidence of the triumph of realism (whose duplicities the twentieth century is busy exposing) and of empirical science (whose basis in language and fictions is being asserted with equal energy and complexity) was no accident. Trollope and Darwin (despite Darwin's own deviousness on the possibility of objective perception) are very much brothers in their almost obsessive need for precise observation, precise recording.

In a remarkable appreciation of Trollope's art, Henry James argues that Trollope's genius was in his "happy, instinctive perception of human varieties," his "love of reality." But, he goes on, Trollope "never attempted to take the so-called scientific view . . . He had no airs of being able to tell you *why* people in a given situation would conduct themselves in a particular way."[15] This is arguable, but partly true insofar as science entails answering *why*. Yet Trollope's whole narrative method (as he describes it) is to achieve a kind of Darwinian position in relation to his materials, to see how they will evolve, to observe meticulously, to let the accumulated evidence provide the primary authority.

Trollope prided himself on Hawthorne's famous comment that his novels were "just as real as if some giant had hewn a great lump out of the earth and put it under a glass case, with all its inhabitants going about their daily business" (*Autobiography*, p. 125). Christopher Herbert suggests that seeing "Trollope's characters as specimens lifted unknowingly out of their daily lives and placed under glass for purposes of observation" implies the techniques of scientific experiment. Herbert claims, moreover, that "it is the scientist's voracious appetite for data that underlies the amazing volume of Trollope's fiction; and it is exactly the impulse of scientific study that governs his passionate insistence on fidelity to nature as well as his studiously flat, unpoetic, 'objective' writing style."[16] The rhetorical strategies of the two autobiographies reflect a mutual determination to get at the real akin to that of the novels and scientific studies. The manner reflects the dominant empiricist ideal: all mere literary ornateness, all obstructions from myth or metaphor must be dismissed, as the Royal Society dismissed them two centuries before, for the plain style that makes possible an unmediated apprehension of the world.

Darwin laments the *consequences* of his obsession with scientific method because the intense commitment to observation and the explanation of what is observed seems to exclude the possibility of that joy in the real which, on James's account, Trollope had—which, indeed, is often manifest in Darwin's own writing. In an alienated and dispassionate way, Darwin characterizes himself, his powers, his emotional inadequacies, and sees himself as "a machine for grinding general laws out of large collections of facts" (*Autobiography,* p. 139). Surely, he underestimated the "grinding." The particular disrupts easy generalizations, and the power to handle the particular—especially in the construction of a large encompassing theory—is a mark of genius. In his *Autobiography* Darwin seems content simply to lay down the facts, to avoid introspection and theorizing, and he creates what we might call a small, loose baggy monster, as, in the *Origin*, he created a large one. But apparent looseness disguises a new kind of structure built from an obsessive cultural need to examine empirically the myriad of particulars that would, in the Platonic scheme of things, be thought of as mere "accidents." The same might be said of the great Victorian realist novels, and of Trollope's work.

Except, of course, that pure mimesis is a fiction. We have seen that Darwin did not quite believe in the empiricism he espoused; Trollope created worlds which in particularities imply a representativeness that verges on the modes he refused, allegory and symbol.[17] The autobiographies understate and evade their tendencies toward speculation and symbolization, just as they minimize the genius of their authors. The narratives they unfold celebrate the virtues of doggedness and application, talk about their authors as "other," with a kind of scientific detachment, demonstrating that such detachment is possible even about the self. Moreover, they imply, in their chronological tracking of apparently unmerited achievements, that anybody else could have done the same thing. Except for the matter of will and doggedness, their particular successes are mere chance. Making minimal claims for the self, they protect themselves and their work from attack; at the same time, in their insistence that no fuss is to be made, they stick with iron determination to their achievements.

The personal, the professional, the social, and the economic run together in this complex. Earlier, I invoked Stephen Jay Gould's argument that "the theory of natural selection should be viewed as an extended analogy . . . to the laissez faire economics of Adam Smith."[18] Smith's theory, like Darwin's, and like Trollope's novels, depends upon the assumption that humans and animals move freely and without artificial interference from government, scientist, breeder, or novelist. The consequence of observing free movement—goes each of the three fictions—will be the discovery that apparent disorder is transformed by natural means, sometimes apparently cruel, into a present order. Although this crude formulation is compatible with the notion of a designer, there is no need for one. The whole process works by itself. A major difference between Smith, on the one side, and Darwin and Trollope, on the other, is that the latter two find no perfection in "order," and no possibility of it. Indeed, the order is not something permanent and stable but part of a continuing process that will shape into succeeding new orders: as Darwin says, "natural selection will not produce absolute perfection" (*Origin,* p. 229). (The question of whether it is, in any case, possible to observe without disrupting the ostensibly "free" movements of the subjects is another critical issue, to be taken up in the next chapter.)

Together, Trollope and Darwin partake of a romantic inheritance: while assuming the primacy of naturalistic explanation, they link "reality" to ostensible disorder, to multiplicity, to minute particulars, to sprawling spaces and long stretches of time, to the blurring of classification and categorical distinctions, to the rejection of extremes by explaining conventions of romance or of catastrophism in terms of the most ordinary quotidian activities.

Trollope's art partakes of inherited conventions of narrative but moves through them to a "new reality." This movement is implied by his comments in *An Autobiography* on his favorite novels. When he began work in the post office, he says, "I had already made up my mind that *Pride and Prejudice* was the best novel in the English language,—a palm which I only partially withdrew after a second reading of *Ivanhoe*, and did not completely bestow elsewhere till *Esmond* was written" (p. 35). Trollope retained his affection for all three, but the progress of his approval might be charted on a graph describing increasing degrees of Trollopeanism—or, perhaps, decreasing degrees of non-Trollopeanism—as they mark a progression away from the primacy of plotting and symmetry of structure. It is clear in all three cases what Trollope admired, but it is also clear that for *his* art, Trollope developed a mode that is, in effect, a sustained critique of the assumptions about the world implicit in the forms of these earlier novels.

The pattern of distancing seems particularly true about *Pride and Prejudice*. Ruth apRoberts has shown how close Trollope's art is to Jane Austen's,[19] but there is an almost complete difference in texture and form, as in Austen's brevity, precision, and epigrammatic wit. Her ironies, while often implying attitudes very Trollopean, are more concentrated and acerbic than his. The narratives, too, are un-Trollopean in their clearly defined progress toward conventionally inevitable yet surprising conclusions, and in their implicit affirmation, with whatever reservations, of the possibilities of stability in society and fiction. The evidence of conscious design is strong and satisfying. Thus, while the social subject of *Pride and Prejudice*, its clear-eyed ironies, its impatience with pomposity, are close to Trollope, its formal virtues, its ways of composing its world, are almost as distant as possible. Like *Mansfield Park*, which threatens to become Trollopean in its resistance to design and to the stabilities it ultimately affirms, *Pride and Prejudice* belongs to a cultural moment in which natural theology could thrive.

Ivanhoe is ostensibly far less Trollopean than *Pride and Prejudice*. Good romance that it is, it is plot-ridden. Yet the actual execution of the narrative is un-Austenian and implies a loosening of formal restrictions and an implicit difference about the way the world is organized. The dominance of plot does not inhibit Scott's penchant for antiquarian irrelevance, or, to put it another way, for focusing on minutiae of ordinary life not relevant to plot. The often balanced subplots, in the manner of renaissance drama, reflect the kind of patterning (deliberately Shakespearean) to be found in Scott's novels.[20] In *Ivanhoe*, as elsewhere, the patterning is implicitly denied by Scott's historicism and feeling for the instability of social traditions and artificiality of formal patterns. The conventional happy ending of romance can provide in historical fiction no timelessly satisfying confirmation of order. Thackeray's *Rebecca and Rowena*, which marries Ivanhoe at last to the dark-haired Jewish (but now converted) Rebecca, suggests something of the inadequacy of Scott's resolution *as* a happy ending. History is not fair; rewards are not distributed according to a rational scheme of justice. Moral distinctions are harder to make than our biases would lead us to believe, and species, as it were, of the good and the bad turn out to be descended from the same human line and are not obviously distinguishable. Moreover, bathetically, time denies closure, as Scott himself certainly understood in his playful exploitation of the conventional happy ending. The big bow-wow Scott was an incipient realist. His historicism is one of the vital elements in most subsequent fiction, in Darwinian theory, and in the way modern cultures imagine themselves.

Thackeray's *Esmond* moves to the edge of Trollope's world, even though it is, formally, the most rigorously controlled of Thackeray's works. Trollope saw Thackeray as his own true precursor, and in the assumed worldly wisdom of their narrators, their "ironic temper,"[21] they share a detached, sometimes scientific vision of experience. Scott, Trollope, and Thackeray were also "almost ideologically careless about plots."[22] All three refused, both overtly, and through various narrative strategies, to take "Art" with full seriousness.[23] The real is always the priority, and art, the means to the representation of life, always obstructs its own object. It gets in the way. But this denial of art is the obverse side of taking it very seriously, for only the artist who has completely tamed art, disciplined it to submission to the real, expelling from it all ornament and "artistry," can claim to be a realist. Ironically, that is to say, the realist who minimizes art requires the greatest discipline in control of it. The struggle is parallel to that required of Darwin, who had to expel from language all those idealist, intentionalist, and anthropocentric elements that language needs to do its work. To break out of the constraints of language, he needed to give his utmost care to it.

For Trollope, in certain respects an ideologue of the "gentleman" and of the individualist bourgeois culture he represents, a deep commitment to any idea or ideology, to any partisan position, becomes a contradiction of the realist enterprise.[24] The narrator may often pause with conventional moralisms or worldly-wise speculations about human nature and society, but the narrative that provokes such pauses often, as in *The Claverings*, undercuts the expressed ideas. This partly accounts for the huge distance between Trollope and Dickens. Everything must finally be translated into the quotidian and removed from the intensities of belief and feeling that are implicit in highly designed, highly figurative, highly engaged art. (Note even the indefinite article in the title *An Autobiography*, implicitly just one of many, not therefore terribly significant.) Thus, although Trollope admires the "tragic" quality of Lady Castlewood's life, it is not so much the tragic that charms him as that "we feel that men and women with flesh and blood, creatures with whom we can sympathize, are struggling amidst their woes."[25]

While *Esmond* is highly plotted, it incorporates a historicism learned from Scott, and a movement—from one historical cause to another—recognizably Scott-like, that begins to break down moral categories. There is no question that Esmond sees the immorality of the Catholic cause, and can incorporate much of the Victorian prejudice against Catholicism in the eighteenth-century wrapping. Yet dramatically, except for some ineffectual conniving by Father Holt, Catholicism as it is embodied in the characters seems not much different from any human commitment. Eliot professed to have learned disbelief from Scott, in whose novels she discovered that goodness can exist on either side of opposed causes. So too with *Esmond*, where ironies that might emerge in Trollope as unembittered disenchantment often seem merely cynical. It is unlike Trollope, both in its muted cynicism and in its virtuoso craft, by which the cynicism is disguised. Reality in *Esmond* is radically compromising, and the radically compromised hero emigrates away from the problems implied by the narrative; the happy ending is surely a strange second best. Perhaps most important, the refusal of resolution opens the book out into other books. Henry enters other narratives, which cut off the possibility of regarding a narrative as ending at all, even though *Esmond* itself is narrated retrospectively, the action ostensibly over.

Trollope inherited and rejected conventions implicit in this rapid overview of these three novels, which might be taken as marking stages in a transformation of attitudes not only about how novels should be written but about how the world and art actually work. All of these novels he so much admired are very different from his own. Even more than Thackeray, he withdrew from the intensities of romantic art and from the demands of high style. Thackeray's elegiac quality, registering sadness at the loss, through experience and knowledge, of the romantic dream, does not reappear in Trollope.

There is a parallel movement in Darwin. In his *Autobiography* he notoriously confesses that as he pursued his scientific interests, he lost his capacity to read and enjoy serious literature. Shakespeare, he admitted, positively nauseated him. "On the other hand," he says, "novels, which are works of the imagination, though not of a very high order, have been for years a wonderful relief and pleasure to me, and I often bless all novelists. A surprising number have been read aloud to me, and I like all if moderately good, and if they do not end unhappily—against which a law ought to be passed. A novel, according to my taste, does not come into the first class unless it contains some person whom one can thoroughly love, and if a pretty woman all the better" (p. 54). Trollope would have agreed. He always urged upon us a lovable, lovely, and (as he put it) little brown heroine; he always claimed that novels were of a less high order than poetry, or than George Eliot. He always aspired to happy endings.

Darwin had trouble with great art not merely because it strained his mind—reserved for his scientific work—but because it included much pain in the world even while it gave evidence of design. To imagine a deity in this world is to imagine a malevolent deity. "Natural selection," says Donald Fleming, "was precisely the denial of nature as a planned work of art and an effort to dissipate the affective tone that natural theologians tried to lend it. It would be tempting to say that Darwin turned against works of art because he had determined to smash the greatest of all."[26] But Fleming does not sufficiently emphasize the elegiac (almost Thackerayan) tone of loss in Darwin's confession of his developing anaesthesia. The loss of art mattered much to a man who took Milton with him around the world. Yet design came to seem increasingly appalling to him as his studies revealed that war of nature out of which the higher organisms were to emerge. Chance and randomness are preferable to the idea of a designer cruel enough to kill humans, animals, and insects, with such abandon as we see in nature. Only a literary form that engages the individual and eschews premature generalization and ornamental disguises can cope adequately with such a world.

Trollope does provide those happy endings that Darwin sought as an escape from the pain of everyday experience, but with a quiet skepticism that makes them very different from Austen's happy endings. He minimizes the evidences of design and implies that the romantic dream of ideal endings cannot be enacted in realistic narrative. His autobiography may even be read—as James seems to have read it—as an attack on art itself. Thus, while Trollope appeared to be writing the kind of "moderately good" novels Darwin sought, he was actually describing Darwin's world and mimicking its form. His equable detachment manifests itself in a style that aspires to be transparent and to leave little mark of its creator.

In one of his letters Darwin notes that he did not very much like George Eliot's *The Mill on the Floss*, but he found *Silas Marner* "charming" (*Life and Letters,* III, 40). The tragic ending of *The Mill on the Floss* would seem to have required a law to be passed against it. Moreover, it was preoccupied with problems of "descent," inheritance, and adaptation; and with great stylistic intensity it showed the world to be a very painful one.

It is a Darwinian world, presented with un-Darwinian intensity and formal urgency. *Silas Marner*, that most un-Trollopean performance, is by contrast formally precise, infusing all its language with pattern or allegorical import. It projects a world so clearly designed that injustice in it must imply a heartless designer. But *Silas Marner* gives us Eliot's most exquisitely distributed justice. Such a world Darwin could fully enjoy, almost certainly because he could not believe it for a moment. Darwin regarded *Adam Bede* as "excellent," but Eliot should always have been worrisome for him because although she probably understood and incorporated more of Darwin's ideas into her fiction than Trollope would have cared to think about (here Gillian Beer's study is much to the point), she was committed to a strenuous and necessarily unpleasant moral exploration of the implications of a Darwinian world, and to an ideal of art that could not for a moment allow the evidences of artistic design to slip away.

Ironically, then, Trollope's world is probably more compatible with Darwin's than Eliot's. Good realist that he was, Trollope had to resist or exploit conventions of ordering from literature and language: the objective is the appearance of life not the shapeliness of form. As James Kincaid has finely put it, "We are urged [by Trollope] to find full meaning in pattern suggested by the action, but there is a concurrent sense of the artificiality, even falseness of that pattern, a sense that genuine life is to be found only outside all pattern."[27]

In what follows I discuss those qualities of Trollope's fiction that are resistant to the traditions of storytelling he inherited, and that point therefore to what has been regarded as most distinctly realistic and lifelike about his craft. In E. S. Dallas's interesting overview of Trollope's career up to 1859, he describes certain aspects of the novels in which Trollope later announced he took great pride. Dallas notes Trollope's aversion to melodrama but complains that he carries it to an extreme. "Mr. Trollope," he says, "has vowed that there shall be no surprises in his novels. The characters shall be naturally evolved; the incidents shall grow out of each other; the passion shall not be exaggerated, and the sentiment shall veritably belong to the event."[28] To a realist this would read like praise. For Dallas, these qualities minimize the pleasures and excitements of reading. "Mr. Trollope," says Dallas, "throws away, needlessly, we think, some of the resources of his art." But in doing so, "he wins our respect." In other words, legitimate assumptions about reality were unartful; but they were at the same time popular. In fact, Dallas says all this in an essay expressing some slight dismay at how very popular Trollope had become. Trollope's dogged commitment to the transparent rendering of reality leads him away from romance conventions but apparently helped make him popular. What in Darwin emerged as revolutionary and threatening to religious tradition of order and teleology, in Trollope is taken as comfortable and convincing.

The popularity suggests that these qualities are not exclusively Trollopean. In resisting melodramatic conventions, absolute distinctions between good and evil, and the idea of a just world in which conclusions once and for all settle things, Trollope was simply a Victorian novelist. But no Victorian novelists succeed as consistently as Trollope in this resistance or are, therefore, more exemplary in projecting a world that parallels Darwin's own.

Let us take, for example, Trollope's ostensibly casual way with plot. "When I sit down to write a novel," he says, "I do not at all know, and I do not very much care, how it is to end" (*Autobiography,* p. 220). Obviously, he did know, at least in the sense that readers themselves usually know. There is no need to read very far in *Dr. Thorne* to know that Mary Thorne will marry Frank Gresham, or that both Roger Scatcherd and his son will die in time to make Mary an heiress and save the Gresham estate. Focus, as in Darwinian narrative, is shifting from the satisfactions of teleology to the pleasures of process.

Trollope was, in fact, concerned with form and order, and well understood that realism did not entail splashing facts randomly against a canvas. One of his complaints about Thackeray is that he did not sufficiently think out the shape of his narratives. "To think of a story is much harder work than to write it . . . To think it over as you lie in bed, or walk about, or sit cosily over your fire, to turn it all in your thoughts, and make the things fit—that requires elbow-grease of the mind. The arrangement of the words is as though you were walking simply along a road. The arrangement of your story is as though you were carrying a sack of flour while you walked."[29] The arrangement of the story is not quite plot, for the design of "plot" that determines the fate of characters is antithetical to Trollope's objectives. Although Trollope's criticism of Thackeray tends to give the lie to the *Autobiography*'s self-caricature of a writer who does not think or plan, it does not imply that Trollope falsified in describing his weakness at plotting or thinking and his preference for characterization.

Dr. Thorne provides an interesting example of the way Trollope's actual work denies the strategies of transparency and spontaneity he claims to use. So the deaths of Scatcherd and, even more, of his son, betray the author's hand. The most interesting tension in the novel is only artificially resolved by the deaths: how Frank and Mary would behave against the unpleasant but sensible pressures of the Greshams to force Frank to marry money. The uncompromised Trollopean method suggests other possibilities that *Dr. Thorne* representatively compromises, just as Darwin's pure empiricism is (privately but self-consciously) compromised by the theoretical underpinning that supports the discovery and interpretation of facts. In the big Victorian novels complex issues are frequently settled by the materials of conventional storytelling—deaths, wills, chance meetings, coincidences.

Dr. Thorne, for example, is rather overestimated in the Trollope canon because of its relatively unified plot. Its greatest strengths reside in its wanderings. Thorne's various duels with Dr. Fillgrave (whose name, with very many others in Trollope, belongs to a different comic tradition) are both funny and significant in establishing the structure of value according to which we are being taught to understand the citizens of Barchester and environs. Miss Dunstable's wit, bluntness, maturity (and the hinted possibility of her corruption by money) are far more important than her brief role in the narrative as alternative to Mary in Frank's affections. Scatcherd, too, is hardly central to the plot, but he is complexly there as a murderer, an alcoholic, a captain of industry, a politician, a fundamentally honest and morally defeated man. Trollope's strategies of delay suggest other novels that don't quite get written and keep off the unwelcome ending in which everybody will be presumed (until the next novel) to live if not happily at least appropriately ever after.

The minimizing of plot is obviously consonant with the Darwinian expulsion of teleology, and thus of the conventional happy ending, from the world. Darwin's "science," moreover, was handicapped as science because it was retrospective, without the power of prediction. Darwin the observer did not know what the plot would be, only that given a set of variations, adaptation of some kind would take place. Trollope is the Darwinian rather than the Herschelian scientist; nor do his plots always curve downward toward loss of energy, in the way of thermodynamics. Rather, their multiplicity and variousness guarantee renewed bursts of compromised but flourishing life.

The attitude toward plot is accompanied by other Darwinian qualities. Trollope is caught as completely as Eliot in the assumption that the world is only comprehensible in terms of cause and effect. The processes are not merely random; they are stochastic. Narratively, Trollope was always troubled by the need for a great deal of what he regarded as dull description before he could get under way. *Dr. Thorne*'s second chapter opens with an apology "for beginning with two long dull chapters full of description." But he needs them because, as he says, "I find that I cannot make poor Mr. Gresham hem and haw and turn himself uneasily in his arm-chair in a natural manner till I have said why he is uneasy. I cannot bring in my doctor speaking his mind freely among the big-wigs till I have explained that it is in accordance with his usual character to do so" (p. 19). There can be no secrets, no surprises in Trollope's world. Like Darwin, he wants to explain all phenomena, and all phenomena can be explained in terms of clusters of causes within a complex environment. And the faith in causes is itself evidence of a primary assumption of continuities, not of Foucauldian disruptions and leaps.

Compare this strategy and these assumptions with Dickens's manner. Rigaud in *Little Dorrit*, Tulkinghorn in *Bleak House*, remain unexplained in Trollope's sense. Superficial, secondary explanations are possible: but Dickens's world is ultimately significant not because it reads experience in terms of cause and effect, but because the various mysterious forces that move through it become part of complex moral and psychological patterns of meaning. Lying behind Dickens's manner is a religious tradition, but Trollope's attitude toward meaning and explanation is profoundly secular. Irrational behavior—which he takes to be the norm—can be explained rationally; and even if it can't, that's not a terrible problem. Explanation, as James suggests, is less important to Trollope than representation.

Both Trollope and Darwin attempt to account rationally for a world full of incongruities and anomalies that traditional literature and science have tried to squeeze into inappropriately regular forms. Darwin, as I have argued, describes a world full of aberrations and maladaptations, "frigate-birds with webbed feet . . . long-toed corncrakes living in meadows instead of in swamps . . . woodpeckers where not a tree grows . . . diving thrushes and petrels with the habits of auks" (*Origin,* p. 217). Trollope, too, fills the world with misfits better adapted for life in different places or times. With quiet daring—not so anxious or self-conscious as Thackeray, not so desperate for another mode of explanation as Eliot—Trollope gives us, rationally, a world not only impelled by irrational energies, but irrational in structure.

He is instinctively distant from the tradition of natural theology, in which meaning inheres in nature, and all particulars can be assimilated to design. He sees with Darwin that the beautiful singing birds are preyed upon and preying, with Eliot that behind the apple blossoms or among the golden corn there is an image of agony. Nobody was more conscious than Trollope of the Darwinian struggle. His later novels are filled with a sense that all of society is dominated by it.

In Trollope, the incongruity between plot (which implies the romance conventions that express the fully rational or intentionally designed world) and the actual energies and concerns of the novels is particularly striking. "I have indeed," he says, "for many years almost abandoned the effort to think, trusting myself, with the narrowest thread of a plot, to work the matter out when the pen is in my hand" (*Autobiography,* p. 134). The abandonment, while it implies confidence in his own powers and disguises a deep engagement in what unfolds in his world, also might suggest an undesigning God, or a cold scientific dispassion surrendering to the play of causes. In Trollope, as in Darwin, the large scale and the small scale are on the whole presented with the same dispassionate voice. In both, the ostensibly trivial figures displace the large romantic ones of traditional narrative, and by virtue of the attention they get become significant. Those Darwinian seeds build to a theory of evolution; the very ordinary people who fill so much of Trollope's deliberately antiheroic narrative become the subjects of the realistic romance.

The strategy of transforming the small to the large is the strategy of scientific disinterest, the refusal to give dramatic priority. Trollope's *Autobiography* provides considerable evidence of this way of perceiving: "I can tell the story of my own father and mother, of my own brother and sister, almost as coldly as I have often done some scene of intended pathos in fiction" (p. 29). Contemporary critics complained of this dispassion, but for Trollope the effect of reality depended upon it. It freed him for his thrusts against romantic form and for his quietly daring refusal to see the world "bright with gladness." There are losses, extinctions, compromises, sacrifices—the most famous of which is in the story of Lily Dale and Johnny Eames. But one might imagine Mary Thorne herself as a sacrifice had not Trollope intruded his compromised, designing hand. Trollope's method of composition parallels Darwin's own way of seeing pain and loss: "It has always appeared to me more satisfactory," Darwin wrote, "to look at the immense amount of pain and suffering in the world as the inevitable result of the natural sequence of events, i.e., general laws, rather than from the direct intervention of God" (*Life and Letters,* III, 64).

Trollope's novels are "scientific experiments" only as Eliot's are what she called "experiments in life." As author, he is a designer, however loose or disguised the designing hands. But his Darwinian vision leads him to test the same materials repeatedly in different ways over long stretches of time. He points out in the *Autobiography* that "novelists have [not] often set before themselves the state of progressive change" (p. 273). And he says that he himself might have failed to think about it were he not so "allured" back to his "old friends." That is, thinking about characters not as inhabiting a plot with beginning, middle, and end, but as people, he was forced to incorporate the conception of time relentlessly refusing to stop a story until the organism was dead; and beyond that, to discover that the cessation of one story in no way implied the cessation of story itself. The same story, or a variation on it, was likely to be played out endlessly, by other characters.

The characteristic Trollopean manner emerges in this confluence of concern with the individual character and awareness of the artificiality of plot. A pattern in one novel may give way to a variation on it in the next, the consequence of which is that together his novels suggest that all patterns are impermanent, that no narrative ends, that few of the many possibilities survive, but for those remaining the proliferation, the variation, the subtle incremental changes stretch toward the infinite. Trollope's world might almost be seen as an extended experiment on the human species, not complete until all the variations are played, and consequently always falling short of satisfying scientific generalization despite the worldly wisdom of the novelist's stance. Disinterested, empirical, obsessively observing particulars, conscious of time and change, Trollope is, in short, Darwinian in his approach to fiction.

One of the patterns that runs through several of the novels is a love triangle in which a young man in love with the good young heroine is attracted or persuaded to another, wealthier woman. Christopher Herbert describes this as an "experimental situation": "In his imaginative laboratory the novelist . . . runs through this situation over and over in successive stories."[30] Three of the Barsetshire chronicles provide an excellent example of such patterning. The relationships normally are structured around any of three possibilities—that the potential lovers are interested primarily in money, that they are genuinely in love, or that they take pleasure in flirting. Frequently, they are controlled by a strong parent, who stands as obstacle or, perhaps, matchmaker. So in *Dr. Thorne* (1858), there is the triangle of Mary Thorne, Frank Gresham, and Martha Dunstable, with Dr. Thorne himself as parental guide; in *Framley Parsonage* (1861), the triangle is Lucy Robarts, Lord Lufton, and Griselda Grantly, with Lady Lufton an interested and powerful mother; in *The Small House at Allington* (1864), the triangle is Lily Dale, Adolphus Crosbie, and Alexandrina DeCourcy.

Since the triangle is such a staple of romantic literature, these sets may seem inevitable. Each is involved with questions of family, class, and money so as to suggest they are indeed variations, self-consciously commenting on each other and refusing each other priority. In an experiment, after all, the isolated fact can tell nothing without being seen in relation to other relevant facts. In science, presumably, such juxtaposition should lead to generalization. In Darwin, as when he shows how various and complicated the interbreeding of separate varieties and species actually is, it often leads to the exploding of former generalizations that were based on inadequate knowledge of particulars. In Trollope, generalization isn't the point.

Frank Gresham is perhaps the most solid and reliable of the three male figures, which is to say, not very solid and reliable. Although his parents, particularly his mother, try to force him to marry so as to help recoup their finances, Frank's loyalty to the penniless Mary seems only slightly threatened. Through Frank, Martha Dunstable enters the Barsetshire chronicles, and she brings Frank up short in his casual slide into lovemaking. But the fact that Frank was capable of such a slide, that he might well have trapped himself into a loveless but wealthy marriage if circumstances had been right, is a possibility of the plot that subverts Frank's position as romantic hero. Frank's weakness is clear, and the narrative treats it with appropriate worldly-wise dispassion and comedy as his suit develops: "Let his inward resolution to abjure the heiress be ever so strong, he was now in a position which allowed him no choice in the matter. Even Mary Thorne could hardly have blamed him for saying, that so far as his own prowess went, it was quite at Miss Dunstable's service. Had Mary been looking on, she, perhaps, might have thought that he could have done so with less of that look of devotion which he threw into his eyes" (p. 172). *Dr. Thorne*, however, will not indulge the potential cynicism. Determinedly romantic, it plays this particular variation so that Frank's weakness disappears in a flood of money and marriage, the consequence of which is that the thematic stakes—relating to the tensions between class and money, individual desire and social restraints—are comically reduced. There is to be sure a further and more painful and unconventional reading—that Frank's weakness is very much part of the "happily ever after" myth, and that the novel countenances it cynically, unworried.

Frank's kind of weakness and the possibilities of corruption are exposed relentlessly in *Framley Parsonage*. Not that Lord Lufton is imagined as any less honorable than Frank. Consistent with Trollope's refusal of conventional romantic heroism, Lufton is no more hero than Frank: "I may as well confess," says the narrator, "that of absolute, true heroism there was only a moderate admixture in Lord Lufton's composition; but what would the world come to if none but absolute true heroes were to be thought worthy of women's love?"[31] Lufton moves immediately from a proposal to the true heroine to a semivoluntary lovemaking with the woman chosen for him by Lady Lufton, his mother. Here the parallel with Frank is precise. But Lufton backs away from Griselda Grantly quickly only when *she* senses that he prefers Lucy to her. And he has also flirted financially with the corrupting Sowerby. Moreover, Lucy's brother, Mark, although outside the love triangle, is almost as thoroughly seduced into the financial triangle as if he had betrayed his wife sexually.

One of the major variations in *Framley Parsonage* is that the parent who is attempting to persuade her son to reject the good heroine for money and class is *not* one of the corrupt aristocracy but rather the strongest representative of the true-blue, Thorne-like aristocratic family. Lady Lufton participates in the very corruption she is fighting when she tries to keep her protegé, Mark, away from the world of the Duke of Omnium. This complication of the narrative pattern suggests that the world is changing from its true-blue traditions so that moral action becomes secondary to flexibility and adaptiveness. This second experiment comments strongly on *Dr. Thorne* and implicates the triangle situation in major social issues. The inevitable Trollopean consequence is compromise, blurring of moral and class categories, and a very Darwinian world.

In *The Small House at Allington* the romance pattern of *Dr. Thorne* is entirely shattered, the inevitability of change (and, from Trollope's developing point of view, corruption) becomes more pronounced. For here, the male center of the triangle is Adolphus Crosbie, who, unlike Lufton or Gresham, has no parents to persuade or bully him. The absence of traditional inheritance, whether corrupt or true-blue, marks a significant change in the exploration of the experimental situation. Neither tradition nor external pressure influences Crosbie's choice. Recognizing Lily's superiority, as little brown heroine, he jilts her for the DeCourcy world, although as it turns out, he did not *need* the DeCourcys to help him. Chance plays its Hardyesque and Darwinian role, for Crosbie might have been loyal to Lily had he known what the higher circles already knew, that he would be promoted in any case. His former chief is delighted to "surprise" him, but surprise in Trollope remains a moral defect, either in narrator or in character. Straightforward knowledge is a condition of right storytelling, and of right action.

Although he is called villain by the true-blue Christopher Dale and even by his not very true-blue business friend, Pratt, and although the narrative certainly endorses moral disapproval of his action, Crosbie is nevertheless presented from the inside, and sympathetically. At the end, his life is no more justly punished than Lily's, and he even regains his ascendancy at his office. The narrator records dispassionately as Trollope's world leaves no space for the happy ending, no grounds for natural theology. Moral being has no correlation with the way things are.

In *Dr. Thorne* another figure, Mr. Moffat, had jilted Frank's sister, Augusta. This leads to an event significantly alluded to in *The Small House at Allington*, and one of the more unpleasant moments in Trollope, when Frank corners Moffat and almost whips him to death. Nobody, certainly not the dispassionate Trollopean narrator, seems to judge Frank's action severely any more than they judge Roger Scatcherd for murdering (in vengeance for his sister) Dr. Thorne's seducing brother, Henry. When Amelia Gazebee discusses Crosbie's defection with Alexandrina, Crosbie's new fiancée, she reminds her of "what Frank Gresham did to Mr. Moffat when he behaved so badly to poor Augusta." The characters have remained essentially the same; indeed many of them carry over from novel to novel. But neither Bernard, Lily's cousin, nor Christopher Dale, the true-blue counterpart to Lufton and Thorne, respond with anything like Frank's decisiveness. And Mr. Dale, ineffectual where Lady Lufton and Dr. Thorne have been powerful, can only repeat, "On my honour, I do not understand it." "It makes me feel," he says, "that the world is changed, and that it is no longer worth a man's while to live in it."[32]

The validity of the horse-whipping treatment is confirmed by another variation. The hobbledehoy innocent, Johnny Eames, takes up the parental position for Lily, and the whipping of Moffat by Gresham has its comic counterpart in Johnny's laying Crosbie prostrate near "Mr. Smith's bookstall" on the railroad platform, "falling himself into the yellow shilling-novel depot by the over fury of his own energy" (p. 322). The fall into cheap fiction is finely appropriate to the revised pattern of *The Small*

House at Allington. It is perhaps more appropriate that the police carry him away—as they did not Frank Gresham. But there is further irony here, how much under Trollope's control is not entirely clear. Johnny is himself a jilter. Although the narrative insists on the guilt of Amelia Roper, who ropes in the naive Johnny, Johnny did in fact make a declaration of love and is as honor-bound to sustain it as Crosbie. A comic variation on the Lily-Crosbie-Alexandrina triangle, the Lily-Eames-Roper triangle implicates even the most innocent in the pervasive corruption intimated by this single motif. More important, it denies the possibility of the simple classifications of good and evil, villain and hero. The judgments ostensibly endorsed in the main voice and the direction of the narrative are undercut by the fact that nobody seems exempt from the guilts implied. There are yet other variations on the motif, even lower on the comic scale, with Johnny's friend, Cradell, who flirts with Mrs. Lupex, and is threatened with a thrashing by her husband. It is Cradell who gets Johnny off the hook by marrying Amelia Roper. And there is another major variation, to which I shall return shortly.

If the inconstancy of Crosbie and the others is demonstrated to be destructive and immoral, the tangled and proliferating complications of this apparently simple experimental situation make its alternative equally unattractive. All of the protagonists—even Johnny Eames—have their inconstant moments, but constancy, as the further fortunes of Lily Dale suggest, may be yet more destructive. Christopher Dale is an inadequate true-blue precisely because he is constant. In his stubbornness, the male equivalent of the passive stubbornness of another Dale by marriage, Lily's mother, he perpetrates a terrible injustice both against Mrs. Dale and, implicitly, against her daughters. He has made up his mind that Lily will get no money and that Bell should marry Bernard, his nephew. These decisions lead to the waste of Lily's life and threaten the happiness of Bell's—for Bernard is incapable of love, or of strong feeling. After Lily learns of the jilting, when Mr. Dale commiserates with her and curses Crosbie, Lily cries, " 'Uncle! Uncle! I will not have that! I will not listen to a word against him from any human—not a word! Remember that!' And her eyes flashed as she spoke. He did not answer her, but took her hand and pressed it, and then she left him. 'The Dales were ever constant!' he said to himself as he walked up and down the terrace before his house. 'Ever constant' " (p. 289). Moral or immoral, constancy implies inability to adapt, as of a Darwinian organism doomed to extinction. The variations through all these novels suggest on the contrary a narrative richness and flexibility that the Dales themselves are incapable of enacting. The intimation of Lily's later fate, on the one hand, and of the extinction of the values for which the Dales and the Luftons and Thornes stand, on the other, are all quietly implicit here. The heroine gradually loses her charm through two novels, and the Barsetshire novels give way, among other things, to the Palliser series.

Perhaps the most famous variation on the experimental situation is that of Plantaganet Palliser. Here the center of the triangle shifts to the woman, Griselda Grantly, who has already played her part in the pattern in *Dr. Thorne*, and who reappears to flirt in her own stony way with Plantaganet. But now the protagonist is on the DeCourcy-Dumbello-Omnium side rather than on the true-blue side. And the Duke of Omnium almost parodically reverses the parental role played out so poorly by Lady Lufton and Christopher Dale. Appropriately enough, the Duke wishes to discourage the flirtation, but only because it would interfere with his own flirtation with Dumbello's mother; and he threatens to cut off Plantaganet's inheritance. Instead of the honorable hero refusing to breach decorum, Plantaganet is positively inspired to wickedness by his uncle's threat and his uncle's model: "The full amount of this threat Mr. Palliser understood, and, as he thought of it, he acknowledged to himself that he had never felt for Lady Dumbello anything like love . . . Lady Dumbello had been nothing to him. But now,—now that the matter had been put before him in this way, might it not become him as a gentleman, to fall in love with so very beautiful a woman whose name had already been linked with his own?" (p. 410). The inversion, the dispassionately implied cynicism, is complete, and it is difficult to read the rest of the Palliser saga (assuming that one accepts the continuity of the narratives and that it is legitimate to recall an incident from one novel in the midst of another) without remembering that this truest of gentlemen is both weak and silly in his first appearance. The episode changes the way *Can You Forgive Her?* can be read. But the whole affair in *The Small House at Allington* is turned into a comic routine, the worldly Griselda entirely besting the innocently lecherous Plantaganet.

Nevertheless, the episode initiates the Palliser series from the already rich life of Barchester. And the Palliser series, too, works variations, with somewhat higher stakes, on the same triangle pattern. In *Can You Forgive Her?* (1865) Burgo Fitzgerald, whom Glencora loves, becomes the potential adulterous lover Plantaganet has aspired clumsily to be in *The Small House at Allington*. Among the large-scale differences in the pattern in the two series is an important thematic one. In the Barchester series Whig aristocrats, moving toward new values of capitalist culture, tend to be the corrupt attractions in the triangles. Parents and surrogate parents of heroes and heroines tend to reject new money for old values. In the Palliser series Whig aristocrats tend to be central, their blending with the commercial classes less obviously corrupt, and the margins, already blurred in the Barchester series, even more difficult to discern.

There is no need to work out particular details further. The various triangles of the political novels are inextricable from the large social and political issues with which the characters are concerned: they become experimental conveniences for the exploration of these issues. Phineas Finn himself begins disloyal to the Irish Mary, whom he leaves for parliament in the same way (and for the same kind of reasons) Frank and Lord Lufton are potentially disloyal to their heroines. Lady Laura moves through two novels deeply though not literally disloyal to her too constant and ultimately maddened husband. Here, traditional moral markers seem almost beside the point. Trollope's prodigious capacity to fill the world with variations and modifications, and to sense the minutiae of difference that keep all these from being ultimately the same gives to his work a breadth of range within apparently narrow plot conventions that allows for an astonishing openness of observation. With uniformitarian plodding movement, Trollope experiments so that his triangles can imply everything from farcical comedy to near tragic intensity; while paying lip service to morality, he withdraws from the apparently essential Victorian activity of moral placement. His animus against certain kinds of corruption is clear enough; yet the moral placement seems to have nothing to do with the narrative sympathy for Laura, or Phineas, or Glencora, or even, finally, Plantaganet himself. Character is fluid, in process, and from novel to novel, as one narrative fades into another, as one series fades into another, Trollope creates a Darwinian world.

It happens again with an entirely different set of characters in another novel of this period. Following closely on *Can You Forgive Her?* in which Glencora Palliser and Alice Vavasor hover between prudence and romance, *The Claverings* (1866) raises the problem yet once more, with a man as a focus of the main variation, and his romantic alternative (who has herself married "prudently") as an important secondary one. The novel clearly manifests many of the qualities of Trollope's art that I have been discussing. But *The Claverings* is of particular interest for its obvious failures to cope with the questions it raises. The Trollopean strategies of close observation, of uniformitarian movement, of characterization lead to some rather unpleasant consequences that are similar to the consequences of Darwinian theory, are incompatible with the strategies of narrative that we have seen in *Mansfield Park*, and are characteristic of romance plotting. I will treat the unpleasant consequences more directly and fully in the next chapter, in relation to several other writers, but it is important to note here how the quasi-scientific detachment of the Trollopean narrator allows for a distanced treatment of the protagonist, a treatment that seems to run counter to the conventions of narrative within which the protagonist operates. This is one of the more striking examples of how the Austenian plot, with its natural-theological implications, is disrupted by Darwinian elements.

The central figure is Harry Clavering, the young man who engages himself to the poor, "plain," but intelligent and sensitive Florence Burton, and finds himself shortly back in love with his first love, Julia Brabazon, who had jilted him for a wealthy peer and who returns, widowed, by the fifth chapter. Like Frank Gresham and Lord Lufton, Harry moves smoothly from the good and poor to the rich and dubious, but the focus is different, for here it is on moral paralysis. Indeed, the novel almost grinds to a halt with Harry's hopelessly indecisive wavering, and moves forward primarily with comic subplots for long pages. The narrator, meanwhile, slowly and incrementally calls into question the status of Harry as hero: he "will not, I fear, have hitherto presented himself to the reader as having much of the heroic nature in his character. It will, perhaps, be complained of him that he is fickle, vain, easily led, and almost as easily led to evil as to good."[33] *The Claverings* is much more overt about the unattractiveness of the wavering protagonist than the other novels in the series, and ultimately much more cynical about it.

The novel links heroism and knowledge: knowledge deromanticizes; and within the conventions of realism, close scrutiny entails the revelation of flaws, and disenchantment. The typical realist narrator is, as Trollope has said, a close observer, and scrutiny with the intensity of the natural historian exposes weaknesses: "There may be a question," the narrator says, "whether as much evil would not be known of most men, let them be heroes or not be heroes, if their characters were, so to say, turned inside out before our eyes" (p. 79). Later he makes the point again, invoking his function as novelist: "Perhaps no terms have been so injurious to the profession of the novelist as those two words, hero and heroine. In spite of the latitude which is allowed to the writer in putting his own interpretation upon these words, something heroic is still expected; whereas, if he attempt to paint from Nature, how little that is heroic should he describe!" The Trollopean narrator speaks with the detachment and distance of the uniformitarian scientist, allowing in addition only a reminder of the disparity between the value expected and the value actually perceived. In one sense, this move is very similar to Eliot's when she denigrates literary heroism in the interests of her moral program to increase compassion for ordinary people. But Trollope's narrator, while offering compromises and excuses for his hero, manages to make him peculiarly, persistently unattractive. His weakness as dramatized makes him different not merely from the protagonists of romance, but from the flawed heroes of realist fiction. "He should have been chivalric, manly, full of high duty. He should have been all this, and full also of love, and then he would have been a hero. But men as I see them are not often heroic" (p. 240). What makes many of Trollope's narratives more interesting than their ostensible conventionality would suggest is his quiet refusal to replace that unheroic hero with a new kind of realist heroism, as Eliot does from *Adam Bede*, through *Felix Holt*, to *Daniel Deronda*. Trollope keeps the "hero" in place; in *The Claverings* he gives the "hero" his rewards without making him heroic or deserving.

The distance of narrator from protagonist is both created and disguised by the detailed scrutiny to which Harry is subjected. The quality of attention given him would seem to guarantee compassion. In novels by Austen or Eliot, intense observation of the movements of consciousness and of moral waverings implies sympathy even for the potentially weak and immoral. But despite occasional pauses, and despite the implicit compassion for comprehensible human weakness that becomes explicit in narrative intrusions, Harry is almost a specimen. Without the self-conscious scientism of French naturalists, Trollope has a kind of "scientific" toughness about his exposure of Harry. Whatever the gentility of the context and the quiet prose with which he is presented, whatever the refusal on Trollope's part to accentuate the weaknesses or even dramatize them, Harry's behavior is consistently weak, vacillating, and even hypocritical. He is a "hero" only in that the novel is primarily about him. And only because we watch him closely can we avoid the judgments that, with the pretense of offering excuses, the narrator prods us into by pausing from observation for interpretation. In the actual conduct of the narrative that leads us through Harry's illicit commitment to Julia, Trollope is mutedly effective. Instead of handling the material melodramatically, or catastrophically, as I would call it, Trollope handles it gradually, incrementally, and scrupulously avoids moral commentary. Harry himself, of course, thinks of the potentially villainous implications of his action—"he told himself that he would not become a villain" (p. 178)—and his consciousness is subject to the same scrutiny as his actions. His character is "turned inside out before our eyes."

There is more than a hint of Harry's active sexuality (manifested in the attraction to him of most of the women he meets), and Trollope is almost clinical in his rendering of the movement by which Harry is seduced into declaring himself to Julia. As Julia describes her isolation and her pain, Harry sits down beside her in silence, "looking away from her at the fire, swearing to himself that he would not become a villain, and yet wishing, almost wishing, that he had the courage to throw his honour overboard. At last, half turning round towards her he took her hand, or rather took her first by the wrist till he could possess himself of her hand. As he did so he touched her hair and her cheek, and she let her hand drop till it rested in his. 'Julia,' he said, 'what can I do to comfort you?' " (p. 178). Note the sharply observed details, particularly the grasping of the wrist, intimating a clumsy sexual attraction. The movement into "villainy" seems merely natural, and only the narrative reminder reintrudes moral considerations: "He had not intended to be cruel. He had drifted into treachery unawares." In the tradition I have been associating with Darwinism, it is not the extraordinary that produces what seems to be extraordinary, but small, incremental, natural causes.

The observation of the narrator is parodied and given dramatic implication by the intrusion of the "Russian spy," Madame Gordeloup, as Harry takes Julia in his arms and "pressed his lip to hers" (p. 179). Gordeloup, to be sure, eventually loses her power to hurt either Julia or Harry, but her presence as spy suggests something of the possible consequences of a Darwinian view of experience. In turning the human into subject, Darwin helped open the way for unscrupulous uses of knowledge. The effect is certainly reductive, or antiheroic, since a uniformitarian reading of human nature minimizes both genius and heroism, as Darwin minimized his own genius in his autobiography. More than that, however, it depersonalizes the individual, exposes what is private to public scrutiny (turns it inside out, as it were). As Alexander Welsh has suggested in his remarkable recent study of George Eliot, the social pathology of scientific information is blackmail.[34] The novelist-narrator has the power of the blackmailer, with access to information that ought to be private but that would be worth something to the blackmailed if it were withheld. Throughout *The Claverings* the morally vulnerable characters, particularly Harry and Julia, are threatened with the possibility of blackmail, literal or metaphorical. And the book is thematically preoccupied with the need for openness, implying preference for figures who reject secrecy. Madame Gordeloup trades on secrets; her brother Pateroff tries to create and exploit them.

The novel's further variation on the triangle situation pushes it to an extreme that at once affirms the Darwinian openness so characteristic of realistic fiction and paralyzes the narrative. Harry wavers without control between his two commitments and betrays both women in his powerlessness either to speak or to act. An almost Dostoyevskian psychological tension builds an almost actionless drama. Florence begins to act. Harry's mother begins to act. Cecilia, Florence's sister-in-law, even calls on Julia Brabazon, Lady Ongar. Harry does nothing. His one virtue is the frankness of his self-hatred. Subject to the fluctuations of his desires, Harry fails to plan—either what he will say to Florence's sister, on the evening of his confession to her, or what he will write to her afterward, even at the moment of his writing: "At last the words came. I can hardly say that they were the product of any fixed resolve made before he commenced the writing. I think that his mind worked more fully when the pen was in his hand than it had done during the hour through which he sat listless" (p. 263). (Harry's relation to his own will echoes Trollope's professed relation to his novels.) He is at the mercy of whichever woman happens to attract him at the moment and is virtually powerless to extricate himself. The novel too reaches an impasse here, and to extricate himself (and Harry) Trollope changes modes.

For finally, much against the principles laid down in the autobiography and implicit in the narrative itself, Trollope turns to the kinds of providential intrusions that operate so precisely and self-consciously in *Mansfield Park* to reward Fanny and to imply a world ordered around the values implicit in the heroine. And yet the conventions that work so admirably and interestingly in Austen emerge in *The Claverings* as demonstrations of the impossibility of the natural-theological model they mimic. Austen's narrative implies a teleology she is happy to work out; we may not like Fanny, but we understand fully the power Fanny accrues from the manipulations of plot. While on the most literal level her successes may seem like mere chance, formally they follow from the moral structure of the book, and thematically they derive from her powers of selfless observation. In Trollope, "chance" seems far more likely to be the explanation of the turns of plot than providence. There is no real coherence between the way the plot turns and what, morally, the characters seem to have earned.

At the critical point of Harry's paralysis, when he has written to Cecilia Burton and tells her that he must see Julia before he can finally return to Florence, it is clear that once in Julia's presence he will flip over again, to begin another cycle of fluctuations without resolve. At that point he is called back to his home to help—ironically—resolve another marriage question, that of his sister to Mr. Saul. His hypocritical lecture to Mr. Saul is succeeded by his descent into a serious illness recognizable in a Victorian pattern of purgation to be found in Dickens, particularly in Pip and David Copperfield, but also in Thackeray's Pendennis. And yet with all the potential Dostoyevskian implications of this illness, it purges nothing but makes plot resolution possible. While in his weakened state, Harry easily succumbs to his mother's pressure to make him return to Florence. The whole determination comes in a slight scene: Mrs. Clavering speaks to Harry in the evening:

> "Have you slept, dear?" she said.
> "A little before my father came in."
> "My darling," she said,—"you will be true to Florence; will you not?" Then there was a pause. "My own Harry, tell me that you will be true where your truth is due."
> "I will, mother," he said. (p. 293)

There the issue is resolved. Harry manages to lie in a postscript to a letter to Florence, telling her that he has never loved anyone half so much as he loves her. But the resolution results from the chance of his being called back to his home to force a decision in another matter when he is totally incapable of deciding anything.

At this stage "chance" becomes an overt subject. When Harry does see Julia again, he remarks, with an appropriateness that all the preceding narrative confirms, "These things, I fear, go very much by chance." Julia, feeling the implications of this heretical remark, both for fiction and for biology, exclaims, "You do not mean me to suppose that you are taking Miss Burton by chance. That would be uncomplimentary to her as to yourself." "Chance at any rate," he replies, "has been very good to me in this instance" (p. 366). Harry does not avoid the uncomplimentary implications. It is not providence that has been good to him. His reward—if that is what it is—is not commensurate with his actions. It is mere accident. Just as in Darwin the teleology of natural theology is transformed into the chance-initiated developments of natural selection, so in Trollope the providential function of chance in narrative is transformed into meaningless accident.

To play out the game more fully, Trollope introduces a strikingly un-Trollopean event to make the novel end quickly and happily. He kills off both Hugh and Archie Clavering (having already killed off Hugh's son) so that suddenly, without any preparation, Harry becomes heir, can give up his attempts at earning a living, can marry Florence immediately, and can give the living of his estate to Mr. Saul so that the austere curate can marry Fanny Clavering. It is like the developments at the end of *Mansfield Park*, which give Fanny all she wants. But again, it is merely gratuitous. Trollope as narrator does not intrude with Austen's worldly- and novel-wise shortcutting to what readers will expect. Obviously, the outward event has nothing to do with the drama of moral wavering. It is mere chance, and all the worse because in certain respects it seems to confirm the conventions of poetic justice. It brings home for Julia, for example, the ironic waste of her prudential marriage to Lord Ongar, for the man she really loved would soon have come into money and she could have had both prudence and romance. Her punishment (and that is the term often used in the novel) is complete. Moreover, since Hugh was villainously unfeeling, it is quite satisfying to find him killed. And since his wife, Julia's sister, herself married for prudence, she is punished in losing the estate. And since Harry's sister Fanny has agreed to marry a man who could not afford to keep her, she is rewarded by his suddenly acquiring a living that is at Harry's disposal. All of this sounds very Victorianly conventional, and insofar as we can take seriously the moral judgments implicit in this working out, it is just that. But this merely conventional narrative is not the one Trollope was writing and belongs to Austen's not Darwin's world.

The unsatisfactoriness of the ending accurately reflects what the Darwinian elements in mid-century narrative have done to the conventions of fiction. Chance, not providence, here determines rewards and punishments and thereby invalidates the conception of rewards and punishments. The distanced and dispassionate look at the triangles, at the issues of romance versus prudence (spelled out not merely in Harry's story, but in Julia's, in her sister's, in Fanny Clavering's relation to Mr. Saul, in Archie Clavering's comic attempt to marry Julia, in Pateroff's attempt to blackmail her) cannot, in their variety or their chanciness, be correlated to any moral reading. Harry gets the rewards of Fanny Price without her powers of self-discipline, of will, or of observation. And instead of a conclusion that emphasizes the rewards and their moral implications, the last line of the novel confirms the inadequacy of the idea of providence and the weakness of Harry. "Providence," says Florence's brother, "has done very well for Florence. And Providence has done very well for him also—but Providence was making a great mistake when she expected him to earn his bread" (p. 412).

The implicit amorality of this is akin to the implicit amorality of Darwin's argument. Darwin's language itself at the end of the first edition of the *Origin* seems almost a commentary on some aspects of this Trollopean pattern. The most remote types within these triangles turn out to be more closely related to each other than any of the conventional modes of classification, social or moral, could detect. Harry jilts Julia as Julia has jilted him, deserts her as Hugh Clavering deserts his wife, is snobbish to Mr. Saul as Julia was to him when she called him an "usher," in the opening jilting scene. Johnnie Eames is more like Crosbie than even Trollope admits; Lady Lufton is more like the Duke of Omnium than she could conceive; the great Plantaganet Palliser has a touch of Crosbie and Johnnie in him—or even of Johnnie's friend Cradell. And here is Darwin concluding his arguments about the arbitrariness of the conventional separation of species: "Hereafter we shall be compelled to acknowledge that the only distinction between species and well-marked varieties is, that the latter are known or believed, to be connected at the present day by intermediate gradations, whereas species were formerly thus connected. Hence, without quite rejecting the consideration of the present existence of intermediate gradations between any two forms, we shall be led to weigh more carefully and to value higher the actual amount of difference between them" (*Origin,* p. 45). Darwin talks of species; Trollope talks of individuals. But the thrust toward fine individual discriminations, away from the fictions of generality, is the same.

According to Darwin, we must "regard every production of nature as one which has had a history" (*Origin,* p. 456). Trollope, we recall, had written, "I do not think that novelists have often set before themselves the state of progressive change" (*Autobiography,* p. 273). Both writers pursue their subjects through the intricacies of pasts that have shaped the present, recognizing that change is the only constant. Their worlds are scattered with anomalies, vestigial figures—old Harding, who cannot talk to his granddaughter Griselda; Christopher Dale, who cannot understand the new generation; Josiah Crawley, who has shaped his austerity into a distortion of life itself. Each of these figures is linked to traditions that no longer sustain them. History reveals itself through them and leaves them in pain. So in Darwin, history is the clue; and anomalies are the best evidence of history. In his last thrust at the creationists, he says, "On the view of each organic being and each separate organ having been specially created, how utterly inexplicable it is that parts, like the teeth in the embryonic calf or like the shrivelled wings under the soldered wingcovers of some beetles, should thus so frequently bear the plain stamp of inutility" (*Origin*, p. 452).

For both Trollope and Darwin, then, the world, however irrational in moral terms, is explicable in time, in history. To quote Darwin once more, "To my mind it accords better with what we know of the laws impressed on matter by the Creator, that the production and extinction of the past and present inhabitants of the world should have been due to secondary causes" (*Origin*, p. 458). The creator invoked here is the one discussed in Hillis Miller's *The Disappearance of God.* To make sense in terms of "secondary causes" entails delaying, perhaps permanently, the satisfaction of design. So it was in those large loose baggy monsters; for the world simply would not yield to human observation evidence of a perfect intelligence. Nor would it resolve its stories so that event corresponds to moral condition. Characters like the inflexible Lily Dale, Josiah Crawley, and Louis Trevelyan attempt to impose design upon the world, in a sense write narratives in the tradition of natural theology. Too constant, even when—as in each case—they are *morally* correct, they slip toward alienation and death.

Darwin, the great observer of aberrations, struggled to find new kinds of biological laws that might account for them. Trollope registered, in the mode of the great Victorian compromisers, such aberrations. Together, they offer narratives of a world impermanent, prolifically populated, interdependent and interrelated, but pervasively imperfect. Great observers, they create worlds that force confrontation between the empirical ideal of observation and the possibilities of order and meaning. Attention to particulars is a move toward disanalogy, a breakdown of the Western means to overcome that radical Western dualism of mind and spirit. A little observation is a dangerous thing. A lot is even more dangerous.

8

The Perils of Observation

I TAKE the Darwinian aspects of Trollope's art as representative of the mainstream of Victorian realist narrative, most particularly in his uniformitarian insistence on continuities and change, the casual antiteleological abundance of his worlds, and his antimetaphorical and antitheoretical commitment to literal recording of precisely observed minutiae. But before concluding the argument of this book with an exploration of the fate of this kind of Darwinian narrative, I want to consider in more detail some of the contradictory implications of the scientific theme of the authority of observation itself, complications intimated but not developed by Trollope. This aspect of the scientific/Darwinian interaction with narrative leads to fundamental questions of how narratives establish their authority, how they get told. To treat this problem adequately, it will be necessary to look at the pervasiveness and implications of the culture's reverence for observation and then, if only briefly, at some writers who *had* read Darwin and who self-consciously worried out the problems he raised.

Although scientists and theorists of science from Bacon on have insisted on the primacy of observation for any true scientific knowledge, the empirical ideal to which observation is connected can have what seem like *antiscientific* implications. Trollope and Darwin's dependence on the authority of observation is part of the armory with which they vanquish natural-theological views of the world. Close observation tends to de-animate nature: the determination to value the experience of the here and now over any earlier traditions of interpretation is already a step toward secularization. Observation privileges the particular over the general because the general is an induction from accumulated experiences—at least within the empiricist argument. One of the ways that Darwinism shattered the British harmony between science and religion (not a logically inevitable development) was precisely by privileging the individual observation.

But it was not only religion that had trouble with Darwin. As we have seen, one of the reasons that Newtonian scientists, like Herschel, hesitated to accept Darwin's theory was that it seemed to escape the rule of law. If facts only become scientific when they can be fit into general laws, Darwin's arguments were apparently unscientific. Empiricism threatens both theology and the laws of nature themselves.

Biological, physiological, and medical science sought ways to reconcile empirical observation with scientific generality; indeed, this was part of Darwin's project. To do that, however, he had to transform his peculiar subject, organic life, including—especially—human life, into material for scientific observation and investigation. The power science exercised over nature, by virtue of its extension of knowledge, was to extend over human beings themselves. In the great democracy of disinterested empirical investigation, the human subject becomes equivalent to the planetary or the geological. And the final twist, of particular interest in my argument, is that as the observer gains power over the observed, he or she *becomes* the observed, and is made peculiarly vulnerable. It is this complex development from the simple concept of observation, as it manifests itself in narrative practice, that I trace and explore in this chapter.

The famous second book of Locke's *Essay Concerning Human Understanding*, one of the foundation stones of British empiricism, opens with a consideration of where the understanding gets its ideas. Appealing appropriately to "every one's own observation and experience," Locke puts the question this way: "Whence has it all the materials of reason and knowledge? To this I answer, in one word, from EXPERIENCE." And the medium for experience is, he says, "observation"—either of "sensible objects" or of the "internal operations of our minds."[1] Almost three centuries later, Herschel adopts vigorously the Lockean formulation, identifying "as the great, and indeed only ultimate source of our knowledge of nature and its laws, EXPERIENCE."[2] And experience, he says, can be acquired in two ways, through "observation" and "experiment."

Observation had become a critical part of a continuing intellectual, religious, and social revolution, stretching back beyond Galileo; it implied an alternative to traditional authority, to the dominance of ancient institutions, but particularly to the dominance of the word, of texts. "Genuine knowledge," Maurice Mandelbaum has noted, was believed in the nineteenth century to involve "some form of immediate apprehension, in which what we know must be both directly present and grasped in concrete detail."[3] And in the continuing Baconian struggle against traditional authority, T. H. Huxley would announce that "the improver of natural knowledge absolutely refuses to recognize authority as such."[4] It is not stretching things to find the same impulses at work in the Pre-Raphaelites' declared "entire adherence to the simplicity of nature,"[5] which implied powers of unimpeded observation of the natural world. Darwin's authority was based in this pervasive romantic ideal of direct access to nature.

But this ideal is compromised by the limitations of the human condition. Science attempts to clear the human difficulties away, and for it the empiricist ideal entails, or is entailed by, the Cartesian dualism of matter and spirit, and of subject and object. Modern scientific discourse, as the most powerful intellectual heir of this tradition, was built (correctly or not) on a structure of alienation from the object of study. Banishing the various "idols" of traditional mythologizing and of inadequate knowledge, scientific discourse authorizes, and is authorized by, objectivity and disinterest. Herschel immediately qualifies his commitment to "experience" with an analysis of the kinds of "prejudices" that threaten to compromise it. Matthew Arnold, in what would seem a very different and very antiscientific sphere, appeals to the ideal of "disinterest," to be achieved by moving beneath our ordinary selves to those best selves not governed by self-interest, desire, or political and social biases.

Observation of any kind, but particularly what might be taken as objective and disinterested observation, becomes, as Foucault understood it and as nineteenth-century narratives seem frequently to testify, an institutional or socially sanctioned act of power and aggression, by which the limitations of the individual self are transcended and the full weight of a larger authority is asserted. In effect, the consequence and the condition of this move to achieve objective and disinterested observation is, at least officially, that to observe a thing carefully, one must not care about it. The self must be distanced from the act of knowing and from the thing known. Darwin intensified the problematic nature of that distancing, and the difficult consequences that followed from the authority it pretended to achieve. It is one thing to remain detached about the distant planets and stars, or even about the rocks and the rivers; it is quite another when the subject becomes human life itself. Darwin's extraordinary success in making a case for the unity of nature, and thus in convincing his contemporaries that all organic life could and should be subjected to naturalistic explanation, raised enormous problems that manifest themselves in the way nineteenth-century narratives got written. Writers, self-conscious about the impossibility of finding a perspective that could reveal the whole truth, not only reflected on the adequacy of their invented narrators and on their narratives as structures of observation, but with increasing frequency made observation itself a theme and a principle of plotting.

Foucault emphasizes the relation of the growth of observation to a new kind of power, adapted to a culture in which individuation had become primary. To a certain extent, he argues, individualism is both created and controlled by observation, or by what he calls the development of the "examination."[6] Darwinian biology, with its emphasis on the individual and the unique, led increasingly to the consideration of what Foucault calls "the case," an objectification and distancing of the individual as it is subjected to close scrutiny:

> The examination is at the centre of the procedures that constitute the individual as effect and object of power, as effect and object of knowledge. It is the examination which, by combining hierarchical surveillance and normalizing judgement, assures the great disciplinary functions of distribution and classification, maximum extraction of forces and time, continuous genetic accumulation, optimum combination of aptitudes and, thereby, the fabrication of cellular, organic, genetic and combinatory individuality. With it are ritualized those disciplines that may be characterized in a word by saying that they are a modality of power for which individual difference is relevant.[7]

Foucault's concern here is with discipline in both senses—the attempt to control behavior, and the establishment of disciplines, as in the current structures of knowledge. Their relation is critical to his argument, and to mine. Foucault talks about the historical shift from "the individuality of the memorable man" to that of "the calculable man,"[8] and the development of "scientifico-disciplinary mechanisms" that define and control individuality. He even links this development with the cultural shift reflected in literature from a focus on the adventures of a noble, "memorable," figure to a concentration on the ostensibly ordinary protagonist associated with realism. Darwinian theory sanctions this shift of perspective, which is fundamental to the development of the human sciences; these are interested not in the historically exceptional but in the nature of the ordinary human being within race and society. Predictability becomes essential for insurance companies and the new knowledge.

A brief look at the implications of the rhetoric of the first two sentences of the *Origin* may help suggest how much work is going on in the attempt to establish observational authority:

> When we look to the individuals of the same variety or subvariety of our old cultivated plants and animals, one of the first points that strikes us is, that they generally differ much more from each other, than do the individuals of any one species or variety in a state of nature. When we reflect on the vast diversity of the plants and animals that have been cultivated and have varied during all ages under the most different climates and treatment, I think we are driven to conclude that the great variability is simply due to our domestic productions having been raised under conditions of life not so uniform as and somewhat different from, those to which the parent-species have been exposed under nature. (p. 71)

This is no act of pure observation. "We look" and "we reflect," as a consequence of which we are "struck" and "driven." The implication is that simple observation overwhelms us so that there is no escape from the coming conclusions. But the formulation disguises preliminary questions, themata that precede the observation—what are we looking for and why? And it disguises a fairly complicated chain of reasoning and a whole buried argument that actually turns common sense on its head: life is more uniform in nature than under domestic conditions! And there is an interesting tension between the word "think" and the word "driven," as though Darwin wants at the same time to project that rhetorical modesty and fair-mindedness that so characterizes his style and to insist on the all but physical pressure of observed fact that obliterates the thinking self. Observation transforms the active if inadequate observer into the passive and totally accurate receiver. It is the power of reality not the scientist that determines the conclusion.

In the empiricist scheme, the observer contemplates but does not affect and is not affected by what is observed. The heart, in Hopkins's phrase, is like Fanny Price's, "in hiding." Darwin's rather mild opening passage describes in small the move to self-annihilation that protects the scientist from the consequences of knowledge, preserves the convention of the separation of observer from observed, and thus authenticates knowledge. Strategically, the beginning is in the observing. But the beginning disguises a hypothesis, a predisposition to find something in what is observed. And the verbs investing authority in what is observed disguise and diminish the activity of the observer and leave him blameless for the consequences of the vision. The case is similar for the writer, and Victorian literature is full of escapes and investments of this kind: Hopkins's "heart in hiding stirred for the bird, the achieve of, the mastery of the thing." These famous ambiguities emphasize the ineffectiveness of the hiding, however, for the heart is exposed, stirring, not merely admiring achievement and mastery, but being mastered. In Hopkins as in Herschel, the observation is of a reality so forceful that, as Herschel notes, "we may follow out its traces and recognize its features through the mist of interest or in the storm of emotion."[9]

The quiet conviction of Darwin's rhetoric was essential to his cause. Whereas, for example, Baden Powell, that otherwise surprisingly Darwinian clergyman, could insist that moral and human phenomena lay outside scientific study, belonging to a "DIFFERENT ORDER OF THINGS,"[10] Darwin's move to put the human within the purview of scientific study demolished that preciously held difference. As George Stocking has recently put it, "natural selection threatened to take all ethical meaning out of temporal process, not simply by postulating evolutionary change independent of God's direction, but also by premising it implicitly on vice rather than virtue."[11]

There was, in any case, a flourishing ethnological tradition before Darwin, based, as Stocking has shown, on the study of "dark-skinned savages." Such study allowed for an objectification and a distancing that became much less easy once it was understood that the "dark-skinned savages" were no more subject to scientific observation than white skinned Europeans, who, it turned out, were descended from hairy quadrupeds, even if they had traveled a longer distance from their ancestors than savages.

Early-century ethnology was actually built on the evangelical principle of human unity, and evangelical interest in the humanity of "dark-skinned savages." The name "anthropology" came to describe the scientific activity of study of humanity through various compromises with what became a profoundly—"objectively"—racist organization, the Anthropological Society of London, established in 1866 and led by James Hunt.[12] The Darwinians were opposed to Hunt's position, which was based in the transcendental biology I have identified as a kind of natural theology. Hunt was committed to physical anthropology, the measuring of brains and skulls and bones and organs, by which the differences among races could be objectively confirmed, the Negro shown to be a different species, closer to the apes than to Europeans. The initial Darwinian naturalistic studies of the human were not compatible with this blatantly racist science.

It is worth pausing just a moment longer on the issue, however, because it suggests how deeply implicated in social and political issues the observational sciences were (and continue to be). Pre-Darwinian ethnology—as opposed to this Huntian physicalism—might be fairly represented by this passage, which concludes John Hall's preface to Pickering's *The Races of Man*: "We are fully satisfied, that all the races of man are, as the Gospel clearly expresses it, 'of one blood'—THAT THE BLACK MAN, RED MAN, AND THE WHITE MAN, ARE LINKS IN ONE GREAT CHAIN OF RELATIONSHIP, AND ALIKE CHILDREN WHICH HAVE DESCENDED FROM ONE COMMON PARENT." The opinion that the Negro is a link between man and the brute creation is, Hall tries to show, "altogether opposed to the facts," that is, it is unscientific. He laments that people who are "altogether ignorant of the anatomical structure of the human body" should have likened the Negro "to a brute and endeavoured to sink him below the level of the human species, for the purpose of degrading him, thereby to palliate the cruel hardships he still suffers."[13] The appeal is not to God but to nature, and Hall identifies those physical qualities that all humans share. The ethnologists drew the moral line between human and animals, but not among human races. And if their starting point was often evangelical activism, they were ready to appeal to the close observation natural history required.

Much then was at stake in the quest for objectification in ethnology and anthropology, and the seriousness of the issue intensified the questioning to which observational authority was to be subjected. But Darwin, when he turned at last to the subject of the human was relatively unself-conscious about the way his arguments had compromised the possibility of objectivity in such study. His cool doggedness did not finally counteract the damages to the empiricist model. The investigator is himself being investigated, the observer observed. To get around the difficulty, Darwin adopted Herschel's principle by which to compensate for limits of perception, and he ransacked the work of biologists all over the world. The vastness of his correspondence and his scientific reading testifies to his unwillingess to settle for the evidence of his own perceptions.

For *The Expression of the Emotions in Man and Animals*, he circulated numerous copies of an elaborate questionnaire about the various ways humans from many different parts of the world express their feelings. For example, the second question (out of sixteen) is: "Does shame excite a blush when the colour of the skin allows it to be visible? and especially how low down the body does the blush extend?" The whole questionnaire marks an extraordinary invasion of privacy and must presume to turn the individual into a distanced object of study, a case. However much Darwin's theory, too, points to the derivation of all humans from one source (the "hairy quadruped" of *The Descent of Man*), the perspective is very much that of a Westerner regarding the dark-skinned people as something "other." (The development within Darwinian anthropology that would circumvent the belief in the "unity of man" was the belief that the dark-skinned people had developed less far from their primate origins. Science can be used for *any* political position.) The "other" may be subjected to such investigations, such invasions of privacy. But the awareness that the other is conspecific with ourselves makes any such investigation reflexive, and Darwin was not at all uneasy about this in *The Expression of the Emotions*. Aware that prejudice and defensiveness might operate in the study of humans, even of "the most distinct and savage races of man," Darwin attends closely to the expression of passions "in some of the commoner animals," since they afford "the safest basis for generalisation on the causes, or origin, of the various movements of Expression. In observing animals, we are not so likely to be biassed by our imagination; and we may feel safe that their expressions are not conventional."[14]

Darwin was ready to invade the territory of personal intimacy in his role as professional scientist because he was willing to have the answers apply to himself. But how does one get answers to questions like this? Darwin got thirty-six answers, which allowed him to conclude confidently that the uniformity of nature he had found in geology and botany is operative on human behavior as well: "It follows from information thus acquired, that the same state of mind is expressed throughout the world with remarkable uniformity."[15]

Darwin is a witness who must apply the answers to his own behavior, since he is extrapolating out from the questionnaires to the whole human species. The principle of uniformity requires that there be no leaps from one race to another, and when Darwin is "struck" it will be with the force of self-exposure. The cherished detachment of natural philosophy is preserved only by the refusal of the writer to allow his feelings space in the prose. Shy, unaggressive, serious, full of curiosity, Darwin goes to zoos and thrusts his face toward a cage to elicit reactions from the animals. He plays with his dog and is sensitive to every twist of the ear, arch of the back, wag of the tail. He reads literature with a remarkable eye for physical detail, and the Shakespeare whom he supposedly came to find nauseating is everywhere as evidence; and poetry of all sorts, fiction, autobiography, appear in the text as part of the accumulated testimony of universal observation. Shylock says he has borne Antonio's complaints "with a patient shrug," and is enlisted by Darwin in a long argument showing that shrugging is part of a pattern of antithesis, to be observed in animals as well. It is the opposite of the movement made when humans are asserting determination, aggression, power.

One of the consequences of the application of Darwinian theory to the observation of human behavior is that all observation is ultimately observation of the self. Darwin's naturalism feeds Victorian positivism, and the banishment of metaphysical and theological explanation for physical effects leads logically to a kind of solipsism. W. K. Clifford, the brilliant mathematician and iconoclast argues the point lucidly: "Either I have some source of knowledge other than experience, and I must admit the existence of *a priori* truths, independent of experience; or I cannot know that any universal statement is true. Now the doctrine of evolution itself forbids me to admit any transcendental source of knowledge; so that I am driven to conclude in regard to every apparent universal statement, either that it is not really universal, but a particular statement about my nervous system, about my apparatus of thought; or that I do not know that it is true."[16] Clifford's skepticism is the obverse side of positivism. Confronting the same epistemological problem Whewell contended with, he rejects, in the name of Darwin, transcendental explanation. The universal is first reduced to the empirical, which cannot be universal; the empirical no longer means direct access to the thing in itself but mediated access, through one's own perceptual equipment; observation, then, is observation of the self perceiving, not of the thing perceived. All truth is ultimately about the self formulating it. All universal laws are laws about oneself.

Darwinism was quickly absorbed by the developments in empiricist theory that entailed a culture-wide critical examination of the primary act of knowledge, observation. So Karl Pearson would shift the idea of observation to that of "sense impressions." "The mind," he says "is entirely limited to the one source, sense-impression, for its contents."[17]

These forces, leading to the internalization of knowledge in science, are reflected in the famous difference between Arnold's desire to "see the object as in itself it really is," and Pater's, "to know one's impression as it really is." By facing directly the relativist and solipsist implications of the empiricist emphasis on "observation" of the external world, Pearson makes the move which has striking parallels not only in Pater's essays, but notoriously in the fiction of James and Conrad. Science, says Pearson, "deals not with the external world, but with the contents of the mind."[18]

George Eliot's preoccupation with consciousness, says Alexander Welsh, was a "European phenomenon."[19] We have seen it in a minor way already in Trollope, whose study of Harry Clavering shifts from narrative action to the drama of psychological paralysis. Welsh is not linking this new preoccupation with Pearson's extensions of empiricism, although I believe the links are there. The scientific objectification of the human develops parallel to a new emphasis on subjectivism because it withdraws all external authority for knowledge and leaves it to observation. Welsh discusses these scientific developments in the context of the objectifications entailed in the movement from small to urban industrial communities. The pressures of observation by alien eyes are on almost all citizens of modern society; the examination of individuated human "cases" that Foucault describes intensifies the self-consciousness of the individual. It makes each aware of impersonal forces of evaluation and judgment that help foster shame and guilt. The study of Harry Clavering's guilt registers the degree to which he seeks to avoid being observed and the way in which he internalizes the judgments he knows the society will make. Thus, the consciousness Welsh describes is "the inward awareness of the outward forces of information."[20]

The scientific and epistemological history of the reflexivity of observation has its parallel in a poetic tradition, suggesting again the mutuality of scientific and literary enterprise. Frank Kermode's description of the romantic poet's quest for unmediated vision of the real echoes the Herschelian ideal of uncorrupted observation.[21] One of Kermode's examples is Keats's *Fall of Hyperion*, with its opening image of the poet's numbing climb, almost to the point of total extinction of self, to Moneta's altar:

> "None can usurp this height," returned the shade,
> "But those to whom the miseries of the world
> Are misery . . ."

To attain the height from which to observe and understand the nature of the real itself requires—as on Baconian principles—a refusal to disguise or diminish the pain of the truth. No prior ameliorating structures are to be imposed on direct observation (as, for example, natural theology entailed).

But to achieve this scientific ideal, Keats's narrative implies, is all but self-annihilating. Put less dramatically, clear vision requires selflessness, the cleansing from the mind of Baconian idols and Herschelian prejudices and Arnoldian interests. Natural theology promised the consonance of truth and meaningfulness; the rejection of natural theology led inevitably to the possibility that the truth would not be humanly satisfying. Fanny Price's narrative dramatizes the natural-theological pattern of observation, but her story also, through machinations of plot rather than through Fanny's poetic or active power, connects selfless observation and clear vision with power. In her case as in Keats's only the initial denial of self that is the condition for clear vision can protect the self from the anguish of the "miseries of the world." "If we had a keen vision and feeling of all ordinary human life," George Eliot notes in one of her most famous—and distinctly post-Darwinian—passages, "it would be like hearing the grass grow and the squirrel's heart beat, and we should die of that roar which lies on the other side of silence."[22]

These are other versions of the self-minimizing strategies of Darwin's autobiography. Selflessness become a protection of the self; and yet achieving it opens the threat to selfhood posed by a reality too overwhelming, too out of harmony with human needs to bear. As Kermode writes, Moneta's face "is the emblem of the cost as well as of the benefits of knowledge and immortality."[23] Nineteenth-century science unambivalently aspires to the selflessness that both allows the unmediated experience and distances such knowledge. (This selflessness is the other side of the "discipline" Foucault discusses. Foucault points out that increase in individuality among those observed is a reflex of the diminution of individuality among the empowered, in the bureaucratic structures of the culture. The denial of the self frees one from the responsibility of oppression and exempts one from the objectification and scrutiny accorded to "cases.") Nineteenth-century literature, emblematically and powerfully in the figure of Keats aspiring to poetic vision, tends to explore the costs of vision.

In the Keatsian tradition, much English writing of the century imagines the cost for normal human intercourse: Teufelsdröckh in his tower in Weissnichtwo, seeing the arterial flow of that city—in deliberately material terms of bodily circulation—and the roofless houses exposing their inhabitants; the Lady of Shalott, or, an example Kermode cites, Empedocles on Aetna. All, like Keats's poet, are above the world and alone. Darwin lived at Down, and the price to him was also great. There is no need to speculate on his mysterious and almost life-long illness; but he armed himself against the social and cultural repercussions of his vision by isolating himself (although un-Keatsianly, within a warm and loving Victorian family) and avoided combat with his opponents at all costs. Artist and scientist risk human isolation as they climb the steps toward Moneta. And they often end celebrating artistry and science, unimpeded observation and extraordinary Jamesian sensitivity, as the saving condition of truth in a culture made uneasy by it. Impersonality and isolation are the price they must pay for clarity of vision and the power that accompanies it, power over the subjects of information, who find themselves publicly exposed to a Teufelsdröckhian gaze as they think themselves private and secluded from view.

For the rest of this chapter I will look at some examples of how this epistemological, poetic, and scientific cultural myth manifests itself in Victorian narrative. Welsh, developing Foucault's argument, states that "the spirit of surveillance and reporting inhabited the project of the nineteenth-century novel."[24] Narrators see themselves as "reporters," but what is required of them is "surveillance"; and a broad range of Victorian fiction strikingly centers on "observing" figures—characters whose primary actions seem to be observing others, or whose fate is tightly woven with their observing or having been observed (Trollope's scientific enterprise as narrator has provided one example). But there are also ways in which "observation" is problematized both as theme and as structural principle, and its dangers and necessity become the obsessive preoccupation of novels. Repeatedly, the novels turn on moments in which characters are detected unawares, revealing to another character or to a distant narrator, some truth or accident that will determine their fates. The narrators themselves are preoccupied with authenticating their positions, testing their own ability to withstand the mists of bias, storms of emotion, and finding strategies by which to overcome the limits of consciousness. Plots emphasize the traditional narrative concern with what happens next by concentrating on the search to discover what has already happened, making that concern their subject and suggesting that close observation would provide information for resolving narrative complications. (Welsh offers a sustained and valuable argument about the relation between such narrative practices and Freudian analysis.)

A familiar place to start would be *Bleak House*, in which the dual narrations underline the limitations of knowledge and the relation between what is known and observed and how one suffers or exercises power. The presence of that professional observer, Bucket, who succeeds best when least noticed and who pursues knowledge with cold efficiency, following upon the cold-blooded acuity of Tulkinghorn, makes observation thematically central to the novel. No longer the province of common sense and good intention, observation becomes almost mysterious, for in the urban development of modern England, professional qualifications are required to penetrate the multiplicity and impersonality of social phenomena and social agents. Bucket, surely, is scientific as Sherlock Holmes would be scientific, in his capacity to read clues, to construct the fossil organism from fragments and traces. Bucket's later warmth and the narrative's obvious admiration of him suggest how important to Dickens is the power of observation not only for the materials of his fiction but, particularly, in the exercise of discipline, surveillance, or control.

But the bitterness of the third-person narrator suggests already the pain inseparable from accurate observation. The voice expresses the experience of taking a good look at the worst; implicitly, it is a voice withdrawing in its ironies and bitterness from the world, and that withdrawal is a condition of its clarity. Ironically, the narrator in his very un-Jarndycean harshness adopts a strategy rather like that of Jarndyce, and both of them anticipate Thomas Hardy. That is, they attempt to keep themselves (often unsuccessfully, to be sure) outside the experience they see. Jarndyce tries *not* to see, but his assumptions about the reality out there are rather similar to those of the narrator, who does see. He protects himself from Jarndyce and Jarndyce; but he is Jarndyce. And yet *Bleak House* begins a series of novels that play ambivalently with the necessity—moral, scientific, aesthetic—for quite professional investigation and observation. (I do not think it a coincidence that these are the very novels I earlier described as exploring the possibilities of self-obliteration.) The world is too abundant, complicated, mazelike for all but a kind of Christlike innocence, like Little Dorrit's. *Bleak House*, *Little Dorrit*, *Our Mutual Friend*, and obviously *The Mystery of Edwin Drood*, all hang upon secrets and an extraordinary collection of observers trying to discover them.

The attempt to celebrate the ideal in Esther Summerson, to whom much of the truth is revealed, is complicated by the almost anti-intellectual implications of her character. On the whole—unlike the morally similar Fanny Price—Esther cannot observe very well. Or, rather, she is predisposed to see innocently, so that what she carefully observes strikes her as attractive and the reader as ironic. She is dependent in such matters on more authoritative, even professional observers, like Bucket. (One remembers, for example, her idyllic description of the barefooted Hortense, which captures the scene vividly but misses the ominousness the reader is meant to pick up.) Yet like Fanny, Esther gets the rewards that come with selflessness in nineteenth-century fiction and in scientific experiments.

Our Mutual Friend, a novel like *Little Dorrit* very much about secrets, is more directly about the problem of distanced observation. Its narrative might be taken as a kind of mythic expression of the ideal of scientific observation and as an implicit critique of it, reflecting its dangers, self-contradictions, and uses. The figure of John Harmon broods over the entire narrative; retrospectively, we see that he has been observing almost everything, but he is both observer and experimenter. Harmon is free to observe human behavior by, in effect, dying. Dead, he is invulnerable to other people's observations (and thus he has to avoid being observed by the inspector, another "professional"); dead, he can observe without distorting. Although not usually looked at in this way, *Our Mutual Friend* is one of those interesting Victorian novels that thematize the transformation of the human into scientific subject by turning their plots on experiments with humans—novels like *Hard Times* and *The Ordeal of Richard Feverel* and James's *Portrait of a Lady*. Almost like Ralph Touchett with Isabel Archer, Harmon devises an experiment that allows him to observe his subject, Bella Wilfer, as in herself she really is. (We needn't explore the complexities of the way the apparently separate observer is, inevitably, shaping and therefore changing the observed.)

Although the rhetoric of the novel suggests that we are to take Harmon's withdrawal favorably, it comments indirectly on his kind of observation, which is a form of spying. It does not, like James's novel, overtly raise the issue of whether the experiment takes unnecessary risks with the life of the heroine; indeed, it implies that only by not knowing of the experiment can Bella be morally saved. But Harmon's distanced observation, like Touchett's, has something of the novelist's nature about it. All set up "experimental" situations and create the illusion that the subjects of the experiment are free to act out their real natures. One can find out about "real natures" only by observing when the objects think they are not being observed.

Our Mutual Friend seems at times to be one long sequence of spying. Wegg spies on Boffin, Venus on Wegg, Boffin on Venus and Wegg, Headstone on Wrayburn, Riderhood on Headstone. All of these figures, like Harmon, believe that the truth is hidden from social view. Peace comes only outside the world of spying and being spied upon, with Jennie Wren, on the roof. And yet from that perspective, earned by Jennie as by Keats's poet, from self-obliterating acquisitions of knowledge, Jennie gets a fuller perspective on human experience. Away from the corruptions below, she calls down, "come up and be dead." Everyone in this world of spying becomes immensely vulnerable, and the only safety is an obliteration of selfhood so complete as to become a literal death.

In their most self-conscious form, later in the century, these concerns reemerge in Wilde's *Picture of Dorian Gray*, a Gothic version of the human experiment story, where the scientific notion of experiment threads its way through the narrative. "It often happened," says the narrator, "that when we thought we were experimenting on others we were really experimenting on ourselves."[25] Lord Henry manages to survive the experiment because he insists on the distancing and detachment requisite for accurate observation. But the close alliance of human experimenting with death—either metaphorical or literal—of experimenter or subject, is no coincidence. Dickens's story resolves the experiment successfully, but the problems it raises, representatively for late nineteenth-century fiction, become in the philosophy of science epistemological problems. In art, they become the center of late-century tragedy—the incompatibility of consciousness with what consciousness knows.

Of all English novelists of the century, George Eliot was the most sensitive to the complications of perspective. She was equally alert to the reflexive dangers of observation, which often takes the form—as in "The Lifted Veil"—of a rejection of knowledge. In *Adam Bede* the subject is not quite central, yet the narrative turns on a moment of detached observation. Adam stands staring at the beech tree to "convince himself that it was not two trees wedded together but only one," observing the object, which seems quite different from himself. But trained novel readers will feel the reflexivity of the question of the "wedded" trees on Adam. And that Hetty will turn up momentarily confirms that what is perceived is conditioned by the perceiver. Adam's life, at this point, is severed in two: "For the rest of his life he remembered that moment when he was calmly examining the beech, as a man remembers his last glimpse of the home where his youth was passed, before the road turned, and he saw it no more."[26] One step more and Adam sees before him, twenty yards away, Arthur and Hetty, holding hands, he bending to kiss her. The force of this moment lies in the inescapability of knowledge. Observation "strikes" Adam, and no further interpretation can change its effect. The observation and the interpretation are simultaneous so that the work of the interpreting and interested mind is entirely absorbed in the fact. And to this fact, the unintentional observer is peculiarly vulnerable.

Seeing in *Adam Bede* is a necessary precondition for the moral life. But it is also a painful one. Adam, whose clarity of vision, precision of workmanship, and sensitivity to the nuances of the natural world, are superior to those of any character in the book, is an isolated figure, isolated in part by his knowledge. Only in his relation to Hetty is Adam's mind seriously clouded (and this parallels the ostensibly aberrant story of "The Lifted Veil," in which Latimer has second sight for all except Bertha, whom he therefore loves). As an observer, Adam has been hampered only by his unself-conscious centrality, by his pride in his own mastery, his imposing on the world his own imaginative structure of values. But the imaginatively masterful observer is, in *Adam Bede*, peculiarly vulnerable. Thematically, the realistic novel is concerned to dramatize that vulnerability, and the imaginative dreaming protagonists of so many Victorian novels are, in a sense, pre-Darwinian scientists, that is, natural theologians. Their imaginative lives are like Bridgewater Treatises in that, while they often are wonderfully perceptive, they invariably read their observation within the theme of teleology, divine intention, grace, and human centrality. In *Adam Bede*, the scene burned into Adam's consciousness is of the fall—a fall into knowledge; it is not redeemed by nature and does not redeem nature. A truth once discovered is, for Eliot, irrevocably there. Redemption is possible only through social act, which does not change and soften the truth, but changes the perceiver.

Adam, the strong observer, is vulnerable because he is *driven* to conclude the truth, and truth is not only painful; it destroys his sense of the way the world is. He is not disinterested. The truth is about himself. Although this might be a scientific disqualification, the novel itself admits nobody as disinterested. And this is one of the most powerful developing insights of nineteenth-century fiction—that there is no pure scientist, not even apparently omniscient narrators. In *Adam Bede* everyone is implicated in the moment of discovery—the unself-conscious lovers, the unself-conscious observer. There remains only the privileged author, but he (or she) has testified that her testimony is itself "defective." She is committed to being a witness to the truth, to avoid reshaping nature. The price of this avoidance is vulnerability. The metaphor of the "witness-box" makes truthfulness essential since perjury is a punishable offense; the truth may well be painful, and it extends beyond the margins of the book into the lives of writer and readers.

The more intensely recognized vulnerability of the observer leads Eliot to increasingly complicated manipulations of perspective, culminating with *Middlemarch*, that most scientifically knowledgeable of novels. Observers in *Middlemarch* are vulnerable, like the extreme case, Raffles, or the more attractive but extremely prudent Mary Garth, in that what they know exposes them to the hostility of interested parties, like Bulstrode or the Featherstone family. Mary's story suggests the possible value of not knowing. But observers are also, obviously, unreliable, and every observation needs the supplement of others both because of limitations of perspective (what Herschel called "prejudices of sense") and limitations of interest ("prejudices of opinion"). And the complexities of perspective in *Middlemarch* lead Eliot even further, to the intellectual complexities of *Daniel Deronda*, with its attempt to imagine a fully authoritative way of seeing that will reinstate both the Wordsworthian mastery and control of the investigator, and the pre-Darwinian significance of the object. The most acute observers in *Daniel Deronda* are Grandcourt and Deronda, and their functions are, as Welsh has brilliantly shown, very similar. Although Grandcourt uses his sharpness of observation to exercise power, and Deronda seems benevolent in intention, in fact his first act of observation is an act of power. He intimidates Gwendolen at the gambling table simply by watching her. More important for Welsh's argument, Deronda seeks to instill shame in Gwendolen, just as Grandcourt tries to instill fear—and the source of both shame and fear is the possible exposure to communal observation and consequent judgment.[27] The price of observation to Grandcourt is, of course, his death. Deronda ultimately severs his connection with Gwendolen and does not use his knowledge. Eliot's exploration of the problems of observation here is an attempt to reimagine a fully authoritative way of seeing, but one that will circumvent the crises of power and vulnerability with which observation seems inevitably attended.

Victorian narratives consistently suggest that the full authority dreamed of always has about it something dangerous and unpleasant. Authoritative observation brings with it political implications (as with Grandcourt's stunning assertions of power). The observer is a kind of power broker, or a tyrant, and we can find this motif strongly announced in pre-Darwinian fiction. In Dickens, Tulkinghorn and Jaggers are two strong expressions of it. Other narrators, who are themselves very precise observers, often describe observant characters whom they do not like. Full authority of observation is precisely what the omniscient narrator, and the novelist, seek. So when Thackeray introduces a novelist in *Pendennis*, he names him Mr. Wagg and makes him a rather nasty, comic purveyor of gossip, capable of doing damage by telling what he uncannily observes. Wagg "eyed and noted everything." He seized minutiae "in spite of himself."[28] In Thackeray's scheme of things, these essential virtues of the novelist seem to produce knowledge and information that can compromise the objects. Harry Clavering is peculiarly sharp in his observation of vulgarity and immediately dismisses Florence's brother because he frequently dusts his shoes with a cloth. But the narrator observing Harry's observation thoroughly places Harry. The realist novel itself is a kind of continuing exposure, using the modern paleontological skills of imagining a complete monster from mere fragments carefully observed (and then articulated). The novel is implicated in the observations it uses and exposes and is, therefore, in the hands of self-conscious practitioners, incapable of avoiding those complications of power and vulnerability Eliot tried to avert. The observing narrator is, by the condition of narration, being observed, and subject consequently to the same kinds of exposures that threaten the characters.

Hardy, of all novelists, seems most sensitive to this kind of problem. His is the observer's art; much of its strength derives from almost uncannily detailed, deeply moving encounters with phenomena that ordinary attention would miss. The narrators, like many of the protagonists, are aware of how easily they might be exposed, shamed, defeated. Consequently in extreme ways they confront, or, rather, attempt to evade, the perils of observation. They keep their distance from their subjects, adopting an archaic and noncolloquial language, seeking an almost deathlike disguise that might exempt them from the assaults of observation to which all characters in the novels are subjected, like subjects of biological experiments.[29] The observer must be aware that observation may not merely reveal the subject; it also threatens to transform it. And watching the living creature surreptitiously exposed confirms the possibility of the observer's exposure, as well. To see without inflicting violence—for the antivivisectionist movement observing seemed utterly inhumane—is thus an ideal designed to provide at least temporary protection not only for the observed, but for the observer. The novelist performing his Zolaesque experiments will often project the "experiment" inside the fiction, where it enacts the novelist's own vulnerability and participation in the organism, social or biological, being observed.

But even as Hardy was using the Herschelian model, with its strong emphasis on the importance of "observation," he was writing a fiction that increasingly undermined "observation" as an adequate authority for knowledge. Like many of the scientists in the last third of the century, he saw that the materials of knowledge are too complex to be handled empirically, that what is required, minimally, is what Tyndall had called "the scientific use of the imagination." Or, in another sense, he was reverting to Whewell's Kantian notion that observation always contains within it an element of thought, and cannot, independently, become the source of knowledge. Thus, he wrote, "As, in looking at a carpet, by following one colour a certain pattern is suggested, by following another colour, another; so in life the seer should watch that pattern among general things which his idiosyncrasy moves him to observe, and describe that alone. This is, quite accurately, a going to Nature; yet the result is no mere photography, but purely the product of the writer's own mind."[30]

Welsh has suggested convincingly the connection between the decline of realism, as practiced by mid-century novelists, and the developments in science that emphasized the "creative, constructive aspects of scientific imagination in the foundation of hypotheses and of intellectual models."[31] Hardy's fiction, in its loving registration of particulars, seems to be affirming the primacy of observation. But those particulars, as Beer has shown, serve as counterpoint in fictions whose large and intricately careful patterns disrupt and destroy them. Hardy at once affirms the priority of observation, and in exploring its effects, undercuts it.

The sense of enormous vulnerability in the necessary act of observation is the major force in the shaping of Hardy's novels—not only in the various, often painfully conventional plots, in their landscapes, and in the moral and psychological dramas, but in their narrative mode as well. The novels expose private consciousness to communal observation, to stray but interested observers bringing the weight of public values to bear on private longings. Almost all characters in his novels seem subject to the assaults of observation, some particularly so because they act out what must have been Hardy's own self-thwarted passion to be careless, not to be respectable, not to be responsible to the conventions of the community: they pay the consequences he was ultimately unwilling to pay. The narrator is intensely conscious of the *unwariness* of the characters and sees them within contexts they could not perceive; he becomes, in a way, one of the innumerable voyeurs who populate his fictions.

Within each novel there are usually dozens of critical moments in which various characters observe other characters who are unaware of being observed as they conduct their daily business. Observation of this sort always feels like a violation. At the beginning of *The Woodlanders*, a novel obsessively preoccupied with observation, the narrator places the barber, Percomb, in the midst of a deserted highway as the evening darkens. Percomb's sudden alertness to his solitude intensifies the reader's sense not of reading about him but of observing him, vulnerable and nervous, from a precisely defined perspective. The narrator pauses to read his character from his "finical style of dress," and then follows him on his way to buy Marty South's hair.

The image of Mrs. Dollery's wagon, taking Percomb toward Marty's house, superbly implies the vision of the novel as a whole, and of much of Hardy's work. The wagon is described with a detached precision striking in its technical detail: "The vehicle had a square black tilt which nodded with the motion of the wheels, and at a point in it over the driver's head was a hook to which the reins were hitched at times, forming a catenary curve from the horse's shoulder."[32] There is much more, but the apparent detachment has already done its work of implying the authority of the observation. So precise an attention gives to the object—or seems to—a significance that is not made explicit. It has become an object of attention, important somehow in its own right, and not merely as a narrative "vehicle." Indeed, from the point of view of narrative business, one wonders what the wagon is doing there at all. There is, to be sure, business to be done—the suggestion of the remoteness of Marty's cottage, the antiquated style of the community, and possibly mere antiquarian pleasure. But it is attention itself that is primary here. The author's voice as detached scientific spectator increases, not diminishes, mystery and the ominous awareness that there is much watching and being watched going on here.

The fullest expression of this comes at the close of the description, by way of the glass window that Mrs. Dollery "cleaned with her pocket-handkerchief every market day before starting": "Looking at the van from the back the spectator could thus see, through its interior, a square piece of the same sky and landscape that he saw without, but intruded on by the profiles of the seated passengers, who, as they rumbled onward their lips moving and heads nodding in animated private converse, remained in cheerful unconsciousness that their mannerisms and facial peculiarities were sharply defined to the public eye" (p. 43). The "spectator," the "public eye" are quintessential Hardy. The extraordinary force of this image lies not simply in its precision, but in the way it registers both the large outer world ("the square piece of the . . . sky and landscape") and the dark interior of the wagon with the entirely unself-conscious passengers, enclosed, unaware either of the large world before them or the "spectator" behind. The passengers forgetting "the sorrows of the world" are not exempt from the world. The privacy is an illusion. And Hardy's novels tend to be about all that is implicit in that illusion and in this image; the strategies of their construction and of their language equally derive from this vision—this vision of a spectator envisioning.

Percomb himself becomes the unobserved observer that the narrator has been as he peers through the darkness toward Marty's cottage. It is a moment perfectly representative of the perils of being observed, for Marty has left her door ajar to allow the smoke from the fire out and has thus left herself unself-consciously vulnerable to observation. The reader sees Marty for the first time as Percomb sees her, that is, from the position of an intruder on her privacy. Such imagery is characteristic of thriller movies, which frequently move the camera in from some distance on an innocent potential victim and expose her (usually a "her") through her own unself-consciousness. "In the still water of privacy," the narrator says, in order to account for the "fullness of expression" on Marty's face, "every feeling and sentiment unfolds in visible luxuriance" (p. 48). The assumption here, as in Dickens and Eliot, is that a character's full nature can be detected only when it is out of social context. A character needs to be perceived by an unseen observer—like the novelist.

The rural settings of Hardy novels emphasize the solitude that allows such fullness, and almost invariably exposes it to the trained control over feeling of more cosmopolitan people. The destruction of the traditional countryside has much to do with its openness to observation. Modern urban life like modern science intensifies the need to hide, for the possibility of having one's most private secrets exposed to observation is everywhere. Darwinism was, in a way, a massive invasion of the privacy of the individual, a transformation of the individual into a "case." In a statement from *The Woodlanders* that echoes implicitly throughout Hardy and suggests much about the nature of his narrators and narratives, Percomb tells Marty, "You should shut your door—then you'd hear folk open it" (p. 49). Observation threatens exposure and at least metaphorical violence, here something like rape. It is not, I believe, merely gratuitous that Percomb has come for Marty's hair, for a rape of the lock.

The narrator, with his archaisms, his strained and distant locutions, remains impersonal and distant yet somehow capable of the most minute attention. He keeps us, as observers, scrutinizing details almost hypersensitively, aware of our otherness, our apartness; and he thus protects himself from the shame of public exposure. Shame is as central to Hardy's narratives as Allon White describes it to be to George Meredith's.[33]

Hardy's preoccupation with precise observation is partly a consequence of deep internalization of the values of the society he exposed. The observations acquire full significance for him and in the fictions themselves within the context of moral and social codes that provide the terms of selectivity. That is, the novelist observes intensely, as the scientist asks questions and investigates, what the society is most concerned about. The Darwinian abundance and multiplicity that Gillian Beer accurately locates in Hardy's narratives is one of the great dangers to society, and one of the strongest evidences of the novels' resistance to the social codes that make respectability so important to the characters, and to Hardy himself. In a similar way, Darwin's observing eye, straining to bring the multiplicity of nature within the rule of law, sees more than law can contain; and his vision of the world is in excess of the theory he can formulate to express it.

Hardy's observed world is constantly transgressing the narrow limits of respectability within which the narrator wants to remain and out of which his most attractive protagonists are cast. His narrators observe life in such a way that the human center simply cannot hold, and the full implications of the Darwinian rejection of natural theology and teleology are dramatized. It is hard to keep the single character in central focus because so much natural detail incessantly struggles to displace it. Even the lesser novels, often wonderfully rich within the constraints of conventional and contrived narrative, depend upon such struggles for displacement. In *A Pair of Blue Eyes*, for example, the jilted lover Stephen Smith observes surreptitiously Elfride Swancourt with her new lover, Henry Knight. Smith spies them entering the summer house: "The scratch of a striking light was heard, and a glow radiated from the interior of the building. The light gave birth to dancing leaf-shadows, lustrous streaks, dots, sparkles, and threads of silver sheen of all imaginable variety and transience. It awakened gnats, which flew towards it, revealed shiny gossamer threads, disturbed earthworms. Stephen gave but little attention to these phenomena, and less time. He saw in the summer-house a strongly illuminated picture."[34] Stephen has discovered the lovers by accident, but his observations become deeply interested. What he sees hurts him, not the lovers. But he is only a small part of the complex of life and observation. While the readers' feelings and his own are concentrated on the revelation, he ignores what the narrator will not allow the reader to ignore—the spiders, gnats, earthworms, who are also disturbed by the lovers. Their sudden birth under the illumination of the match reveals them as Marty South was revealed. The intrusion of their narratively irrelevant life on the stories of the protagonists is part of what those stories are about. The scene ramifies further because beyond the spying Smith there is another observer, Mrs. Jethway, whose grudge against Elfride leads *her* to spy also. She has seen the lovers *and* Stephen spying. Whereas Stephen is "shattered in spirit and sick to his heart's centre," Mrs. Jethway has already felt the displacement that Stephen now feels: my heart is desolate, and nobody cares about it, she correctly explains.

All of these displacements imply a world in which proper perspective is almost impossible to achieve. The Darwinian themes of abundance and entanglement alter the question of observation because they make anthropocentric interpretation impossible. What observation reveals is one's own marginality and vulnerability. The narrator makes any perspective ironic by reminders that there are many others, even in the same scene, but beyond the margins of awareness of any single character. In observing Elfride and Henry, Stephen disturbs the insects and wounds himself, while he cannot be aware of Mrs. Jethway's pain, which also derives from her observations. Each is locked into an "interest" that makes the observation inadequate. The narrator, watching all, exploits the distance to protect himself as he turns the spying and double spying into something like comedy, except for the pain. In a moment less comically conceived, Hardy exposes even his pure Tess to the ironies of alternative points of view. He notes, with a Darwinian sense of the abundance and indifference of life, how Tess's gauzy skirt, on an idyllic Sunday morning, "had brushed up from the grass innumerable flies and butterflies which unable to escape, remained caged in the transparent tissue as in an aviary."[35] In the very act of compassionate observation, the narrator at once affirms its value and, by the strangeness of his focus, suggests its limits. Like the butterflies, the observer is driven by a power larger than his, but now to the recognition that observation is meaningless outside humanly imagined context.

Thus, it is not only the observed character, like Marty South, who is likely to suffer because she is innocent and unguarded from prying eyes, but the observer as well. Often the observer is more likely to suffer: the most extreme and moving example of this comes in Giles Winterbourne's quiet protective observation of Grace Melbury, while he slowly dies of fever and exposure to the cold and the rain. Giles dies for respectability; he dies acting out the idea of disinterested observation. The preliminary moral act to such observation is self-annihilation, and in this sequence the association of disinterest and death is complete. The association connects observation with a social order. Here, Giles surrenders his own interests because any attempt on his part to enter the shelter with Grace would be, to his sensitive woodland eyes, a violation of propriety. The "fact" is only a fact because it is alive with social significance.

Hardy's narrator works like the post-Darwinian scientist. He needs to achieve the fullest possible distance from his subjects compatible with the closest possible scrutiny of them. He needs to discover the large general patterns, to understand the interconnections among widely diverse and apparently isolated scenes, characters, actions, to recognize, as he says of Marty South and Giles Winterbourne, that "their lonely courses formed no detached design at all, but were part of the pattern in the great web of human doings then weaving in both hemispheres from the White Sea to Cape Horn" (p. 59). And he needs, finally, to do all this without exposing himself to the pain the interweaving entails and without imposing upon the observed world the passion for centrality and dominance to which his own characters so frequently succumb.

In his *Problems of Life and Mind*, G. H. Lewes defines "Science" as "seeing with other eyes." Knowledge, he says, is "a presentation of feeling."[36] The language Lewes uses is an elaboration of the language of empiricism, and it leads to some of the problems and attitudes I have been sketching here. Since all experience, which is the source of all knowledge, is "feeling," it is inevitable that the attempt to separate observation from feeling is doomed. Although Lewes tries to mean something different by feeling than we mean when we use the word in ordinary language, the kinship of knowledge to experience and feeling remains. The strategy of the scientist is to turn the feeling into an abstraction, so as to be able to manipulate it. Hardy's narratives make such abstraction impossible. In *Two on a Tower*, at times an almost comic expression of the implications of scientific observation, Hardy makes the young astronomer—isolated, like Keats's poet, on the inaccessible tower—bleakly pessimistic, and the science of astronomy a serio-comic commentary on the perils of observation: "The interest of their sidereal observation led them on, till the knowledge that scarce any other human vision was travelling within a hundred million miles of their own gave them such a sense of the isolation of that faculty as almost to be a sense of isolation in respect of their whole personality, causing a shudder at its absoluteness."[37]

That all this is happening while Swithin and Lady Constantine are falling in love, ironically undercuts the abstractness of their pain. Yet the option is shown to be either Keatsian vision or sexual love; and in fact, they are inseparable. Observers, Swithin and Lady Constantine spend most of the novel trying to avoid being observed. However much Hardy may play with the theme, as here, it is an overriding concern of his fiction. Swithin and Lady Constantine are very much victims of their fear of being observed, although far less so of their observation of the cosmos. "At crossings, and other occasional pauses, pedestrians turned their faces and looked at the pair (for no reason but that, among so many, there were naturally a few who have eyes to note what incidents come in their way as they plod on); but the two in the vehicle could not but fear that these innocent beholders had special detective designs on them."[38] Hardy's narrators seek scientific distance so that they can achieve imaginative and compassionate freedom. Observation held out Moneta's threat—knowing that the misery of the world is misery—and her promise, of true vision. No writer of the nineteenth century, except perhaps Darwin himself, more fully demonstrates the reward or explores the threat than Hardy.

Victorian narratives find their authority in their solidity of specification, in the sort of observation James, Darwin, Trollope, and Hardy all affirmed. Yet in working out their narratives and the theme of observation itself they dramatize its problematic nature, its dependence on something other than what is normally called "experience," its tendency to induce passivity, to deny identity itself, to cut off creative or free action, to react upon the observer to inflict pain, to impose upon the observed to distort or even to destroy. For the scientists, always officially committed to discovering, not creating, experiment—with its necessary concommitant, hypothesis—allowed for the imaginative freedom that deference to an external real would have inhibited. And as science moved in directions that called into question the laws that Herschel and his successors had depended upon to make particular observations intelligible, nineteenth- and early twentieth-century narratives also broke from the conventions of the real as Eliot had announced them in *Adam Bede*, and as Trollope consistently practiced them. The revolution against the word, through experience, finds itself with the novel betrayed by the words required to make the observations knowable.

The attacks on science that dominated the last part of the century can perhaps suggest something of this development. Arthur Balfour was only arguing what the scientists themselves had been demonstrating: "the world as science describes it" is in "flagrant" contradiction of "the world as it appears."[39] This position is consistent with the full empiricist argument. But Balfour believed in addition that despite its empiricist claims, science had surrendered its faith in experience and was only creating another mythology. The move to a focus on consciousness was, too, an empiricist move. Nevertheless, Balfour had well understood a critical aspect both of empiricism and of Darwin's argument. Darwin could not through observation prove his position. To make his case he had to show that what the world had seemed to be was not at all what the world was. Moreover the empiricists, recognizing that what they observed were "sensations," that is, impressions on consciousness rather than the immediately apprehended object, had, like the novelists, moved all the drama inside. Faith in observation, on the terms the scientists were then accepting, would be precisely that, "faith." "We claim," Balfour says,

> to found all our scientific opinions on experience; and the experience on which we found our theories of the physical universe is our *sense-perception* of that universe. That *is* experience; and in the region of belief there is no other. Yet the conclusions which thus profess to be entirely founded upon experience are to all appearances fundamentally opposed to it; our knowledge of reality is based upon illusion, and the very conceptions we use in describing it to others, or in thinking of it ourselves, are abstracted from anthropomorphic fancies, which science forbids us to believe and Nature compels us to entertain.[40]

Taking the scientists' assumption that there is an experiential ground for knowledge—an assumption so radically challenged in the twentieth century that it is difficult to make a fuss about it any more—Balfour holds them to it, and finds none even on their accounting. Science as a system, he says, "is wholly without proof." It is, finally, "incoherent."

One way to think about the break—arbitrary as it is, and imprecise—between Victorianism and modernism is as a shift from belief in observation as authority to deep distrust of it. The Darwinism that made its mark under an empiricist label and that affirmed the unity, continuity, and coherence of nature could easily become an instrument in the disruption of these fundamental ideas. Darwin's complex argument, while adapted forcefully by Huxley and the scientific naturalists, and by the culture as a whole in the renewed faith in Nature, naturalism and "things," was in fact terribly disruptive in its implications; and I believe that a Darwinism fully consistent with the vision of the *Origin* would in fact have led directly to a revolutionary sense of the basis for authority and the nature of knowledge. It would have obliterated without resurrection in different form in social Darwinism and the like the metaphorical link between nature and human social order.

The undermining of empiricism, even from within, was significant both inside and outside science. Trollopean narratives as part of the great tradition of Victorian realism gradually lost their authority. The peculiar sanction that "observation" had through most of the century as the almost sacred source of all knowledge was disappearing. Not that the English could ever surrender empiricism or totally devalue "observation." But certainly, turn-of-the-century fiction reflects a shift of value, in its complex experiments with point of view, with stream of consciousness, with the sort of seamless prose George Moore tried to borrow on an analogy with Wagnerian music, with increasing interest not only in Dickensian mystery that was, in the end, presumably resolvable, but also in "impenetrable mystery."

One need only recall Oscar Wilde's outrageous play with "lying," and his at least half-serious argument that what we see is determined by what art has made us see; this, of course, is an aesthetic version of the anti-empiricist argument, already discussed by nineteenth-century scientists, that observation is never neutral. There is no such thing as an innocent eye. What eye less innocent than Dorian Gray's, as he steps cynically back from his own life and quotes the artist figure, Sir Henry, to the artist, Basil: "To become the spectator of one's own life, as Harry says, is to escape the suffering of life."[41] The dominance of observation, with its implication of an absolute reality beyond consciousness that the human mind progressively uncovers, was implicit in the determinist vision with which George Eliot, for example, struggled, or against which the butterflies caught in Tess's gauzy dress flapped their wings in vain.

Conrad, who always manages to put the worst face on things, may provide a good stopping point for these speculations. His Marlow seems to continue the quest for the long lost absolute object, for that presence will make sense of "things"; but his narratives reflect only absence. With

Paterian intensity, he tries to register the "rescued fragment" of experience, to show "its vibration, its colour, its form."[42] But he stays with his impenetrable mysteries because, at the tail end of the great empiricist-scientific tradition, he has fully accepted the inaccessibility of the thing in itself. And in the famous passage in *Heart of Darkness*, which has peculiar relevance to my exploration of observation, Marlow is described as telling his story not to get at the meaning as though it were the kernel of the nut, but to find meaning as "a glow brings out a haze."[43] That is, what is observed remains important, but what signifies is not the observed.

Marlow enters the darkness to find Kurtz, an imperialist Moneta, one might say, and pursues him with an obsessive, almost scientific desire to know. But the price is as high as it was for Keats, and the results less satisfying. With Darwinian inevitability, what he observes turns out to be himself; the journey of observation turns out to be an act of self-annihilation. And if we can make anything out of the compromises and evasions in the conclusion of *Heart of Darkness*, it may be that Conrad and Marlow together find what Dickens and Eliot allowed themselves tentatively to think, that it is best after all not to try to know, that observation of nature in the traditional sense is really impossible, the attempt to achieve it, too perilous.

9

From Scott to Darwin to Conrad: Revolution Not Evolution

TROLLOPEAN REALISM appears to be on the side of common sense. It makes no fuss and appears to stir no waters. But with Darwinian tolerance of aberrations and persistent secularity, it hazardously wrests value from God and therefore from nature and puts it into the hands of those who compose the community that constitutes realism's world. In his battle with natural theology, Darwin had in effect done the same, despite his scientific refusal to enter the ethical wars or to seek "origins." His theory of evolution by natural selection had led him to posit a naturalistic basis for the distinctively human attributes of intelligence and morality, although, as he said at the outset of his discussion of "instinct," "I have nothing to do with the origin of the mental powers, any more than I have with that of life itself" (*Origin,* p. 234). The last stage of Darwin's argument was to prove that human species, too, developed through natural selection despite the fact that man "manifestly owes [his] immense superiority to his intellectual faculties, his social habits, which lead him to aid and defend his fellows, and to his corporeal structure" (*Descent,* I, 137). In the *Descent* Darwin tries unabashedly to connect the higher faculties with original "low motives" (I, 163–164). The source of human value is ultimately the self. Here again, Darwin's argument is counter-intuitive. And, as I have been suggesting, the implications of Trollope's apparently commonsensical worlds are similarly so.

In Trollope almost as much as in Hardy, the ideal is at odds with the real, and is deeply destructive—as we have seen in Lily Dale, for example. It produces an inflexibility resulting from the imposition of idea on "reality."[1] Trollope's world is not rational, not designed. Like Darwin's, it separates mind from nature, denying the latter any normative power or derivation from superhuman authority. The more Trollope observes the world, the more obviously does its order derive from the observer. Trollope tends to respond by building into his narratives an ostensibly decisive cultural tradition of continuity that reabsorbs the aberrant into the uniformitarian movement or expels it for its refusal to surrender its personal ideal for the sake of an obviously irrational and arbitrary tradition called society or history.

In Hardy and Conrad this whole movement is much more self-conscious, and the arbitrariness of the "social" and "natural" is built into the structure of their narratives. The discontinuities between thing and value are obvious, and the tenuousness of the "natural" self, really constructed out of arbitrary social conventions, becomes thematically central. In making the scientific and evolutionary subject self-consciously part of the struggle over ethics and social order, Hardy and Conrad emphasize the disruptive elements of the Darwinian vision. They bring forward elements in Darwinism resistant to the conforming and uniformitarian elements of continuity that were so easily enlisted in defense of established power; in so doing, they reveal contradictory elements in the project of realism itself.

This last chapter focuses on just those elements that made Darwin's theory potentially disruptive. His deep faith that nothing in nature makes leaps was absorbed by George Eliot and most others into the political idea that human society cannot make leaps either: evolution not revolution. Using nature as political model surreptitiously reinvests it with the values it had lost when God disappeared; but latent if repressed in Darwinian nature were great gaps requiring the great leaps Darwin needed to deny.

Near the end of the *Descent* Darwin writes as though the logic of his argument has led him to social prescription in a Galtonian mode: "Both sexes," he says, "ought to refrain from marriage if in any marked degree inferior in body or mind; but such hope is Utopian and will never be even partially realised until the laws of inheritance are thoroughly known." Darwin is plainly worried that if the poor or the intellectually and physically weak marry recklessly while the prudent avoid marriage, "inferior members will tend to supplant the better members of society" (II, 403). It was not then a long step to enlist Darwinism on the side of modern capitalist society, although Darwin's own humanity led him to a contradictory unease about the ruthlessness toward the weak and outcast that this position implies.

One of the great ironies of nineteenth-century thought, although a perfectly comprehensible one, is Marx's interest in Darwin's work, his belief that Darwin's theory provided naturalist ground for his own social theories. Robert Young remarks that Marx's use of Darwin for his own purposes "marked the beginning of a complex and fraught history, wherein not only Marxists but also the proponents of practically every conceivable political position sought to ground it in Darwinism."[2] That ideology can interpret and use science is particularly clear in Darwin's relation to Marxist thought. The connections with Malthusian economics have long led to associations between Darwinian and capitalist theory. Yet not only did Marx view the *Origin* as powerful support for his own ideas; Engels saw evolutionism as central to the dialectics of nature. Engels attempted, with remarkable hubris and even more remarkable knowledge and intelligence, to construct a scientific view of nature that would see it as dialectical in structure. Here is how he reads Darwin: "Darwin did not know what a bitter satire he wrote on mankind, and especially on his countrymen, when he showed that free competition, the struggle for existence, which the economists celebrate as the highest historical achievement, is the normal state of the *animal kingdom*. Only conscious organisation of social production, in which production and distribution are carried on in a planned way, can lift mankind above the rest of the animal world as regards the social aspects, in the same way that production in general has done this for men in their aspect as species."[3]

Taking the same "fact," Darwin's relation to political economy, Engels is arguing precisely against the reductionist naturalistic fallacy that has dominated most modern social thought in relation to science and Darwinism. (In trying, however, to show that nature itself is dialectical, Engels is, of course, creating another version of that fallacy.) The move is similar to Huxley's in his "Prolegomena" to *Evolution and Ethics*, in which he argues that you cannot, with Spencer, take nature as moral model. Whatever nature's constraints on human action, men and women must construct society on grounds other than their biology: "There is this vast and fundamental difference between bee society and human society. In the former, the members of the society are each organically predestined to the performance of one particular class of functions only . . . Among mankind, on the contrary, there is no such predestination to a sharply defined place in the social organism."[4] Huxley is not, to be sure, free of the tendency to apply nature's lessons to the human. But he begins by rejecting the reductionism—based on unspoken analogy—that takes all organic behavior as a model for human behavior. And he argues that human "nature" can survive only by denying the normal processes of nature: "That which lies before the human race is a constant struggle to maintain and improve, in opposition to the State of Nature, the State of Art of an organized polity; in which, and by which, man may develop a worthy civilization, capable of maintaining and constantly improving itself, until the evolution of our globe shall have entered so far upon its downward course that the cosmic process resumes its sway; and, once more, the State of Nature prevails over the surface of our planet."[5] Huxley's social project is, at least, frankly ideological: we strive for the "civilization" we desire rather than for the one decreed by "nature." For Huxley, however, the entropic vision closes down the Darwinian one; still, it does not lead him to make the leap from science to prescription.

Engels's ideological project is a different one, but for him, too, the goal is to lift *beyond* the animal source, and he argues precisely the reverse of the political economists' view that Darwinism endorses capitalism. The fact that humans are animals does not mean that they should take animal behavior as their own norm. It is a satire on capitalism that it uses animal behavior as such a norm. The separation of value from fact—which at some point must entail direct recognition of the arbitrarily willed social project—is critical to the whole Darwinian debate.

Obviously this separation complicates the political uses of Darwinism and particularly its use *against* disruptive political action. Its participation in the revolution that displaced natural theology and the social structures, like those described in Austen's novels, that natural theology sustained, should suggest other possibilities. But within the English tradition, Darwinism's potential for revolutionary significance, the very fact that we talk about the "Darwinian revolution," has tended to be minimized. The English novel has been no more receptive to revolutions than England itself. And English realism constructs reality so that ideological antagonism to revolution is implicit in its very forms. This is odd, given our normal association of Marxist writing with realism, and given, too, Georg Lukács's impressive defense of realism and his particular attention to Scott as the formative realistic historical writer of the nineteenth century.[6] I do not want to pause here to look into the ideological sources of realism's resistance to revolution, some of which, in relation to Darwinism, have been at least briefly noted in earlier chapters. The issue here is how the gradualist reading of change in English nineteenth-century realism, which I have associated through the reading of Trollope with Darwin's major ideas, was almost automatically transferred from the realms of science to the realms of society and politics. Once again, fictional narratives should help to expose how the guiding assumptions about what is real, usually also implying an ethical imperative to behave "naturally," concealed their own incoherence.

I will frame my argument with two political assasinations—quasi-fictional ones that span the century. Scott's *Old Mortality* (1816) projects its action from the moment of the assassination of the Archbishop of St. Andrews, by Balfour of Burley, and Conrad's entire *Under Western Eyes* (1911) depends upon the fact of the assassination by the student, Haldin, of Mr. de P——, president of the "notorious Repressive Commission" in pre-Revolutionary Russia. When Balfour, somberly preoccupied with his cause, asks Henry Morton how he could have allowed himself to participate in the government-endorsed wappenschaw, Morton replies: "I do my duty as a subject, and pursue my harmless recreations according to my own pleasure."[7] And when Razumov, thinking only of the prize essay he is preparing, suddenly finds Haldin in his room and hears from him of the assassination, his mind turns immediately to private affairs: "There goes my silver medal!"[8]

The protagonists of these novels are obviously not political animals. They are what Lukács might have called "maintaining" rather than "world-historical" figures, giving the novelist opportunity, through their ordinariness, to reflect the historical "disturbances" with which the characters prefer not to engage themselves. Lukács sees treatment of such figures as distinctive of social realism, allowing, he believes, for rich depiction of the way historical reality works itself into the texture of all private life. Yet there are paradoxes here. While working within a tradition that Lukács argues is largely invented by Scott, Conrad makes that tradition an element in the kind of modernist novel Lukács abjured,[9] for it can offer no decisive reading of the "objective historical situation." The sign and the interpretation are, in this kind of art, severed. Moreover, the protagonists are not only unpolitical. The very pressures that a cruel, even absurd history places upon their ignobly decent ambitions lead to a strongly antirevolutionary fiction.

These romantic and modernist versions of the revolutionary situation, while differing implicitly on how typical revolutionary action may be, share a deep discomfort with abrupt change. The protagonists' first moments in each book are moments of abrupt and unexpected change, leading to a transformation of life so radical that they are threatened with the loss not only of their places within the community but of their identities as well. (Note, by the way, that the threat to identity comes with a burst of unwanted information.) The excitement of the narratives implies an uneasiness, and the novels quickly displace world-historical events with the feelings and behavior of characters who want nothing to do with them. Revolution is represented as something that happens while we are not looking (although, of course, Scott exploits the romantic antiquarianism of historic battles, described nevertheless from the perspective of the personal and unhistorical interests of various characters). What matters about revolution here is that it threatens to disrupt what has been complacently taken as the natural growth and continuity of normal lives. In Scott revolution, which seems to be his constant subject, is the consequence of moral excesses and produces them; but it is absorbed into an ultimately intelligible and continuous movement of time. In Conrad revolution is gradually perceived as the monstrously normal condition of time and human experience. Razumov, resisting his author's gloomy vision of the discontinuity and absurdity of experience, looks back to the romantic vision: after Haldin finally leaves his rooms, he notoriously scrawls, amidst a set of slogans, "evolution not revolution" (p. 54).

Although there are many English novels that touch on revolution, literal or metaphorical, the abrupt and radical nature of Scott and Conrad's revolutionary violence at the start of these novels, and the suddenness of their intrusion on the protagonists, as they mark the chronological beginning and end of my argument, is at least on the face of it rather unusual. Yet they are only overt expressions of conditions implicit in the practice of mainstream English nineteenth-century realism. They help locate some of the central concerns and latent contradictions in the stubbornly domestic English novel between 1816 and 1910. The similarity of subject and the ultimate difference of implication suggest the limits of the great tradition of compromise that marks the romantic-Victorian novel, which is preoccupied with Darwinian (which is "natural") change. With an almost Carlylean revulsion from paper constitutions, the tradition abjures violence, while implying its possibility as a threat, occasionally allowing it as a terror and an outrage. The attempt at radical change is perceived as inevitably violent, and as a disruption of normal human and social relations: the disruption of the ideal of the organic community or the natural human bond, as in Hetty Sorrel's murder of her illegitimate infant, or Sikes's murder of Nancy.

The aesthetic, moral, and political problem revolves around the nature of the "natural"; for the Victorian novel above all things, even in its closest examinations of manners and of social conditions, assumes the model of the natural for ethical as well as descriptive purposes. Its truest heroes and heroines grow into the condition of the natural as they discover more fully what it is and learn, as Maggie Tulliver tries to, "patient obedience to its teachings." But the natural was being radically redefined throughout the century, and patient obedience to it could be confusing or criminal, depending on which nature you believed in. The ostensibly "unnatural" and disruptive acts of assassination put to the test, then, the assumptions about reality described in this book. These assumptions obviously underlay the tradition of Victorian realism, but to challenge them was to challenge its characteristic aesthetic, political, and even epistemological conditions. Yet the disruptive actions of Haldin and Burley do not clearly imply a narrative reality very different from that of realism, for the crisis within the novels is not in the assassinations themselves, but in the moral demands by the assassins upon the protagonists. The assassination is immediately absorbed into a community and puts the values of that community to the test.

The assassinations immediately question the conventions of normality and civilization that govern ordinary life, and the way realist narratives unfold. In particular, they strain the ideal of the refusal of violence and disruption and force strategies of reconciliation that are not always convincing. If the assassin has acted the part of Cain, the protagonist who might refuse to give aid to the assassin is asking, "Am I my brother's keeper?" The protagonist himself, in obeying the moral sanctions of society, also betrays them and becomes an assassin: the important barrier between normal and abnormal, natural and unnatural, good and evil, breaks down. The realist novel, which depends so importantly on the conception of character and continuous development, is threatened both by the blurring of the margins of identity and by the leaps in time and action required to avoid the inevitable implication of the most ordinary figures in the brutalities revolutionary activity entails on either side. As Conrad develops his version of the story, the attempt to sustain the conventions of normality is entirely relinquished, and breakdowns in identity, gaps in narrative, disruptions of all kinds become the norm.

Opening narratives with such disruptive situations can suggest some general implications about the way narrative is constructed. First, they imply two streams of narrative: one the narrative of history, a macro-narrative, as it were; the other a narrative of private life, a micronarrative. (The typicality of this bifurcation for Victorian realism might be suggested by considering the role the Reform Bill plays in *Middlemarch*.) This "stream" conforms to the Darwinian conventions of continuous change and the organic interrelation of all individuals within an organism or a social unit. Second, the movement of the two narratives strikingly dramatizes the way the micro is determined by the macro, and the way world-historical events are governed by minutiae ostensibly irrelevant to the large issues that official history would emphasize. In this way, the assassinations radically reaffirm the persistent Victorian-realist theme of disenchantment: the usually slow recognition that the self is constrained by a larger world of which it is only an insignificant part, and that the apparently heroic is built of quotidian detail which, if it were observed closely, would lead to no surprises, no disruptions. Thus, third, what happens is beyond the control of the private hero, the novel's protagonist, who is at the mercy of forces and motives that feel entirely beside the point of his life but that implicate him entirely. The Darwinian sense of chance, which Dickens half resisted, comes into full play here. The events apparently offer the protagonist opportunities for choice, but the choices are on the whole illusory, and almost always are among unsatisfactory alternatives. Neither the protagonist's intention nor his will has power over the movement of the narrative. Fourth, the world implied by that movement is deterministic—that is, all events are implicated in the movement of cause and effect, determined by what has already happened, and at the macro level, insofar as the novelist knows what happened in history, the characters' fate amidst the options they believe available is already sealed: events move in a foreknown or predetermined direction, totally without reference to the desires of any of the characters (unless the novelist becomes a romancer and chooses to make history subject to a protagonist's will). We have here a kind of reverse teleology, the kind that might be inferred from the Darwinian narrative of natural selection despite Darwin's rejection of the happy-ending teleology of Paley. Thus the illusion of freedom normally implied in micronarratives is shattered. Yet, fifth, the experience of that shattering within a deterministic world has the experiential texture of "chance."

With the historicist orientation of both Scott and Conrad, chance operates in the antiteleological sense described in the fourth chapter. What happens follows from a sequence of causes, and there is nothing "chancy" about either of the assassinations and their aftermaths. Causal sequences are satisfactory *in novels* if the events are both likely to have happened in the circumstances described and if there is some connection between the nature of the characters and what happens to them. Psychological explanation of events, however primitive, carries more weight in fiction than it is likely to do in life. Balfour has fought on the side of Morton's father and fully intends to draw Morton himself into the good cause. Morton, because he is honorable and remains loyal to his own inheritance, agrees to protect Balfour from the authorities. Similarly, Haldin has long *planned* to hide in Razumov's room because he has made some interesting (but, we understand, incorrect) guesses about Razumov's character. We have, then, explanations.

Yet in both cases we need to deal with the novelistic experience of intrusion, disruption, discontinuity, and with the sense of the arbitrariness of the meetings. Burley's character explains his behavior, to be sure, as does Morton's, once he is enmeshed in the affair. But after all, Morton has come to participate, for pleasure, in the archery competition, not to make a political statement; Razumov returns to his room with his mind full of his competitive exams and his prospects for success. Their lives do not seem to be connected with the assassins; they become, in effect, pawns of other characters' intentions, and what happens to them has nothing to do with what they have been feeling or what is on their minds. Yet the novel not only puts the protagonists in positions that make them seem revolutionary but rhetorically diminishes their differences from the assassins.

Although the psychological explanations have a realistic plausibility, the coincidences of event are not determined psychologically at all. Moreover, the psychological validity of the coincidences applies to one but not both of the parties. From a wider perspective, Morton and Razumov are in the hands not only of individual characters who misunderstand them, but of the history that governs those other characters' lives as well. Through Balfour and Haldin history erupts into narrative sequences that were moving in other directions. The crossing of narrative lines creates the conditions of "chance" discussed earlier, the convergence of two causal streams that have been moving regardless of each other.[10]

The event is *experienced* as chance. What occurs beyond the range of our possible observation must be felt as meaningless—mere chance—when it touches us. There is a famous passage in *Middlemarch* (chapter 41) in which Eliot must invoke "Uriel watching the progress of planetary history from the Sun," in order to justify the manipulations by which Raffles enters the Bulstrode story. To the distant observing Angel, "one result would be just as much of a coincidence as the other." As scientific, detached observer, Uriel is an inadequate device for sustaining the conventions of uniformitarian realism. The mending operation feels forced. Yet the passage focuses precisely the issue raised by the sequences in Scott and Conrad: what *appears to be* coincidence, a sudden disruption of expected sequence, can be seen to be the inevitable consequence of innumerable small events. Within the tradition of scientific objectivity as the only true authority, the observer can only achieve accuracy by separating himself from the event. Uriel (or a scientist or a Lukácsian—Marxist—critic) can see what such engaged figures as Morton and Razumov cannot. Trollope's narrators may play Uriel. Conrad's, deliberately limited, cannot: the realist's leap of faith that all events are causally connected, that all disruptions and catastrophes can be understood as the accumulation of small effects, and that all events are ultimately comprehensible, is no longer possible. The play with point of view that is so important to the development of the novel in the last half of the century releases the forces of chance and the irrational as it discards Uriel. Evolution turns into revolution.

Yet Conrad, surely, is a convinced Darwinian, and Scott, in his historicist moderation, helped pave the way for the triumph of the Darwinian imagination. It is worth one last brief look at Darwin's argument to consider how it could have been used in both a Trollopean and a Conradian way. The problem is that while the language of *The Origin of Species* is, as we have seen, adaptable as the nineteenth-century's most imaginative and powerful denial of revolutionary change, it was latent with disruptiveness and its persistent naturalizing ate away at conventions of value that quietly determined its conservative uses. Even its remarkable, experientially persuasive rhetoric in favor of the view that natural selection "can act only by the preservation and accumulation of infinitesimally small inherited modifications" and its determination to "banish" belief in "any great and sudden modification" (p. 142) is counterintuitive and makes the whole world strange. Darwin must tell stories and push his counterintuitive arguments against the limits of imagination. Common sense tells us that nature makes enormous leaps, into the majesty of the Alps, through the cataclysmic horror of earthquakes, in the overwhelming differences among living creatures, from the minute infusoria and primitive mollusks to the mind of man.

In effect, Darwin tries to account for all apparent discontinuities across space and time. Where significantly different species occupying similar places occur in juxtaposed spaces, Darwin finds a history of gradations between them that have not survived, even in the fossil record. The question, for instance, of how a land carnivorous animal could have been converted into one with aquatic habits is difficult because we assume that animals in their transitional states would have had trouble subsisting. But Darwin says, "It would be easy to show that within the same group carnivorous animals exist having every intermediate grade between truly aquatic and strictly terrestrial habits" (p. 212). There follows a brilliantly selected metamorphosis, in which Darwin talks of squirrels and lemurs and fish that fly, and birds that don't. Drawing on an enormous range of examples that break down common-sense notions of identity and remind us of how we have, as it were, domesticated the strange, Darwin transforms the world. If, he says, "about a dozen genera of birds had become extinct or were unknown, who would have ventured to have surmised that birds might have existed which used their wings solely as flappers, like the logger-headed duck . . . ; as fins in the water and front legs on the land, like the penguin; as sails, like the ostrich; and functionally for no purpose, like the Apteryx. Yet the structure of each of these birds is good for it" (p. 213).

Characteristically, Darwin turns to apparent anomalies as evidence for his argument. The world makes no sense in precisely the places separate creationists and natural theologians found their justifications. It is neither economical nor efficient—qualities theologians would want to attribute to the creator and scientists would expect of an ordered world—but rather prolific, wasteful, eccentric. On a theory of design it would be difficult to understand a web-footed goose living away from water, a woodpecker away from trees; on Darwin's theory such anomalies are evidence of natural selection. The apparent leaps are in fact the result of innumerable slow transitions: "If any one being vary ever so little, either in habits or structure," it gains an advantage over some other inhabitant, and will seize its place, "however different it may be from its own place" (p. 217).

The world as it exists now becomes a history (at one point, Darwin even calls it a museum): "Every detail of structure in every living creature . . . may be viewed as having been of special use to some ancestral form" (p. 228). Life is continuous and incremental. All details matter and can be read as part of a narrative that makes sense and culminates in a present where every living creature is adapted to its place, however strange and anomalous its structure. Strangeness becomes not evidence of wild aberration, of some monstrous leap of nature, but of continuity itself. The margins of identity blur (as does the notion of stable perfection), as Darwin reveals a world developing constantly through subtle intermediate gradations. This is one part of Darwin's story: it is the model for

Razumov's desperate scrawl. Without overt political intention, Darwin constructs a nature that can make revolution "unnatural."

But gradualism could not be an entirely self-consistent theory, and rhetorically Darwin himself depends on the shock of enormous differences having been influenced by apparently trivial causes. There *are* vast differences, and in the higgledy-piggledy world of natural selection, some of what Darwin described did not seem lawful. We have seen that there is much in natural selection—particularly its focus on the individual and the unique—that justifies John Herschel's first negative description. Beer has importantly emphasized Darwin's resistance to any kind of stability, and finds him constantly reworking his language to emphasize "physical process, not completed idea."[11] The instability this introduced into a world hitherto imagined as designed, and still essentially Aristotelian and essentialist, was radical.

Beginning with the discussion of "Domestic Selection," Darwin is intent to show that individual variations are not the exception, but the norm, that they happen constantly, and are only controlled by domestic breeders, who select some and reject others. "Nature," he says, "gives successive variations; man adds them up in certain directions useful to him" (p. 90). Variability is simply the inexplicable given for Darwin. It is true that for the purposes of his argument, Darwin does not need to explain it, only to demonstrate that it is pervasive. But in his attempt to show that the variations are not the product of "design" he emphasizes how often individual variations are utterly useless to the organism. In this stochastic process, law enters only *after* the variations occur: then the human breeder, or, more potently, natural selection, does its transformative work. We may understand why natural selection allows some variations to survive and others not, but the variation is a sudden intrusion—and, it should be emphasized, a necessary and continuing one—on the lawful processes of nature. The variation appears as "chance," as sudden and unexplained as Haldin's appearance in Razumov's rooms. Into the law-bound system that Darwin was attempting to create, lawlessness immediately thrusts itself.

The conservatism of Darwinian gradualism is thus countered by the conception of the source of new generation in his world. Since it does not come through divine act, it must be through chance variations; an entirely law-bound world of the sort that the uniformitarian view projects leaves no space for the new. My point, of course, is that while the uniformitarian basis of Darwin's arguments, which parallel so closely the methods and themes of realistic fiction, implicitly denies the possibility of successful revolution, the true generating power of Darwin's theory is what cannot be reduced to law, nor accounted for by gradualism. Thus, Darwin's theory can be used to expose by analogy fundamental contradictions in the Victorian realist project. The determination to view all experience from the perspective of the ordinary closes out the possibility of real change and locks all characters into an organic-deterministic system. The conventions of coincidence by which even a novel like *Middlemarch* releases its protagonists from social or psychological imprisonment do not so much represent a retreat from the ideals of realism as a necessary element in any imagination of the possibility of real change and growth in the realist's world. The move, however, is a risky one. (Beer has helpfully pointed out to me that change for Darwin could come from *outside*, that is, by invasion, not by revolution.)

The support that Darwin's theory gave to the realist novelist's program could also threaten the stability it affirmed. Its rigorous exclusion of design and of the divine hand tended toward an unredeemed secularity that made experience meaningful only in the trivial sense that it could be explained by laws. But on the logic of the theory, laws are not universally applicable after all because the generating power of the new is chance. There is no reason, then, to assume that the new will in any way affirm the highest human values, or be in any way relevant to them. Hardy's exploitation of the conventions of coincidence and happenstance to increase not diminish the protagonists' suffering is one entirely legitimate inference from the Darwinian scheme. Sally Shuttleworth points interestingly to this problem in George Eliot's works, where Darwinian organicism and gradualism are theoretically central. She notes that several of Eliot's novels "offer two conflicting models of history, one based on ideas of continuity, moral order, individual responsibility and control, and another which stresses gaps and jumps in historical development, chance, individual powerlessness, and self-division."[12]

I frame my subject with the two assassinations because I want to emphasize how central to the project of the ideologically conservative realism of Victorian novelists are those very disruptive elements of chance which they, like Darwin, attempted to repress. *Old Mortality* is a particularly valuable example, in part because Scott played so important a role in the development of the Victorian novel and, reflecting and capturing a whole culture's interest, was part of that medium that cultivated Darwin's theory. The chance encounter that opens the story of *Old Mortality* marks the crossing of two histories—the personal one of Henry Morton and the history of the Covenanters' rebellion against the Stewarts. The point of intersection is the point at which the novel first dramatizes the dependence of the private life on large social-historical conditions, and ironically it suggests that the realist novel's insistence on that dependence entails narrative methods that come from outside the conventions of realism and that at least give the appearance of chance as I have discussed it—that is, the convergence of two or more narrative lines.

Although most narratives may well require such convergence, when it occurs it almost always has the effect of surprise or even shock. Darwin himself uses it that way, for in order to bring home the interdependence of all organisms, he looks for examples that emphasize the unlikeliness of convergence. There is a quality of wonder in the explanation of how the enclosure acts helped "determine the existence of Scotch fir": careful observation shows that where cattle graze the fir gets no chance to grow; and in the fact that "in several parts of the world insects determine the existence of cattle" (*Origin,* p. 124). Clearly, the enclosure acts were not designed to affect the growth of Scotch fir, nor do the flies of Paraguay act in order to affect the life of cattle. Any "design" must be inferred by Uriel; and in fiction, novelists must devise plausible ways to bring narrative lines not ostensibly related to each other by design or intention into contact. Seen from this perspective, the device of the omniscient narrator is not an accident of nineteenth-century realism but a condition of it. Once we allow not only that limitation of perspective is a condition of all actors, but also that Uriel-like vision is inaccessible to anyone, the realist project of discovering the paths of necesary sequence breaks down. It becomes conceivable that none has the power to discover the paths and, moreover, that the paths are not there in the first place. The realist program of recognizing mutual dependencies entails Darwinian shock tactics since no one in fact is omniscient; it entails also, therefore, narrative techniques incompatible with the gradualist and ordered principles of realism.

This is particularly striking in Scott because Morton's story aspires to Shuttleworth's conditions of "continuity, moral order, individual responsibility and control," while Burley's is revolutionary and disruptive, and, in the cause of Calvinism, ultimately denies individual responsibility. Moreover, while the story begins dramatically with the wappenschaw and the meeting of Burley and Morton shortly afterward, the novel itself begins with Peter Pattieson. In drawing on the anecdotes of "Old Mortality" and on his own research, Pattieson is attempting to make sense of the irrational and violent events and behavior of the past. He wants, indeed, to bring them within the pale of common sense and, in effect, to diminish their abruptness and violence by allowing us to understand, Uriel-like, what seems incomprehensible: "We may safely hope, that the souls of the brave and sincere on either side have long looked down with surprise and pity upon the ill-appreciated motives which caused their mutual hatred and hostility, while in this valley of darkness, blood, and tears. Peace to their memory!" (p. 70). In wanting "to do justice to the merits of both parties," Pattieson transforms the abruptness and violence of revolution into a history that confirms the quotidian reality and values of the present. Through Pattieson, Scott submits the abrupt and revolutionary opening to the normative work of narrative ordering; like natural selection working on chance variations, the ordered processes of history absorb and regulate the disruptions, making them seem the aberration rather than the norm. Open-mindedness, of the sort that Eliot claimed to have learned from Scott, tends within the realist ethos to excuse all by understanding all.

The realistic protagonist, Henry Morton, as critics and Scott himself all agree, is not a particularly interesting figure. Yet he has a power in the novel incommensurate with his own ability to effect change. The secret to this is that Morton, situated between the fanatical historical forces, stands for precisely the values that Peter Pattieson claims govern the narration of the story. Pattieson neatly balances the opponents and judges them against a standard of moderation: "If recollection of former injuries, extra-loyalty, and contempt and hatred of their adversaries, produced rigour and tyranny in the one party, it will hardly be denied, on the other hand, that, if the zeal for God's house did not eat up the conventiclers, it devoured at least . . . no small portion of their loyalty, sober sense, and good breeding" (pp. 69–70). Morton, conceived on the model of the prudential hero so well analyzed by Alexander Welsh, takes precisely the same view:

> He had formed few congenial ties with those who were the objects of persecution, and was disgusted alike by their narrow-minded and selfish party-spirit, their gloomy fanaticism, their abhorrent condemnation of all elegant studies or innocent exercises, and the envenomed rancour of their political hatred. But his mind was still more revolted by the tyrannical and oppressive conduct of the government, the misrule, license, and brutality of the soldiery, the executions on the scaffold, the slaughters in the open field, the free quarters and

> exactions imposed by military law, which placed the lives and fortunes of a free people on a level with Asiatic slaves. Condemning, therefore, each party as its excesses fell under his eyes, disgusted with the sight of evils which he had no means of alleviating, and hearing alternative complaints and exultations with which he could not sympathize, he would long ere this have left Scotland, had it not been for his attachment to Edith Bellenden. (p. 187)

The characteristics of this passage are immediately recognizable as typical of the Scott predicament. The hero who tries to affirm moderation and the virtues of civilization is pressed by opposing fanaticisms into a position that makes immoderate behavior on his part almost essential. And yet, by making so centrist a figure the hero, Scott is obliged to create a form that will somehow, without contradicting its own values, make the nonviolent center triumphant. The device by which this is largely accomplished is in rendering the hero powerless through the convergence of the two streams of history.

The macrohistory belongs to Claverhouse and Burley, both of whom are historical figures, and both of whom get long and interesting antiquarian footnotes in Scott's text. But Morton, caught between the forces of the larger history, is an unhistorical character belonging for the most part to the microhistory. His is the story not of revolution and civil war, but of a love affair, obstructed in the tradition of Romeo and Juliet, by forces totally irrelevant to it. Morton's story can only close when his love for Edith is resolved, one way or the other, and we think we know which way that must be. That is, the political issue raised in the passage is not resolved by directly addressing it. In fact, Morton's political feelings would lead him to abandon politics entirely. What keeps him where politics can hurt him is his personal feeling for another unhistorical character. (A similar imagination of politics, and evasion of it, is at work in *Middlemarch*, where Ladislaw's politics are determined by Dorothea's presence, and in Conrad's *Nostromo*, in which Decoud thrusts himself quite cynically into the revolution in Costaguana beause of his attraction to a woman; the implications for history and narrative practice are similar.)

Within the context of macrohistory, individual identity and desire amount to nothing. Claverhouse sentences Morton to death almost immediately upon meeting him. Later in the novel, when he is captured by fanatics, Morton is once again doomed—and then saved fortuitously by Evandale, the other merely civilized soldier, and contestant for Edith's hand. In order to save Morton from the violent logic and the demand for corrupting compromise of the convergence of separate narrative streams, Scott's narrative must take advantage of various traditional fictional tricks and perpetrate last-minute rescues. The plot is abrupt and discontinuous in order to save continuity and order; it is the narrative version, as it were, of the intrusion of the divine hand. From *Mansfield Park* we are familiar enough with this sort of thing, which has always been part of the storyteller's bag of tricks, and continued to be so through Victorian fiction, though often more than half-embarassedly.

In *Old Mortality*, however, Scott rather complicates matters. The novel is as novel almost powerless to resolve itself. Within the terms of history Scott can realistically allow, Morton is doomed. He is in fact banished into an utter loss of identity. Powerless himself, Morton is confirmed not by action but by history. In the battle between fanaticisms, neither side wins, and with the instauration of William and Mary, the country itself adopts the values that Morton had been trying to live out during the first part of the novel. The logic of the microhistory leads to a dead end. What is required is a leap beyond narrative logic, which Scott produces by allowing a quiet revolution offstage to confirm his protagonist, who nevertheless has had nothing whatever to do with it. In mixing the two streams of history, Scott is true to his vision of the utter dependence of individuals on the movements of macrohistory, but he therefore must make large narrative leaps, and he begins the last section of the novel by announcing one:

> It is fortunate for tale-tellers that they are not tied down like theatrical writers to the unities of time and place, but may conduct their personages to Athens and Thebes at their pleasure, and bring them back at their convenience. Time, to use Rosalind's simile, has hitherto paced with the hero of our tale; for, betwixt Morton's first appearances as a competitor for the popinjay, and his final departure for Holland, hardly two months elapsed. Years, however, glided away ere we find it possible to resume the thread of our narrative, and Time must be held to have galloped over the interval. (pp. 399–400)

The historical transformation has taken place offstage, just as the assassination did at the start of the book. Morton is as much caught up in the later movement of history as in the earlier, and his life in the interim is a great gap in the narrative. Ironically, his return into more unhistorical circumstances places him once again between contending forces; but they wipe themselves out, and he emerges, as we had always expected, with Edith his bride.

Nevertheless, Morton accomplishes all this by doing and being almost nothing. "I am so changed," he thinks, "that no breathing creature that I have known and loved will now acknowledge me!" (p. 435). As he moves rather helplessly through the last sequences of the novel, he manages not to accomplish what he wants, and as a consequence, Evandale is killed and Morton marries Edith. Scott attempts to keep Morton free of the taint of revolution, which is the initial occasion for the action, but he can manage this only by contrivance that allows the macrohistory and Morton's own powerlessness to achieve the conventional ends of narrative design. Yet Morton is himself close to the extreme actions that his whole narrative denies because he must sanction great violence. One telling moment shows him restrained from protest at the torture of MacBriar during the trial of the Covenanters; only this restraint, this requirement that he not protest the violence he despises, keeps him from the death sentence. The betrayal muted here is elaborated in the foreground of Conrad's tale, when Razumov in effect becomes an assassin.

Finally, then, *Old Mortality* impressively lays out the conditions for Victorian realism and some of the problems implicit in the affirmation of its values. Morton remains a silent witness to history. In a pattern we have been tracing through many narratives infected with the nineteenth-century empiricist values, he survives the macrohistorical battles by being ineffectual, and his survival entails loss of identity through acquiescence in the processes of history rather than resistance to them. Before Darwin or Conrad, Scott affirms gradualism and the idea, evolution not revolution, even as his narratives indulge the leaps and the excitement of revolutionary movement, the refusal to acquiesce in "nature," or in history. Without a Darwinian, dysteleological sense of nature, Scott could comfortably believe that gradualism was compatible with design and rather casually call on history to work out its justice and its moral progress and do the divine bidding that Darwin was to deny—except when reading novels for their happy endings. Without a sense of the mindless strenuousness of natural selection, he could allow the moral superior to survive the conflict of two stronger, and cannier opponents. Yet to do even this, he had to employ narrative leaps, discontinuities between intention, motive, and action, evidences of his protagonist's absence of responsibility.[13] Realism attempts to suppress the discontinuities, the violent gaps, the irrationality, and to heroize the ordinary. Scott allows the ordinary to triumph even if he is wildly reckless in the nonrealistic means he uses to achieve the triumph. The Victorians picked up in their own practices the triumph of the center and attempted much more self-consciously to perceive that as the norm. Scott could not somehow achieve a narrative that performed this, although he asserted it often enough. And he was perfectly consistent with his own notions of narrative in finding in what he called Austen's "correct and striking representation of that which is daily taking place" a superb form of art.[14]

The pattern of disenchantment and discovery that characterizes Austen's novels parallels the pattern in Scott's more extravagant fictions. That pattern, in which protagonists must learn to recognize and then make peace with the harsh unaccommodating actual, is as I argued earlier the dominant pattern of Victorian realistic fiction; it is a domesticated version of the revolutionary plot we have just been looking at in Scott; and it confirms the importance for realism of establishing a perspective that can unequivocally affirm invariability of sequence. But as in the classic case of Dorothea Brooke and Edward Casaubon, the convergence of two histories is normally a convergence of microhistories, which is part of the point. The collision must be seen to be "natural," and the natural is normally not catastrophic. Yet in muted domestic and uniformitarian terms it is no less so.

In *Middlemarch* there is, of course, a macrohistory and a revolutionary history; the Reform Bill does in fact influence the little dramas of provincial life, just as the reality of German historical scholarship influences Casaubon's scholarly and marital failures. But *Middlemarch* is the classic working of the integration of converging narratives within a gradualist schema, and the subtle handling of history averts the revolutionary subject, or transforms it into local or banal phenomena—Mr. Brooke's electoral campaign, or Mr. Garth's encounter with the navvies. The Darwinian model, organicist and gradualist, is at work throughout that novel. And the gaps and discontinuities which, on my theory, are inevitable in any gradualist tale, are carefully absorbed so that gradualism does indeed seem the norm of human experience. The narrative voice reminds us, in the very Uriel-like perspective it takes on the reform agitation and its manifestations in Middlemarch, that the political revolution, from a distance of forty years, has not much changed the way things are. And the marvelous "Finale" might be taken as a hymn to gradualism.

As the summa of Victorian realism, *Middlemarch* needs to be checked against fiction that does risk direct confrontation with revolution as subject. Such novels are likely to have many of the characteristics of *Old Mortality*: that is, first, framing or control by a wise voice that sees beyond the conflict, can absorb it into more normal quotidian life, and diminish it by retrospect (note how Charlotte Brontë's *Shirley*, which like everything else she wrote resists the dominant patterning of the time, is still the only one of her novels narrated in the third person); second, protagonists who serve as vicars of compromise, approximating as closely as possible to the condition of the wise narrator, and who, while in the midst of action, are somehow incapable of affecting it and are threatened with loss of identity or compromises so total that they are in danger of becoming like the violent antagonists in the larger conflict; third, elements that seem to be outside the conventions of realism and the values affirmed by the centrist narrator—random intrusions of narrative streams into each other, gaps in time, breakdowns in identity, sudden and emphatic disruptions of ordinary life; yet fourth, the location of value in micro- not macrohistories, in the domestic, not the revolutionary.

To take an obvious example, Thackeray's *The History of Henry Esmond, Esq.* is actually a novel about revolution: revolutionary activity is part of the young Henry's first experiences; a thwarted revolution concludes the story; and the Scott-like question, under which king? deeply influences Henry's career. The muted treatment of revolution is no accident in Thackeray, for he, of all the novelists considered here and perhaps of all Victorian novelists, is most rigorously the novelist of things-as-they-are, the most intensely skeptical about intensities and extreme actions. A Darwinian before Darwin, Thackeray regularly employs ironic distancing on romantic or heroic moments to make them feel illusory or, at best, momentary in a world universally ordinary—uniformitarian. The strategies of *Henry Esmond*, despite its virtuoso historical reproductions of eighteenth-century language and style, are emphatically antirevolutionary, quintessentially "realistic."

The framing of the story, even more than of Scott's, is immediately calculated to minimize the disruptiveness and dangers of history and of revolution itself. First, there is the voice of Esmond's daughter, Rachel, writing at a distance not only of time, but of place. She is interested not in her father's connections with revolutionary activity but in his qualities as father and husband. She is, as it were, the Peter Pattieson of the novel, standing between a republican husband and a royalist father: "I can love them both, whether wearing the King's colours or the Republic's. I am sure that they love me and one another, and him above all, my father and theirs, the dearest friend of their childhood, the noble gentlemen who bred them from their infancy in the practice and knowledge of Truth, and Love and Honour."[15] Although Rachel's voice is not heard again, the pressure of this imagination of the historical narrative is constant throughout the book. Yet another framing chapter, Esmond's own, is strategically constructed to prevent the reader falling for the romance elements before they even enter. Esmond speaks the voice of Uriel-like realist detachment, with a slight eighteenth-century accent. Anachronistically, perhaps, he has learned a great deal from Macaulay and Scott and insists on looking at history from the perspective of the novelist: "Mr. Hogarth and Mr. Fielding will give our children a much better idea of the manners of the present age in England, than the *Court Gazette* and the newspapers which we get thence" (p. 2). But he has a touch of antiheroic irony about him that may seem appropriate to his eighteenth-century disguise and is not in the tone of Scott or Macaulay. Queen Anne at the hunt is "a hot, red-faced woman, not in the least resembling that statue of her which turns its stone back upon St. Paul's" (p. 2). Esmond translates all life into the quotidian and has "seen too much success in life to take off my hat and huzzah to it as it passes in a gilt coach." Success and failure, virtue and villainy are the consequence of prior conditions over which the individual has little real control: "Give me a chain and red gown and a pudding before me, and I could play the part of Alderman very well, and sentence Jack after dinner. Starve me, keep me from books and honest people, educate me to love dice, gin, and pleasure, and put me on Hounslow Heath, with a purse before me, and I will take it" (p. 4). Material causes determine both character and morality. Thackeray could not of course have read Darwin yet, and his interest in science was marginal, if real, but the strategies of his fiction, both in their uniformitarianism and in their reluctance to allow seriously for nonmaterialist argument, were strikingly harmonious with the direction of Darwin's thinking.

But Thackeray's deflationary irony is also protective. His realism implies the ideology of compromise, and it is no accident that Esmond leaves for America as soon as his love problems are resolved. As G. Robert Stange put it in his excellent preface to an edition of the novel, "Esmond's is not a purposeful settlement in the New World so much as a retreat from the impossibilities of life in England" (p. xix). Like Henry Morton, he retreats from political action. Moreover, it is thematically right that the revolutionary subject is translated into a rake's escapade, the whole restoration of young James thwarted by that prince's attempted seduction of Beatrix. The novel leaves little to choose between either king. The logic of gradualism and of realism leads Thackeray to a position that threatens the whole tradition of rational design, which Darwin's theory was to subvert scientifically. What determines the direction of history in Thackeray is not some cosmic plan or the intention of leaders, but the minutiae of ordinary life, the crossing of narrative streams—chance. When macro- and micro-history converge, Thackeray shows that microhistory actually shapes the direction of world-historical events. History is the story of little people and of circumstances, and it doesn't seem to make much sense.

As Barbara Hardy has impressively shown, Thackeray's art is in certain ways radical.[16] His "exposure of luxury" is brilliant, ironic, and ultimately quite serious. But it is precisely the realistic and ironic disenchantment that his protagonists ultimately undergo that initiates and controls most of his narratives. The consequence is that the stunning revelations of corruption, of Queen Anne's red face, of the prince's childish lechery, lead not toward the revolutionary positions that Lukács assumes realism will entail but toward a rejection of the significance, even the possibility of large-scale action. In this respect, although Thackeray is by far the most disenchanted of Victorian realists and takes the techniques of realism to the edge of Darwin's dysteleology and total secularism, he expresses representatively the ideology of Victorian realism. A close look at reality reveals for him that the large is really small, and that the only inevitable and natural sequel to heroism and expectation is domesticity and disappointment. "We have but to change the point of view," Esmond says, reflecting on his relations with the Duke of Marlborough, "and the greatest action looks mean; as we turn the perspective glass, and a giant appears a pygmy" (p. 253).

There is a streak of strong sentiment in both Thackeray and Esmond. Yet the narrative Esmond writes is completed late in his long and happy marriage with Rachel, which is described with very uncharacteristic intensity on the last pages of the novel—perhaps because it is just about over. The ironies and cynicisms of that narrative would seem incompatible with the ideal life in marriage Esmond claims, for on the basis of his language one would have to infer that all romantic intensity is mere illusion.

Revolution itself is reduced to domestic narrative, and Esmond's part in the planned and failed restoration of the Stewarts becomes an occasion for dissolving his relation with Beatrix, who is banished from the book and his affections, and for joining with Rachel. Most important, it is the final occasion for his leaving England entirely, which he does contentedly in the last paragraph. Some of the bathos of Scott's *Redgauntlet* is deliberately implied here, but whereas in Scott there is a touching moment when

Redgauntlet must face the fact that he is no longer dangerous to the government and that dreams of Restoration of the Stewarts are only dreams, here there is no sense of romantic loss at all. The reduction of revolution to domestic bliss is what, in a sense, the book has been all about—except that domestic bliss itself must, in Thackeray, seem suspect.

Different as it is from Thackeray's novel, Dickens's *Tale of Two Cities* also inherits much from Scott and, in its own more melodramatic way, similarly affirms the quotidian over the revolutionary. Dickens does, of course, indulge romantic intensities, and as a consequence he gives greater emphasis to the convergence of narrative streams than Thackeray. That convergence is the center of the narrative, the macrohistory emphasizing the horror and irrationality of revolution, the microhistory affirming the virtue of the quotidian and the domestic. There are, to be sure, no framing voices of the kind we find in both Scott and Thackeray, but the omniscient narrator balances and understands from the very first moment: "It was the best of times, it was the worst of times." As in Scott, the narrator attempts an even-handed view of the horrors and excesses on both sides, and like Morton and Esmond, Darnay responds to that horror by opting out, although he is eventually drawn in. Dickens's metaphors emphasize Darnay's powerlessness even as the narrative aggrandizes him for his willingness to risk his life for a servant: "Like the mariner in the old story, the winds and streams had driven him within the influence of the Loadstone Rock, and it was drawing him to itself, and he must go."[17] The passage echoes the fatalism of *Little Dorrit*, which also emphasizes the superpersonal powers that determine narratives and fate. Just as Esmond, busy as he is, can do nothing to make things happen, to bring the prince to his own revolution, and as Morton seems to have no control over his own fate, so Darnay's liberal values and willingness to take risks are ineffectual: he is imprisoned and powerless for the last third of the novel and unconscious when rescued from the prison. The scientistic determinism inferred from Darwin's materialist science is pervasive in nineteenth-century fiction.

The great heroic final act of the novel is a further diversion from the revolution. Dickens constructs a new kind of heroism out of a domestic rather than a political story, and Carton's self-sacrifice is imagined as a way to avert the catastrophe of the world-historical destroying the private history. Carton dies neither for France nor the people, not even for Darnay. He dies for the woman he loves as, in effect, Evandale dies for Edith. Carton's act makes possible the domestic norm, yet is entirely outside that norm. Violent death is a condition of order and domesticity. This doubleness of intent is characteristic of the Dickens we have examined in previous chapters. His narrative instincts often collide with his ideas and attitudes. And it parallels the doubleness of the Darwinian argument as well.

The abruptness of so much of the narrative is obviously connected with the weekly serial publication form, which always made Dickens uneasy. But the abruptness is not inconsistent with Dickens's handling of the revolutionary theme. The key points of the novel are the points of convergence between the two narratives, beginning with the recovery of Dr. Manette, leaping immediately without explanation from his totally broken and irrational state to his domestic comfort with Lucy. The gap, caused by political repression, cannot within the domestic narrative be treated forthrightly. Lucy, as the embodiment of domestic virtues, provides the thread that knits his life together and that links it with others as well.

But the domestic portion of the life is treated rather generally, and the eruption of Bastille Day marks another convergence of two narrative lines. In a bridge chapter that treats Lucy as a "golden thread," and celebrates the joys of domesticity, the larger history suddenly intrudes: "Headlong, mad, and dangerous footsteps to force their way into anybody's life, footsteps not easily made clean again if once stained red, the footsteps raging in Saint Antoine afar off, as the little circle sat in the dark London window."[18] While the passage that follows is a powerful Dickensian invocation of the wildness and irrationality of the mob, which concludes with Defarge tearing apart the cell in which Manette had been imprisoned, the effect is ultimately to divert attention from the revolution and consider its consequences primarily in relation to domesticity. The storming of the Bastille is the threat of drawing Lucy and her family into a historical horror over which even her infectious lovingness will have no control.

To be sure, Dickens writes with compassion of the suffering and oppression of the French poor and with savage bitterness at the reckless and self-indulgent cruelty of aristocracy. But placing the Defarges at the center of the revolution turns it into a mindless vengeful energy of precisely the sort that will be regardless of all domestic and personal values: they are to be seen as more frightening even than Monseigneur and the corrupt aristocrats. Dickens's narrator knows better than the revolutionaries for he has the Uriel-like distance that allows him to see how things are related. His position in relation to Darnay is roughly like that of Peter Pattieson in relation to Morton, or Thackeray in relation to

Esmond. Invoking a Carlylean rhetoric and perspective, the narrator identifies here the thoughts of Darnay: "It was too much the way of Monseigneur under his reverses as a refugee, and it was much too much the way of native British orthodoxy, to talk of this terrible Revolution as if it were the one only harvest ever known under the skies that had not been sown—as if nothing had ever been done, or omitted to be done, that had led to it—as if observers of the wretched millions in France, and of the misused and perverted resources that should have made them prosperous, had not seen it inevitably coming, years before, and had not in plain words recorded what they saw."[19] Revolution, a radical choice for change within a specific context, is transformed here into a fatalistic consequence of natural gradualist forces. Darnay's desire to do something about it is of course chimerical, and Dickens is entirely consistent in rendering him powerless from the point of his arrival in France.

In the convergence of the two narratives Dickens eagerly affirms the superiority of the domestic, but he does so again in ways that violate the conventions of realism drastically. On the one hand, he almost literally divides Darnay in two. Carton dies in history so that Darnay can live in domesticity. On the other, he brings together the domestic Miss Pross with the revolutionary Madame Defarge, and in a direct melodramatic presentation of a life and death struggle between the two principles, allows the representative of the domestic to act aggressively and demonstrate superior strength. This is only possible, I believe, because Miss Pross is a servant and therefore exempt from the constraints placed upon protagonists, capable of doing the dirty work of violence that domestic narratives normally play offstage.[20] But the point of the superiority of power of the uniform and domestic over the catastrophic and world-historical is implicit. Moreover, the price of this aggression is high: Miss Pross loses her hearing as a result of the struggle (and in this way curiously presages the fate of Razumov). Dickens still wants to heroize the domestic, where Thackeray does not, and in doing so, in giving Carton the last heroic word and Miss Pross the power to outbattle the revolutionary fanatic, he reimposes design and meaning on his world. Within the conventions of narrative, Thackeray must do so too. But Esmond is in exile from a country where the king who won is perhaps only slightly better than the pretender who lost, and where the issues were matters only of abstract loyalty, since the novel has addressed no other. Dickens has given us a world where revolutionary issues are real, where heroism is possible, but where, at last, value is to be found outside world-historical politics.

Although *A Tale of Two Cities* was published in the same year as Darwin's *Origin*, it reflects less of the Darwinian ethos than *Henry Esmond*, which is uniformitarian and scrupulously sequential in its narration and, most interestingly of all, threatens to violate the conventions of teleology—of the happy ending. By making Esmond's love of Beatrix fruitless, Thackeray does indeed thwart the conventional ending, and by marrying Esmond to the woman who is, essentially, his mother, he opens all kinds of disruptive possibilities that I have avoided addressing here. But while Dickens relies on radical disruption of conventions of character, in the breakdown of Manette and the doubleness of Carton/Darnay, and while Thackeray intimates the breakdown of the conception of design and resigns history to chance, neither Dickens nor Thackeray can allow into his fiction a conception of nature that entirely denies design, as writers later in the century try to do. That denial was difficult enough for Darwin, as we have noted. But Dickens and Thackeray, flirting with elements that threaten the ordering power of nature, comfortably work them out within the conventions of Victorian realism that entail, as I have been arguing, the refusal of revolutionary change in the interests of slow progressive movement that assures an ultimate happy ending.

In Conrad, the refusal of revolution takes on some of the qualities of desperation. Conrad constructs his novels out of a sense of the contradiction always implicit in realism—that the very language of which it is constructed is inconsistent with the nature of nature. For nature is ineffable, and language is a late and aberrant outgrowth of the merely material. Razumov quotes Cabanis to Sophia Antonovna: "Man is a digestive tube" (p. 211). Conrad writes in the current of the full acceptance of the disturbing secularist implications of Darwin's theory.

In *Under Western Eyes* Conrad is a true inheritor of the conventions of the Victorian novel and of the Darwinian vision.[21] It has been argued—correctly, I believe—that in *Under Western Eyes* Conrad mutes his interest in science,[22] but the novel is infused with a vision clearly derivable from the developments in Darwinian thought at the end of the century. It struggles with just those problems of the relation of matter to value, of observation to interpretation, of continuity and disruption, chance and necessity that became central to the Darwinian debates. Moreover, even in the "Author's Note" to the novel he self-consciously affirms the authority of detached observation as an aesthetic ideal: "I had never been called before to a greater effort of detachment: detachment from all passions, prejudices and even from personal memories" (p. lx). The strategies of evasion and displacement that mark the narration of this novel reflect the self-annihilating implications of the ideal of scientific objectivity and detachment.

From a late Darwinian perspective, Conrad carries the conventions of realism to such radical extremes that all the latent incoherences become thematically and formally central. His novel challenges not only the notion that nature is compatible with human aspiration, but that language is capable of expressing it. Razumov's desperate affirmation of evolution over revolution is an appeal to George Eliot's conception of evolutionary change, and another of his slogans, "Direction not Disruption," points back to natural theology and the kinds of narrative we have seen in Dickens and Thackeray. Significantly, the structure of the narrative at the point these slogans are recorded denies Razumov's wish: the narrator abruptly interrupts the narrative, and shortly afterward the first part of the book ends with the famous "Where to?"—a question that is not answered for more than a hundred pages. The second section begins with what seems an entirely new narrative, introduced only with the narrator's apology, "this is not a work of imagination" (p. 84).

Under Western Eyes carries out some of the incomplete implications of the novels we have been looking at. In particular, it takes the moment of the crossing of narrative streams, Haldin's intrusion into Razumov's rooms, as the occasion to make chance blur the conventional assumptions that sustain distinctions like evolution/revolution, natural/unnatural, good/bad. "Haldin is disruption," Razumov thinks. The power of chance—hence irrationality—over human preparation is overwhelming: "Events started by human folly," for which read "idealism," "link themselves into a sequence which no sagacity can foresee and no courage can break through. Fatality enters your rooms while your landlady's back is turned; you come home and find it in possession bearing a man's name, clothed in flesh—wearing a brown cloth coat and long boots—lounging against the stove" (pp. 69–70). Echoing the fatalism of a novel like *Little Dorrit*, the passage reverses its implications, pointing not only to the dominance of chance, but to the pervasiveness of chance inside ordinary experience; and in the evocation of trivial details, it rejects melodrama for comedy, and makes the conventions of realism work against themselves. The very banality of the details undercuts assumptions of normality. Disruption and chance come clothed in the garb of realism.

We can say that *Under Western Eyes* is "about" revolution, but more centrally, it is about art and about the novel. Its sustained critique of the conventions of the realist novel follows from Conrad's agreement with Darwin that human ideas have the same status as monkey's instincts, that the mind of man has been developed from a mind as low as that of the lowest animals. His formalist rebellion against the conventions of realism, particularly against uniformitarian and gradualist assumptions, implies a fundamental disbelief in "nature," as the Victorians understood it, and thus in the moral confidence with which they denied revolution. Conrad draws most heavily on those aspects of Darwin that emphasize chance and disruption and the irrational bases of nature from which all spirit has been excluded. It is not that nature can now serve as a moral "model" that somehow endorses revolution, as Victorian nature denied it, but that natural law itself is "irrational," that is, regardless of human interest and intelligence. Human consciousness has no purchase on nature, and those moments of direct contact with it—when the barrier of language is down—are moments of horror. Revolutionary action, in its mad explosions of violence, seems simply to echo the irrationality of all nonverbal reality.

The critique of realism begins at the outset with a new and self-conscious insistence on limited perspective in rejection of the convention of omniscience. Conrad picks up an element of which we have seen important harbingers in Thackeray, the translation of all experience into personal perspective, no part of which can be understood fully because there is no voice in his novels who can speak with Uriel-like authority. Like *Old Mortality* and *Henry Esmond*, *Under Western Eyes* is framed by a narrator with a problematic relation to the action. But the function of the frame, despite the Professsor's protestations, is not to clarify or reconcile or absorb the revolutionary disruptions into a uniformitarian history, but the reverse: the Professor keeps insisting that Razumov and his acquaintances are not comprehensible and that the medium in which they are experienced and in which they are described is untrustworthy. It is Thackerayan skepticism gone mad: "Words, as is well known, are the great foes of reality," the Professor announces on the first page (p. 1). Throughout the narrative, the Professor insists on its unintelligibility, even when it does not seem particularly obscure. The distrust of language throughout is matched by the distrust of ideas (the development of a Trollopean perspective, one might say), which are destructive forces, moving from abstraction into their irrational source, violence. Nathalia Haldin, the idealist sister of the assassin, dreaming of an ideal society, responds to the Professor's skepticism by arguing, "Everything is inconceivable to the strict logic of ideas. And yet the world exists to our sense, and we exist in it" (p. 89). Both Nathalia and the Professor deny the possibility of a Western rationalist understanding of experience. Experience cannot be contained discursively, nor can it be within the confines of conventional realistic narrative.

The ultimate brute reality, analogous to the horror at the heart of darkness, is the vast whiteness of the Russian wilderness, an enormous material solitude, totally beyond the capacity of human consciousness to order or comprehend. Against the backdrop of this vast irrationality, sanity is only possible by simulations of traditional realism, a preoccupation with the trivial facts of experience—like that of the harlequin sailor in *Heart of Darkness*. The setting of the revolutionary cabal in Switzerland almost under the statue of Rousseau is appropriate, for it suggests a contrast between the brute reality of a vast nature and the artificially constructed bourgeois civilization that can sustain itself only in a postage-stamp country.

The arbitrariness of civilization is one of the implications of the radical disruptions of narrative in which the novel indulges, often almost to tedium. The limitation of point of view is oppressive, partly because of the Professor's continuing insistence that his subject is unintelligible, partly also because the last part of the book is rigidly confined to Razumov's consciousness in a series of extraordinarily long discussions with equally limited consciousnesses. Not only is the book a network of written words, various documents that are the source of most of what is known about Razumov, but those words are for the most part the record of spoken words rather than of actions. The translation of words into action is peculiarly dangerous in Conrad's world.

Against this obsessive preoccupation with words, the novel keeps insisting on their contrast with things. The realist valorization of things becomes in Conrad something other than a means to credibility, or representative confirmation of the solidity and recognizability of the world. Things cannot be translated into language, and the central experience of Conrad's novels, the attempt to penetrate into a heart of darkness, while parallel to the realist pattern of disenchantment, will not work itself out. The darkness cannot be expressed in language, nor can Peter Ivanovitch's chain, or Mikulin's hat. Objects are not a sign of the solidity of the world, as one would expect from realist fiction, but of its tenuousness. Hillis Miller captures well this aspect of Conrad's art: "Ordinarily things are assumed into the process of living and so taken for granted that they are scarcely noticed, only used, as a man does not notice the doorknob he turns a dozen times a day. Conrad shows such things wrested from their context in daily life and put before the spectator as mute, static presences. The interpretations ordinarily connecting man to things are broken, and the world is put in parentheses, seen as pure phenomena."[23]

Similarly, realist insistence on the significance of the trivial becomes a kind of phantasmagoric disruption of all assumptions of normality. The trivial detail matters enormously in the novel: "The trivialities of daily existence were an armour for the soul," thinks Razumov (p. 44). And in this respect we are reminded of the harlequin sailor in *Heart of Darkness*, with his marine manual in the midst of the jungle. In every respect Conrad's world and Conrad's fiction are reminders of the total unnaturalness of language, of thought, and, of course, of fiction itself.

Ironically, demystification leads to a deeper mystification: the opacity of language and of fiction, the turn of the novel away from what it "represents" onto its own nature. Adapting the conventions of realism in order to expose their arbitrariness, Conrad nevertheless posits a reality beyond language from which his fictions protect him. That reality is, indeed, revolutionary, in the sense that it is governed not by regularities, but by irrationalities, by forces incomprehensible to human consciousness, and violently threatening. Aware of the artificiality of human constructs, Conrad does not move to a revolutionary displacement: instead he finds himself committed to supporting human constructs in their artificiality against the deep irrationality of phenomena. His letters to Cunningham Grahame are revelatory because in them he often attempts to explain his attitudes toward revolution, and these are bound up with his attitudes toward art.[24]

He writes to Grahame somewhat as the Professor speaks to Nathalia. "I did not seek controversy with you," he writes in one letter,

> I think we do agree. If I've read you aright . . . You are a most hopeless idealist—your aspirations are irrealisable. You want from men faith, honour, fidelity to truth in themselves and others. You want them to have all this, to show it every day, to make out of their words their rule of life. The respectable classes which suspect you of such pernicious longings lock you up and would just as soon have you shot—because your personality counts and you cannot deny that you are a dangerous man. What makes you dangerous is your unwarrantable belief that your desire may be realized. This is the only point of difference between us. I do not believe. And if I desire the very same

> things no one cares. Consequently I am not likely to be locked up or shot. Therein is another difference—this time to your manifest advantage.[25]

It is no wonder then that Conrad develops further than Scott or Dickens the threat of personal annihilation. Razumov is a mere label of a solitary individuality. His identity is sustained by his preoccupation with getting on in society, and this is fatally disrupted by the intrusion of Haldin. As Carton and Darnay fade into each other, as Morton almost dissolves into the oblivion of exile and is saved by sanctioning a brutality equivalent to that of the fanatics he resists on both sides, so Razumov, only tenuously himself, is absorbed into Haldin. In the darkness of Razumov's rooms, Haldin is imagined as a kind of phantom, and in passage after passage, Haldin and Razumov are blurred into each other. When Nathalia first sees Razumov, she manages only to utter her brother's name (p. 145). But the critical turn in the novel is that Razumov, unlike Morton, expressly betrays the assassin, and Haldin is captured and killed. When, at the end, obsessed by the other chance—the attraction to Nathalia—Razumov is driven to confess, he destroys himself as he had done Haldin.

In Conrad, engagement with history means violence and destruction. Art becomes a protector, a way to escape the violence, but the escape means a retreat from personality, even if the art constructs a "reality" whose nature is revolutionary. The antagonism between thought and reality, the terrifying necessity for Conrad to will the thought that is life to him, is enacted in the fiction, which at once engages and separates, which (because it is language) allows the assertion, the negation, the imagined consequences, without the actual consequences. Razumov's is the story of a movement into the violence of a reality extraverbal. When he confesses, which ought to be an act of language, he stops talking. Nathalia pleads with him for the story: "The story, Kirylo Sidorovitch, the story!" But Conrad cannot resolve his story with a story; all the long discussions have gone to create misperceptions, and language has been discovered to be the great foe of reality. "It ends here—on this very spot," says Razumov. "He pressed a denunciatory finger to his breast with force, and became perfectly still" (p. 298).

The dramatized world, in effect, enacts the dualities that Conrad describes—the commitment to the large ideals and their impossibility. Identifying Conrad entirely with the antirevolutionary attitudes implicit in the caricature of the conspirators is inadequate, and there are alternatives:

the growing presence of Haldin as a figure of stature in the book, or Razumov's identification with Haldin and achievement of the heroic integrity Sophia Antonovna ascribes to him at the end of the novel. "There's character" in his fidelity to the truth, in his discovery of the impossibility his life had become through its entanglement with words. His personality counts. Moreover Razumov attempts, unlike Conrad and the Professor, to struggle beyond language; and the consequence is a brutal beating.

Conrad's almost reactionary politics are at odds with the implications of his art. In his critique of realism, he develops techniques that thoroughly disrupt its antirevolutionary assumptions. The Victorian social conservatism that led to the adoption of Darwinism as a sanction for slow, historical change is part of the object of contempt of Conrad's art. His techniques of disruption, discontinuity, of elaborating a radical distrust of language, lead to a vision of the world that undercuts the gradualism in which Darwin and Victorian realists had invested so much. Violence—assassination, self-destruction, brutal repression—is the condition of life not its accident. There can be no Uriel-like narrator because, like Razumov, we are all trapped in a network of chance upon which it is impossible to look disinterestedly. Observation in the Herschelian sense becomes merely chimerical. The consequence, on the Conrad model, is destruction; but with all his deeply conservative hostility to revolutionary ardor, with all his distrust of ideas, with his imagination of a heart of darkness or the whiteness of the Russian interior sustaining the real "nature" of the world, he leaves no space except, as Sophia Antonovna puts it, to rot or to burn. Well, one other—the writing of novels, which allows during the brief space of their writing the conception of rotting or burning without their actuality.

The modernism in which Conrad participates, ideologically conservative as it tended to be, thus developed narrative strategies that, in exploding Victorian conventions of nature and of the possibilities of its representation in language, actually affirm the inevitability of revolution. He takes the other side of Darwin, the one cloaked by Eliot and beside the point of Trollope, finding in chance and an overarching materialism that robs things of their meaning the true nature of the world. His is part of the last step of Western culture in the dismantling of the project of natural theology: he disrupts the Victorian analogy between nature and society by revealing with an almost nauseated disgust the blind arbitrariness of bourgeois, gradualist society. The Darwinian threat to identity goes beyond the rejection of essentialism and the blurring of individuals through time, even beyond the self-annihilation required of the keen observer. Conrad becomes a Darwinian literalist, reducing the human to the animal, and the animal into mere matter subject to the laws of entropy. He leaves no space for the consciousness that allows him to make the reduction, and certainly no reason to model human behavior on nature.

As he describes a revolutionary, chance-ridden, disruptive nature, he sees the antirevolutionary stance of realism itself as a conventional and arbitrary construction of nineteenth-century bourgeois imagination, like the domestic dullness of Geneva, presided over by the statue of Rousseau. He exposes the contradictions latent in that construction in the image of the revolutionary cabal developing in the heart of Geneva. And he retreats from the implications of his vision, as Morton, Esmond, and Darnay retreat, while his protagonist Razumov confronts them and is rewarded in losing contact with the language, that foe of reality, that has implicated him in the first place.

The retreat from language is, in some respects, not inconsistent with Darwin's position. Conrad takes the anti-Kantian position that Darwin and his son took with Max Müller—intellect does not precede sensation, but sensation precedes intellect. With philosophic consistency, one would have to argue that it does not follow from historical priority that feeling is more real than intellect, or that it remains more powerful than the intellect which follows from it. But in turning from the dominant Victorian interpretation of Darwin, and demonstrating that that interpretation is not an inevitable one, Conrad actually continues the tradition that persists into contemporary sociobiology, of extending science's description of the natural world into a description—but *not* a prescription—of the moral world. On the subject of revolution, however, the divergence is clear. Whereas the Victorians took Darwinian reality as a sanction for the rejection of revolutionary action, Conrad takes his Darwinian reality as evidence that revolution is everywhere and inevitable.

The difference, finally, lay in the power exercised by the values and attitudes implicit in the natural theology with which this book began. If nature is not under the divine regulations of a God whose consciousness is compatible with our own, it is nevertheless intelligible. That was the assumption of Victorian science, as it was the assumption of Darwin himself. The latent teleology of Darwin's theory left him ambivalent. The difficulties were great and the movement not inevitable, but it did seem as though natural selection worked for the improvement of the individual and of the species. Conrad's nature has lost its last residuum of natural theology, unless the teleology of entropy can be considered its bleak and material shadow.

Conrad, in any case, exposes in Darwin's theory the elements of its own subversion, particularly in its materialist insistence. Darwin, in Conradian moments, also distrusted human ideals of moral and spiritual perfection. He too saw into the heart of that instinctual animal darkness. Near the end of his life, he wrote to William Graham about that writer's *Creed of Science*, which attempted to turn the materialist argument for the uniformity of nature back toward the divine goodness and authority that science had seemed to replace. It is as though Graham is anticipating and refuting the direction of Conrad's imagination: "The mighty machinery of the universe," as of Conrad's "knitting machine," which knits us in and knits us out, "vast and complicated as it is, permits man, through the knowledge of it, which Science gives, to turn to profit the very invariability of the works of the machinery, the rigid uniformity of the laws, which at first sight seem to crush him under the weight of helpless necessity. By his knowledge of the uniform behaviour of Nature, he regains the practical freedom which the universal reign of invariable law seems at first to take away."[26] Darwin rejects Graham's teleology, which reads late-century science back into the mode of natural theology, although he approves, in a vague way, of Graham's repudiation of chance. But as I have suggested, that tenacious aftershadow of natural theology persists. Darwin wants to believe that the universe is not the result of chance, although he is never quite convincing in his attempts. But what is most striking about Darwin's response to Graham is that he almost subverts entirely his whole intellectual enterprise, and reverts to that radical distrust of reason, that feeling of its anomalousness in nature, which I registered at the end of the last chapter. Once more, near the very end of his life, he reminds himself of the possible implication of his monkey theory. Darwin, the intellectual *malgré lui*, expresses with typical Darwinian tentativeness the possibility that our origins belie all attempts to reach toward an ideal.

At the point when he is ready to urge that the universe is not the result of chance, Darwin exclaims: "But then with me the horrid doubt always arises whether the convictions of man's mind, which has been developed from the mind of the lower animals, are of any value or at all trustworthy. Would any one trust in the convictions of a monkey's mind, if there are any convictions in such a mind?"[27] Darwin arrives unwillingly at the impasse that Conrad explores. It is a revolutionary conception, which uses his own theory against his own theory. If we are as the theory describes then we cannot trust the theory. Conrad couldn't have thought of a tighter bind. *Under Western Eyes* expresses both the vision of the Darwin who theorized and the vision of the Darwin who denied the theory. Against the hopelessness of the contradiction and of the irrational violence latent in it, Conrad (really like Darwin) constructed a narrative that confirms the power of language by demonstrating its failure to correspond to "reality." He affirms the conventions of Western thought while he demonstrates their arbitrariness, invoking them as the only possible resistance to a world beyond language to which everything human is alien.

The Darwinian imagination persists in narrative, but its close connection with gradualism, direction, meaning has been permanently severed. His way of seeing has imprinted itself permanently on the imagination of the West, but his presence is so diffused and various, so much part of the Freudian mythology, of the deconstructive turn, of the largest movements of mind in the twentieth century that efforts to trace it further become futile. Nevertheless, our way of telling stories, of creating meaning, of distrusting both phenomena and language owe much to Darwin and the writers who absorbed, extended, and reacted to his imagination.

Notes

Index

Notes

1. Darwin among the Novelists

1. Gillian Beer, *Darwin's Plots: Evolutionary Narrative in Darwin, George Eliot, and Nineteenth-Century Fiction* (London: Routledge and Kegan Paul, 1983), p. 5. Beer's study is the classic consideration of Darwin as writer and as disseminator of literary myths.

2. Perhaps the best known full-scale general studies are Leo Henkin's 1940 study, *Darwinism in the English Novel, 1860–1910* (repr., New York: Russell and Russell, 1963), and Lionel Stevenson, *Darwin among the Poets* (Chicago: University of Chicago Press, 1932). Henkin is exclusively interested in thematic influence and provides what amounts to an extended catalogue of references to Darwinism ideas among the novelists. Stevenson's book is very different in objective and argument from mine. See also A. E. Jones, *Darwinism and Its Relationship to Realism and Naturalism in American Fiction* (Madison, N.J.: Drew University Press, 1950); G. Roppen, *Evolution and Poetic Belief: A Study in Some Victorian and Modern Writers* (Oslo: Oslo University Press, 1956).

3. See Sally Shuttleworth, *George Eliot and Nineteenth-Century Science* (Cambridge: Cambridge University Press, 1984).

4. Roger Ebbatson, *The Evolutionary Self: Hardy, Foster, Lawrence* (Sussex: Harvester Press, 1982).

5. Redmond O'Hanlon, *Joseph Conrad and Charles Darwin: The Influence of Scientific Thought on Conrad's Fiction* (Atlantic Highlands, N.J.: Humanities Press, 1984).

6. Beer, *Darwin's Plots,* p. 6.

7. See the introduction to Michel Serres, *Hermes: Literature, Science, Philosophy,* ed. Josué V. Harari and David F. Bell (Baltimore: Johns Hopkins University Press, 1982), p. xix.

8. Katherine Hayles, "Information or Noise? Economy of Explanation in Barthes' *S/Z* and Shannon's Information Theory," in *One Culture: Essays in Science*

and Literature, ed. George Levine (Madison: University of Wisconsin Press, 1987), p. 120.

9. I do not want to claim this as a Foucauldian exercise in "Archaeology." Nevertheless, Foucault's sense of "enunciative regularities" is related to Serres's and my own view that the science of a particular era, in this case, Darwinian science, can be viewed as an intersection of various traditions that form a "discursive practice." What I am concerned with here is tracing some of these regularities across ostensibly diverse discourses, narrative and scientific, finding the areas of mutuality, and, less overtly, intimating their dependence on other related discourses, particularly the discourses of social and political power. See Michel Foucault, *The Archaeology of Knowledge and the Discourse on Language,* trans. A. M. Sheridan Smith (New York: Harper and Row, 1972), pp. 143–148. See also Alan Sheridan, *Foucault: The Will to Truth* (London: Tavistock Publications, 1980), pp. 103–110.

10. Among the many volumes of illustrated microscope studies, Philip Gosse's (1859) was a representative success. The preface promises "to open the path to the myriad wonders of creation which, altogether unseen by the unassisted eye, are made cognisable to sight by the aid of the microscope." The author adopts "a colloquial and familiar style" in "a series of imaginary *conversaziones,* or microscopical *soirées*" in which he acts "as the provider of scientific entertainment and instruction" (*Evenings at the Microscope; or, Researches among the Minuter Organisms and Forms of Animal Life,* 2nd ed., London: SPCK, 1877, pp. ii, iv).

11. Beer, *Darwin's Plots,* p. 32.

12. Stephen G. Brush, *The Temperature of History* (New York: Burt Franklin and Co., 1978), pp. 1–2.

13. For a brief survey of some of the major arguments, see George Levine, "Literary Science—Scientific Literature," *Raritan Review,* 6 (Winter 1987): 24–41. This essay, in somewhat different form, serves as an introduction to *One Culture: Essays in Science and Literature.*

14. The question of changes in science is basic to all the arguments among philosophers of science about the rationality, or nonrationality of science. Change implies the possibility of previous error, and the movement from one theory to another does not always seem adequately explained by internal factors. The sociology of science builds on the assumption that internalist explanation is inadequate to account for change (or for the failure to change when, from the perspective of modern science, change was in order). For an excellent account of the nature of the problem and a proffered anti-Kuhnian but modified rationalist solution, see W. H. Newton-Smith, *The Rationality of Science* (London: Routledge and Kegan Paul, 1981).

15. Gerald Holton, *Thematic Origins of Scientific Thought: Kepler to Einstein* (Cambridge, Mass.: Harvard University Press, 1973), pp. 23, 24.

16. Talking of the work of sociologists of knowledge, Newton-Smith accepts

the possibility that "social factors having nothing to do with science can prompt members of the scientific community to reassess their theoretical commitments." But, he argues, it is still necessary to explain why the change "issued in the particular paradigm it did and not another one" (*The Rationality of Science,* pp. 263–264).

17. See Barry Barnes, *Scientific Knowledge and Sociological Theory* (London: Routledge and Kegan Paul, 1974), p. 19.

18. See Stephen Jay Gould, *Time's Arrow, Time's Cycle: Myth and Metaphor in the Discovery of Geological Time* (Cambridge, Mass.: Harvard University Press, 1987), pp. 115–137.

19. Niles Eldredge and Ian Tattersall, *The Myths of Human Evolution* (New York: Columbia University Press, 1982), p. 3.

20. The now classic literary and historical versions of this are in Foucault, *The Archaeology of Knowledge and the Discourse on Language,* and Edward Said, *Beginnings: Intention and Method* (New York: Basic Books, 1975). According to Foucault, "if it is true that these discursive, discontinuous series have their regularity, within certain limits, it is clearly no longer possible to establish mechanically causal links or an ideal necessity among their constitutive elements. We must accept the introduction of chance as a category in the production of events" (p. 231). Said discusses Foucault's attempt to "install chance, discontinuity, and materiality . . . as forces operating in discourse" (pp. 301ff.).

21. Said, *Beginnings,* p. 282.

22. See Walter Cannon, "The Whewell-Darwin Controversy," *Journal of the Geological Society of London,* 132 (1976): 377–384, for a suggestive account of Darwin's rejection of Lyell's anti-Lamarckian argument in *The Principles of Geology;* Cannon's "The Uniformitarian-Catastrophist Debate," *Isis,* 51 (1960): "Lyell's Uniformatarianism was an anti-evolutionary creed, postulating repetition rather than cumulative development as the net result of eons of geological time" (p. 39); and Cannon's "The Problems of Miracles in the 1830s," *Victorian Studies,* 4 (September 1960): 6–32. The literature on this subject has been expanding richly, most recently with Gould's *Time's Arrow, Time's Cycle.* Martin J. S. Rudwick, in *The Great Devonian Controversy: The Shaping of Scientific Knowledge among Gentlemanly Specialists* (Chicago: University of Chicago Press, 1985), analyzes in detail a major early nineteenth-century scientific controversy and demonstrates how much of the development of the argument was related to social and biographical forces. Yet Rudwick argues that "the outcome was indeed constrained in unexpected ways by the evidence" (p. 456). See also Rudwick's *The Meaning of Fossils: Episodes in the History of Palaeontology* (London: MacDonald; New York: American Elsevier, 1972).

23. Stephen Jay Gould, *The Panda's Thumb* (New York: W. W. Norton, 1980), pp. 184–185.

24. James R. Moore, "1859 and All That: Remaking the Story of Evolution

and Religion," in *Charles Darwin, 1809–1882: A Centennial Commemorative* (Wellington, N.Z.: Nova Pacifica, 1982), pp. 169–170.

25. Stephen Toulmin and June Goodfield, among others, also claim that "in social and political theory, the authority of Darwin was invoked only to lend support to positions already adopted for other reasons" (*The Discovery of Time,* Chicago: University of Chicago Press, 1982, p. 238).

26. For innovative and stimulating discussion of the disruptive Darwin, whose "biocentrism" entailed radical subversion not only of anthropocentrism, but also of teleological thought in any of its forms, see Margot Norris, *Beasts of the Modern Imagination: Darwin, Nietzsche, Kafka, Ernst, and Lawrence* (Baltimore: Johns Hopkins University Press, 1985).

27. See Peter Morton, *The Vital Science: Biology and the Literary Imagination* (London: George Allen and Unwin, 1984); Morse Peckham, "Darwinism and Darwinisticism," *Victorian Studies,* 3 (September 1959): 19.

28. The Duke of Argyll, *The Reign of Law* (London: Strahan and Co., 1866; repr., 1868), p. 29.

29. John C. Greene, *Darwin and the Modern World View* (Baton Rouge: Louisiana State University Press, 1961), p. 96.

30. Moore argues that "Darwinism was 'social' from the start," and he protests what he considers the bourgeois enterprise of trying to separate Darwin from social Darwinism, which is generally taken as a term of opprobrium. Rather, he investigates the way the phrase is the product of a professional discourse that pretends to divorce science and ideology ("Socializing Darwinism: Historiography and the Fortunes of a Phrase," in *Science as Politics,* ed. Lee Levidow, London: Free Association Books, 1986, pp. 38–80). See also Robert Young's polemical essay "Darwinism *Is* Social," in *The Darwinian Heritage,* ed. David Kohn (Princeton: Princeton University Press, 1986), pp. 609–638. Young convincingly shows the intimate connection between Darwin's biological and his social thought, and would probably not accept my attempt not to see social Darwinism in the *text* of the *Origin:* "Once it is granted that natural and theological conceptions are, in significant ways, projections of social ones, then important aspects of all of the Darwinian debate are social ones, and the distinction between Darwinism and Social Darwinism is one of level and scope, not of what is social and what is asocial" (p. 610).

31. In their attack on sociobiological arguments, R. C. Lewontin, Steven Rose, and Leon J. Kamin discuss its reductionist tendencies, which were manifest in some of the most interesting Victorian considerations of human behavior: "Reductionists try to explain the properties of complex wholes—molecules, say, or societies—in terms of the units of which those molecules or societies are composed" (*Not in Our Genes,* New York: Pantheon, 1984, p. 5). Victorian materialists and modern sociobiologists might fairly be called reductionists on this definition.

32. To keep this from becoming two books, it is necessary for me to refer to an earlier book of mine, *The Realistic Imagination: English Fiction from Frankenstein to Lady Chatterley* (Chicago: University of Chicago Press, 1981), in which I try to describe the Victorian notion of "realism" in fiction. Self-conscious about the difficulties of representation and fully aware of the medium that at once represented and distanced the real, the Victorians were nevertheless unwilling to sever fictional discourse from the normal discourse of description that offered itself as nonfiction. The realism I discuss there is closely related to the Darwinian vision that is the subject of this book. My point is that the two overlap and infuse each other and that seeing them in relation helps illuminate the whole realist project.

33. Peter Bowler, *Evolution: The History of an Idea* (Berkeley: University of California Press, 1984), p. 160.

34. But in her famous essay "The Natural History of German Life," an important theoretical locus for these ideas, George Eliot is still looking to scientific detachment to provide moral sympathy by making us aware of the other. Here connections among art, social science, and evolutionary science are formulated, and here the ideal of a nontheoretical inductive approach to the phenomena of social life is adumbrated in ways that obviously influenced her novel writing: "The thing for mankind to know is, not what are the motives and influences which the moralist thinks *ought* to act on the labourer or the artisan, but what are the motives and influences which *do* act on him" (*Essays of George Eliot,* ed. Thomas Pinney, New York: Columbia University Press, 1963, p. 271). The tradition of "natural history" in Riehl's sense is part of the ethos that fed into Darwinian theory and assimilated it quickly to anthropology.

35. Michael Ghiselin, *The Triumph of the Darwinian Method* (Chicago: University of Chicago Press, 1969; repr., 1984), p. 4.

36. Charles Darwin, *On the Origin of Species by Means of Natural Selection; or, The Preservation of Favoured Races in the Struggle for Life,* ed. J. W. Burrow (Harmondsworth: Penguin Books, 1959); Burrow uses the text of the first (1859) edition. Henceforth cited in the text as *Origin;* unless otherwise noted, page numbers refer to this edition.

37. J. Hillis Miller, in *The Forms of Victorian Fiction* (South Bend, Ind.: University of Notre Dame, 1968), talks about the way "intersubjectivity" displaces divine ordering in the Victorian novel. Psychology is the secular-scientific alternative to religious explanation of the human social and moral condition.

2. *Natural Theology*

1. Cannon, "The Problems of Miracles," p. 11.

2. "Many of the most prominent scientific geologists were men of acknowledged personal piety . . . theological liberals who were well aware of the critical hermeneutics being developed by German scholars . . . They had not hitched their

religious beliefs onto literalistic modes of biblical exegesis; indeed they were among the most vehement critics of the scriptural geologists" (Rudwick, *The Great Devonian Controversy,* p. 44).

3. John Herschel, *A Preliminary Discourse on the Study of Natural Philosophy* (1830; repr., Philadelphia: Carey, Lea and Blanchard, 1835), p. 6.

4. William Buckland, *Geology and Mineralogy Considered with Reference to Natural Theology* (Philadelphia: Carey, Lea and Blanchard, 1837), p. 19.

5. See, e.g., Walter Cannon, "The Whewell-Darwin Controversy," *Journal of the Geological Society,* 132 (1976): 377–384; Michael Ruse, "Darwin's Debt to Philosophy: An Examination of the Influence of the Philosophical Ideas of John F. W. Herschel and William Whewell on the Development of Charles Darwin's Theory of Evolution," *Studies in the History and Philosophy of Science,* 6 (1975): 159–181; Paul Thagard, "Discussion: Darwin and Whewell," *Studies in the History and Philosophy of Science,* 8 (1977): 353–356; see also Dov Ospovat, *The Development of Darwin's Theory: Natural History, Natural Theology, and Natural Selection, 1838–1859* (Cambridge: Cambridge University Press, 1981), esp. pp. 148–149.

6. See Peter Vorzimmer, "The Darwin Reading Notebooks, 1838–1860," *Journal of the History of Biology,* 10 (1977): 107–152. Darwin noted on February 24, 1840, that he "well skimmed (for second time) Whewell's Bridgewater Treatise" (p. 123).

7. "Extracts from B-C-D-E Transmutation Notebooks," ed. Paul H. Barrett, in Howard Gruber, *Darwin on Man* (New York: E. P. Dutton, 1974), p. 455.

8. When Whewell read Darwin's *Origin of Species,* he wrote to Darwin, "I cannot, yet at least, become a convert." According to Francis Darwin, Whewell was so little converted that for some years he refused to allow a copy of Darwin's book into the library of Trinity College. See Francis Darwin, ed., *The Life and Letters of Charles Darwin,* 3 vols. (London: John Murray, 1887), II, 261. Henceforth cited in the text as *Life and Letters.*

9. Indeed, part of Whewell's enterprise was to challenge Lyell's antihistorical uniformitarianism, which allowed for endless expanses of time by sacrificing any sense of development or real change. Whewell, that is, tried to put back into nature the history that Christianity required and that Darwin would himself later have to add to Lyell's theory in order to create his own (see Cannon, "The Problems of Miracles," p. 8).

10. Ibid., p. 32. For an important discussion of Whewell's position here and, more extensively, of Mill's counterarguments and their ultimate place in the evolutionary debate, see George W. Stocking, Jr., *Victorian Anthropology* (New York: Free Press, 1987), pp. 37–41.

11. See Ospovat, *The Development of Darwin's Theory,* pp. 148–149, and Thagard, "Discussion."

12. See James R. Moore, *The Post-Darwinian Controversies: A Study of the*

Protestant Struggle to Come to Terms with Darwin in Great Britain and America, 1870–1900 (Cambridge: Cambridge University Press, 1979), for an extensive consideration of the compatibility of Darwinian theory with "orthodox theology."

13. I am indebted for much information about the Bridgewater Treatises to a paper by John Robson, "The Finger and Fiat of God: The Bridgewater Treatises," delivered at a conference on the Victorian crisis of faith at the University of Toronto, in November 1984. The eight (nine with Babbage's unofficial one) treatises were described as "On the Power, Wisdom, and Goodness of God, as manifested in the Creation; illustrating such work by all reasonable arguments, as for instance the variety and formation of God's creatures in the animal, vegetable, and mineral kingdoms, the effect of digestion, and thereby of conversion; the construction of the hand of man, and an infinite variety of other arguments; as also by discoveries ancient and modern, in arts, sciences, and the whole extent of literature." This was part of the "prefactory notice" published with each of the treatises.

14. William Whewell, *On Astronomy and General Physics Considered with Reference to Natural Theology* (London: H. G. Bohn, 1852), p. 305. Henceforth cited in the text as *Astronomy*.

15. Charles Darwin, *The Descent of Man and Selection in Relation to Sex* (Princeton: Princeton University Press, 1981), I, 153. Henceforth cited in the text as *Descent*. There is some evidence in the first edition of the *Origin* that Darwin was being a bit disingenuous here. He does, in chapter 6, take into account aberrant and apparently useless structures in animals. At the same time, he does write that in the transition from one form to another, as in the case of flying squirrels, "each grade of structure has been useful to its possessor" (*Origin*, p. 213).

16. See Ospovat, *The Development of Darwin's Theory*, p. 148. A large literature has been developing that attempts to show the intimacy of the connection between Darwin's arguments and natural theology. Ospovat's book is the most extensive and impressive example, but see also John Hedley Brooke, "The Relations between Darwin's Science and His Religion," in *Darwinism and Divinity*, ed. John Durant (Oxford: Basil Blackwell, 1985), pp. 40–75. Brooke demonstrates in some detail the nature of the "structural continuity" between Paley and Darwin's arguments. See also A. Hunter Dupree, "Christianity and the Scientific Community in the Age of Darwin," in *God and Nature: Historical Essays on the Encounter between Christianity and Religion*, ed. David C. Lindberg and Ronald L. Numbers (Berkeley: University of California Press, 1986), pp. 351–367.

17. See *Life and Letters*, I, 314–315: "I see a bird which I want for food, take my gun and kill it, I do this *designedly*. An innocent and good man stands under a tree and is killed by a flash of lightning. Do you believe (and I really should like to hear) that God *designedly* killed this man? Many or most persons do believe this; I can't and don't. If you believe so, do you believe that when a swallow snaps up a gnat that God designed that that particular swallow should snap up that

particular gnat at that particular instant? I believe that the man and the gnat are in the same predicament. If the death of neither man nor gnat are designed, I see no good reason to believe their *first* birth or production should be necessarily designed."

18. Charles Darwin, *The Variations of Animals and Plants under Domestication,* 2 vols. (New York, 1900), II, p. 415.

19. Edward Manier, *The Young Darwin and His Cultural Circle* (Dordrecht: Reidel, 1978), p. 54.

20. T. H. Huxley, in "On the Reception of the 'Origin of Species,' " interpolated into *Life and Letters,* argues violently against the idea that Darwin believed in "Chance." "It is not a little wonderful that such an accusation as this should be brought against a writer who has, over and over again, warned his readers that when he uses the word 'spontaneous,' he merely means that he is ignorant of the cause of that which is so termed; and whose whole theory crumbles to pieces if the uniformity and regularity of natural causation for illimitable past ages is denied" (II, 199). This is too strong, but Darwin himself remained uneasy with the fact that he could not explain the source of variations. The last two pages of *Variations in Animals and Plants under Domestication* explain why the problem does not affect his theory. In a letter to Lord Farrer, Darwin wrote, "if we consider the whole universe, the mind refuses to look at it as the outcome of chance—that is, without design or purpose." See Francis Darwin, ed., *More Letters of Charles Darwin,* 2 vols. (New York: Appleton and Co., 1903), II, 395. Henceforth cited in the text as *More Letters.*

21. J. H. Newman, *Apologia pro vita sua* (New York: W. W. Norton, 1969), p. 6.

22. William Whewell, "On the Fundamental Antithesis of Philosophy," book I, chap. 2, of *Philosophy of the Inductive Sciences,* in which these distinctions are elaborated, is reprinted in *Selected Writings on the History of Science,* ed. Yehuda Elkana (Chicago: University of Chicago Press, 1984), pp. 138–155.

23. Whewell, "Fundamental Antithesis," p. 341.

24. John Holloway, *The Victorian Sage: Studies in Argument* (New York: W. W. Norton, 1965).

25. Ernst Mayr, *The Growth of Biological Thought: Diversity, Evolution, and Inheritance* (Cambridge, Mass.: Harvard University Press, 1982), esp. pp. 32–36.

26. W. B. Carpenter, review of *History of the Inductive Sciences, British and Foreign Medical Review,* 5 (1838): 325.

27. Elkana, Introduction to *Selected Writings in the History of Science,* pp. xvi–xvii.

28. See, for an important discussion of the extension of natural law to the human, Robert Young, "Darwin's Metaphor: Does Nature Select?" *Monist,* 55 (1971): 441–503; slightly revised and reprinted in his *Darwin's Metaphor: Nature's Place in Victorian Culture* (Cambridge: Cambridge University Press, 1985).

29. Herschel, *Preliminary Discourse,* p. 11.

30. William Paley, *Natural Theology; or, Evidences of the Existence and Attributes of the Deity, Collected from the Appearances of Nature* (repr., Houston: St. Thomas Press, 1972), p. 345.

31. Beer, *Darwin's Plots,* p. 48.

32. Ibid.

33. Paley, *Natural Theology,* p. 365.

34. See for a popular example P. B. Medawar and J. S. Medawar, *Aristotle to Zoos: A Philosophical Dictionary of Biology* (Cambridge, Mass.: Harvard University Press, 1983), p. 167. Even Burrow's introduction to the *Origin* concludes with the example of the melanistic moth.

35. Charles Darwin, *On the Origin of Species,* 6th ed., intro. Julian Huxley (New York: New American Library, 1958), p. 181.

36. Dov Ospovat, "Perfect Adaptation and Teleological Explanation: Approaches to the Problem of Life in the Mid-Nineteenth Century," in *Studies in the History of Biology,* ed. William Coleman and Camille Limoges (Baltimore: Johns Hopkins University Press, 1978), pp. 33–37.

37. Ospovat, *The Development of Darwin's Theory,* p. 378.

38. Robert Chambers, *Vestiges of the Natural History of Creation,* ed. Gavin de Beer (New York: Humanities Press, 1969), p. 157.

39. Paley, *Natural Theology,* pp. 46, 52.

40. Thomas McFarland, quoted in Ospovat, *The Development of Darwin's Theory,* p. 38.

41. Charles Coulton Gillispie, *Genesis and Geology: The Impact of Scientific Discoveries upon Religious Beliefs in the Decades before Darwin* (New York: Harper and Row, 1959), p. 39. Gillispie also shows that when reviewing Lyell's *Principles of Geology,* Whewell refused to accept uniformitarian explanation and insisted on the necessity of great leaps, of the sort that deny the naturalistic historicity essential to Darwinian theory and characteristic of the developing historicism of the nineteenth century. True to his natural-theological inclinations, Whewell wrote: "We conceive it undeniable . . . that we see in the transition from an earth peopled by one set of animals, to the same earth swarming with entirely new forms of organic life, a distinct manifestation of creative power, transcending the known laws of nature: and it appears to us, that geology has thus lighted a new lamp along the path of natural theology" (*The British Critic,* 9, 1831: 194; quoted in Gillispie, *Genesis and Geology,* p. 146).

42. Gillispie, *Genesis and Geology,* p. 39.

43. Whewell, "Fundamental Antithesis," p. 358.

3. *Mansfield Park*

1. Charles Coulton Gillispie, *The Edge of Objectivity: An Essay in the History of Scientific Ideas* (Princeton: Princeton University Press, 1960), p. 284.

2. Marilyn Butler, *Jane Austen and the War of Ideas* (Oxford: Clarendon Press, 1975), provides a very impressive case for Jane Austen, the conservative, by placing her in the context of these contemporary ideas. I am much indebted to her reading of Austen.

3. See Toulmin and Goodfield, *The Discovery of Time:* "Dalton was now in a position to do something which Lavoisier had hesitated to do: he could equate the fundamental material units of the chemical elements with Newton's 'hard, massy, impenetrable particles,' and identify the molecules of compound substance with the 'particles of the first composition' formed by the simple combination of these elementary particles" (p. 231). Cf. also Gillespie, *The Edge of Objectivity:* It was Dalton "who brought Newton and Lavoisier face to face" (p. 258).

4. See Toulmin and Goodfield, *The Discovery of Time,* p. 218; cf. Alfred North Whitehead, *Science in the Modern World* (New York: Signet, 1948): "No material is lost or gained in any chemical transformations" (pp. 60–61).

5. Jane Austen, *Mansfield Park* (Harmondsworth: Penguin, 1966), p. 448. Page numbers in the text refer to this edition.

6. Richard F. Patteson, "Truth, Certitude, and Stability in Jane Austen's Fiction," *Philological Quarterly,* 60 (1982): 467.

7. Stuart Tave, *Some Words of Jane Austen* (Chicago: University of Chicago Press, 1973), pp. 18, 19.

8. Michel Foucault, *The Order of Things: An Archaeology of the Human Sciences* (New York: Vintage Books, 1970), pp. 128–130.

9. Quoted in Stephen Toulmin and June Goodfield, *The Architecture of Matter* (New York: Harper and Row, 1962), p. 222.

10. See Davydd Greenwood, *The Taming of Evolution: The Persistence of Nonevolutionary Views in the Study of Humans* (Ithaca: Cornell University Press, 1984), p. 52.

11. Foucault, *The Order of Things,* pp. 128–129.

12. See Raymond Williams, *Keywords: A Vocabulary of Culture and Society* (New York: Oxford University Press, 1976), pp. 263–264.

13. Leo Bersani, *A Future for Astyanax* (Boston: Little, Brown, 1976). See the chapter "Realism and the Fear of Desire," pp. 51–89.

14. Greenwood, *The Taming of Evolution,* p. 65.

15. Tanner's introduction to the Penguin edition, which is a version of his essay ("The Quiet Thing: *Mansfield Park*") in *Critical Essays on Jane Austen*, ed. B. C. Southam (London: Routledge and Kegan Paul, 1968), emphasizes both Fanny's distance as observer and her infallibility.

16. Tave, *Some Words,* p. 195.

17. Julia Prewitt Brown, *Jane Austen's Novels: Social Change and Literary Form* (Cambridge, Mass.: Harvard University Press, 1979), p. 100.

18. A roll call of the arguments for Henry and Mary would be a very long one. Perhaps the most famous is Marvin Mudrick, *Jane Austen: Irony as Defense and*

Discovery (Princeton: Princeton University Press, 1952). Lionel Trilling's classic, "Mansfield Park," reprinted in *Jane Austen: A Collection of Essays,* ed. Ian Watt (Englewood Cliffs, N.J.: Prentice-Hall, 1963), pp. 124–140, concedes the superior charm of Mary and Henry, and essentially blames modern culture for the judgment that he shares. He sees Mary and Henry as the "first" representations of the "specifically modern personality." The modernity of the two, for my purposes, can be associated with the kind of threat to the stable Mansfield ideal they represent. Although I have strong reservations about Trilling's argument and believe his claims to the priority and uniqueness of the two figures are overstated, he seems to me correct in emphasizing the complications they manifest: "Never before has the moral life been shown as [Austen] shows it to be, never before had it been conceived to be so complex and difficult and exhausting . . . She is the first novelist to represent society, the general culture, as playing a part in the moral life, generating the concepts of 'sincerity' and 'vulgarity' which no earlier time would have understood the meaning of, and which for us are so subtle that they defy definition" (p. 138). Mary and Henry shadow forth the ecologically complex, taxonomically difficult structures characteristic of the Darwinian world and of Victorian realism.

19. Choderlos Laclos, *Dangerous Acquaintances,* trans. Richard Aldington (New York: New Directions Books, 1957), p. 29.

20. Laclos, *Dangerous Acquaintances,* p. 215.

21. Tony Tanner indicates that "like many Jamesian figures [Fanny] does not fully participate in the world, but as a result she sees things more clearly and accurately than those who do" ("The Quiet Thing," p. 149).

22. D. D. Devlin, *Jane Austen and Education* (London: Macmillan, 1975), p. 91.

23. The conservative implications of Austen's disapproval of improvement are spelled out in Alistair Duckworth, *The Improvement of the Estate* (Baltimore: Johns Hopkins University Press, 1971).

24. The language used to describe the sailor's life captures something of the sort of "passivity" implied as ideal in Mansfield Park; it matters, of course, that the language is filtered through Henry's consciousness, deliberately used by him against his sort of aggressive activity: "The glory of heroism, of usefulness, of exertion, of endurance, made his own habits of selfish indulgence appear in shameful contrast; and he wished he had been a William Price, distinguishing himself and working his way to fortune and consequence with so much self-respect and happy ardour, instead of what he was!" (p. 179). "Usefulness, exertion, endurance" are all qualities attributable to the apparently passive Fanny.

25. Tave, *Some Words,* pp. 203ff.

26. Duckworth, *The Improvement of the Estate* p. 80.

27. See Barbara Hardy, *A Reading of Jane Austen* (New York: New York University Press, 1976), esp. chap. 1, "The Flexible Medium": "In her narrative

voice, in the free indirect style in which she shares commentary with the characters, and in her habit of quick, vivid summary, she moves lightly and unobtrusively from character to group, close-up to distance. Like all the threads that join her private and public worlds, that of her commentary is so fine that its stitches scarcely show" (p. 36).

4. Darwin's Revolution

1. Beer, *Darwin's Plots,* p. 53.

2. For a discussion of how "the *problems* . . . in Darwinism were put to work in the interests of social thought," see Greta Jones, *Social Darwinism and English Thought* (Brighton: Harvester Press, 1980).

3. For discussion of Darwin's development away from natural theology, see Ospovat, *The Development of Darwin's Theory:* "In constructing theories of transmutation in the period 1837–8 . . . Darwin took for granted that adaptation is perfect and that variation is for the purpose of enabling organisms to accommodate to environmental change . . . Until the 1850's he adhered to assumptions that were deeply embedded in the traditional view, and these gave the theory of natural selection . . . the structure of a mechanism of adjustment to change, a means by which the balance of nature is preserved" (p. 3).

4. See John Dewey's famous essay on the subject in *The Influence of Darwin on Philosophy and Other Essays* (New York: Henry Holt, 1910).

5. Frederick Burkhardt et al., eds., *The Correspondence of Charles Darwin,* vol. 1, *1821–1836* (Cambridge: Cambridge University Press, 1985), p. 122.

6. In *Charles Darwin and the Problem of Creation* (Chicago: University of Chicago Press, 1979), Neal C. Gillespie considers Darwin's use of language about the "Creator" and the "creation," arguing that natural-theological assumptions about special creation and design were deeply embedded in the thought of most scientists and that Darwin, in order to affirm a secular, law-driven scientific practice, needed to expose and undermine these assumptions. His hostility to the intervention of religion in science was, Gillespie argues, total. But "theism" was "a part of his total vision up until the end" (p. 144).

7. Edward Manier has argued that "the manner in which he represents his theory was remarkably effective and the *representation* of the theory—metaphors, allegories, and all—was a crucial and absolutely central aspect of Darwin's *scientific* accomplishment (*The Young Darwin,* p. 16).

8. It would be worth exploring how Darwin's style of self-presentation fits into the conditions feminist critics have been noting about Victorian culture: its division between male and female spheres; the way the male conducts the public action while influenced by the feminine values of compassion and self-denial, while remaining tough, competitive, and perhaps ruthless when necessary. Darwin's gentlemanliness had enormous strategic advantages in the public arena.

9. Charles Darwin, *The Foundations of the Origin of Species: Two Essays Written in 1842 and 1844,* ed. Francis Darwin (Cambridge: Cambridge University Press, 1909).

10. Robert Stauffer, ed., *Charles Darwin's Natural Selection, Being the Second Part of His Big Species Book Written from 1856 to 1858* (Cambridge: Cambridge University Press, 1975); the Wallace paper was published with excerpts from Darwin's work on the subject and excerpts from a letter by Darwin to Hooker, and both were sponsored by Lyell and Hooker at a meeting of the Linnaean Society on July 1, 1858. This established Darwin's priority. These documents were all reprinted in *Science before Darwin,* ed. Howard Mumford Jones and I. Bernard Cohen (London: Andre Deutsch, 1963), pp. 337–365.

11. T. H. Huxley, *Darwiniana: Collected Essays,* 9 vols. (London: Macmillan and Co., 1893), II, 79.

12. Herschel, *Preliminary Discourse,* p. 119.

13. There were only two epigraphs to the first edition. The Butler epigraph appeared in later editions.

14. J. B. Mozley, "The Argument of Design," *Quarterly Review,* 127 (1869): 173.

15. Young, "Darwin's Metaphor," p. 82.

16. Burkhardt et al., eds., *Correspondence,* I, 500. David Kohn, who is now editing Darwin's journals, tells me that there is considerable evidence that he and Herschel discussed what Herschel elsewhere called "the mystery of mysteries," that is, the origin of species.

17. Ernst Mayr, *Evolution and the Diversity of Life: Selected Essays* (Cambridge, Mass: Harvard University Press, 1976), p. 33. I assume that Mayr uses the word "perfection" loosely here, since it is certainly not Darwinian, nor is it, strictly speaking, consistent with Mayr's own view of evolution.

18. Manier, *The Young Darwin,* p. 121.

19. Gillispie, *The Problem of Creation,* p. 56.

20. Darwin, *Origin,* 6th ed., p. 88.

21. W. B. Carpenter, *Principles of General and Comparative Physiology Intended as an Introduction to the Study of Human Physiology and as a Guide to the Philosophical Pursuit of Natural History* (London: John Churchill, 1839), p. 9.

22. Herschel, *Preliminary Discourse,* p. 43.

23. T. H. Huxley, *Science and Christian Tradition,* vol. 5 of *Collected Essays* (London: Macmillan and Co., 1893), p. 77.

24. G. H. Lewes, *Comte's Philosophy of the Sciences* (London: H. Bohn, 1853), p. 52.

25. John Stuart Mill, *A System of Logic, Ratiocinative and Inductive, Being a Connected View of the Principles of Evidence and the Methods of Scientific Investigation,* 2 vols., 4th ed. (London: John Parker, 1856), I, 348.

26. One of the more interesting contests of the period was between Max Müller

and Darwin over the question of the origins of language. Darwin, as he wrote to Müller in 1873, believed that "he who is fully convinced, as I am, that man is descended from some lower animal, is almost forced to believe *a priori* that articulate language had been developed from inarticulate cries" (*More Letters,* II, 45); Müller believed that it was impossible to account for language as some accidental development from unintelligent noises, and thought of language as originating in a pure, unmediated connection with the world and as having fallen, through metaphorical distortions, ever since. See Müller, "My Reply to Mr. Darwin," *Contemporary Review* (January 1875): 305. The Darwin replied to is in fact George Darwin.

27. John Herschel, *Physical Geography* (Edinburg: Adam and Charles Black, 1862), p. 12.

28. See the Duke of Argyll, *The Reign of Law,* pp. 28–29.

29. Manier, *The Young Darwin,* p. 117.

30. C. H. Waddington, "How Much Is Evolution Affected by Chance and Necessity?" in *Beyond Chance and Necessity,* ed. John Lewis (London: Teilhard Center for the Future of Man, 1974), p. 94.

31. Jones, *Social Darwinism,* p. 35.

32. Gregory Bateson, *Mind and Nature* (New York: E. P. Dutton, 1979), p. 147.

33. Beer, *Darwin's Plots,* pp. 26–27.

34. Letter from John Herschel to Charles Lyell, February 20, 1836; printed as Appendix to Charles Babbage, *The Ninth Bridgewater Treatise: A Fragment* (Philadelphia: Lea and Blanchard, 1841), p. 212.

35. Stephen Jay Gould and Niles Eldredge, "Punctuated Equilibrium: The Tempo and the Mode of Evolution Considered," *Paleobiology,* 3 (1977): 115–151. For further discussion of this question, see Eldredge and Tattersall, *The Myths of Evolution,* and Gould's *Time's Arrow,* which traces in brilliant detail Hutton and Lyell's nonempirical development of uniformitarian theory. The view of Gould, Eldredge, and Tattersall entails, among other things, a recognition of species as identifiable units, not subject to change in the way Darwin claimed. The epistemology to be deduced from this theory is radically different from that of Darwin, and in effect requires rejecting the idea that "species" is merely an arbitrary name for a flux of characteristics in time.

36. John Beatty, "What's in a Word? Coming to Terms in the Darwinian Revolution," *Journal of the History of Biology,* 15 (Summer 1982): 235. In a later essay Beatty modifies his position, arguing that Darwin distinguished between "what his fellow naturalists *called* 'species' and the non-evolutionary beliefs in terms of which they *defined* 'species' " ("Speaking of Species: Darwin's Strategies," in *The Darwinian Heritage,* p. 266). Ernst Mayr has several valuable essays on the question of nominalism and the reality of species in *Evolution and the Diversity of Life* (Cambridge, Mass.: Belknap Press of Harvard University Press, 1976). See

especially the sections "Speciation" and "Species Concepts and Definitions": "One of the minor tragedies in the history of biology has been the assumption during the one hundred fifty years after Linnaeus that constancy and clear definition of species are strictly correlated and that one must *either* believe in evolution (the 'inconstancy' of species) and then have to deny the existence of species except as purely subjective, arbitrary figments of the imagination, *or,* as most early naturalists have done, believe in the sharp delimitation of species but think that this necessitates denying evolution" (p. 494). Stephen Jay Gould develops Mayr's argument for the "objective taxonomic" reality of species in *The Panda's Thumb,* pp. 204–213. See also note 25, above.

37. F. Max Müller, *Three Introductory Lectures on the Science of Thought* (London: Kegan, Paul, Trench, Trübner, 1909), pp. 77–78.

38. This passage does not appear in the first edition. I quote from a reprint of the sixth edition (New York: New American Library, 1958), p. 88.

39. A. R. Wallace, "On the Tendency of Varieties to Depart Indefinitely from the Original Type," reprinted in *Science before Darwin,* p. 362.

40. Charles Darwin and Alfred Russel Wallace, *Evolution by Natural Selection,* ed. Gavin de Beer (Cambridge: Cambridge University Press, 1958), p. 114.

41. Darwin, *Origin,* 6th ed., p. 90.

42. Ghiselin, in *The Triumph of the Darwinian Method,* argues that "unless one understands . . . that Darwin applied, rigorously and consistently, the modern, hypothetico-deductive method . . . his accomplishments cannot be appreciated. His entire scientific accomplishment must be attributed not to the collection of facts, but to the development of theory" (p. 4). Ghiselin is polemically rationalist in his arguments, insisting that Darwin's theory was worked through in a "clear-cut series of rational operations" (p. 77). He attempts to squeeze the intuition and nonrational elements from Darwin's thought, and while something of Darwin's flexibility and imaginative power is lost in the analysis, Ghiselin is very good in showing how rigorously Darwin reasoned through his arguments, how much they depended not on crude empiricism but on tight reasoning, in the "one long argument."

43. See the balanced discussion of this and related issues in John Greene, *Science, Ideology, and World View* (Berkeley: University of California Press, 1981), chap. 5, "Darwin as a Social Evolutionist."

44. See Silvan S. Schweber, "Darwin and the Political Economists: Divergence of Character," *Journal of the History of Biology,* 13 (Fall 1980): 195–289.

45. See Holloway, *The Victorian Sage:* "All of these authors insist on how acquiring wisdom is somehow an opening of the eyes, making us see in our experience what we failed to see before" (p. 9).

46. A. Dwight Culler, "The Darwinian Revolution and Literary Form," in *The Art of Victorian Prose,* ed. George Levine and William Madden (New York: Oxford University Press, 1968), p. 227.

47. Beer, *Darwin's Plots,* pp. 80–81.

48. See John Hedley Brooke, "The Relations between Darwin's Science and His Religion," p. 45. Owen's view, which was not incompatible with certain kinds of evolutionary thought, was the most powerful among serious biologists of the time and provided a rational explanation for similarities of structure that Darwin could account for only by construction of a narrative in time. That is, the similarities in structure in all living things are, according to Owen, the result of a divine, archetypal idea. It is therefore not aberrant that what we now call vestigial organs are uselessly present in organisms. Our "tails" are rudimentary now, but they reflect an archetypal design. Similarly, Brooke points out one of Darwin's favorite refutations of natural theology, "mammae in men."

49. Beer, *Darwin's Plots,* p. 42.

50. In *Time's Arrow,* Gould argues that Lyell's resistance to Darwin's theory was a resistance to the implication of progress in evolution. But the *Origin* does not justify Lyell's fears. Gould, too, says that Darwin's theory does *not* imply progress. Rather, he "maintained a very ambiguous attitude" toward it, "accepting it provisionally as part of the fossil record, but denying that the theory of natural selection—a statement about adaptation to shifting local environments—required organic advance" (p. 170).

51. For a discussion of this aspect of realism, the tendency to remind readers that the novel is inadequately mimetic because it cannot render the full complexity or multiplicity of reality, see my *The Realistic Imagination,* esp. pp. 154ff.

52. Paley, *Natural Theology,* pp. 26–27.

53. David Bromwich, "Reflections on the Word *Genius,*" *New Literary History,* 17 (Autumn 1985): 158.

54. Paley, *Natural Theology,* p. 21.

55. Ibid., p. 13.

56. Darwin, *Origin,* 6th ed., p. 171.

57. Max Müller, "Lectures on Mr. Darwin's Philosophy of Language," *Fraser's Magazine,* 7 (May 1873): 525–541; 7 (June 1873): 659–678; quotation on p. 678.

5. Dickens and Darwin

1. Charles Dickens, *Bleak House,* ed. George Ford and Sylvére Monod (New York: W. W. Norton, 1977), p. 197. Page numbers in text refer to this edition.

2. Stephen Jay Gould, in "Flaws in a Victorian Veil" (in *The Panda's Thumb*), discusses Louis Agassiz's resistance to Darwin's theory in this context. Agassiz was appalled that blacks might be related to whites and as a consequence he turned unscientifically to the idea of polygeny (we cannot *all* be descended from the same original sources). Gould quotes Agassiz:

I shudder from the consequences. We have already to struggle, in our

progress, against the influence of universal equality, in consequence of the difficulty of preserving the acquisitions of individual eminence, the wealth of refinement and culture growing out of select association. What would be our condition if to these difficulties were added the far more tenacious influences of physical disability. Improvements in our system of education . . . may sooner or later counterbalance the effects of the apathy of the uncultivated and of the rudeness of the lower classes and raise them to a higher standard. But how shall we eradicate the stigma of a lower race when its blood has once been allowed to flow freely into that of our children. (pp. 174–175)

3. Beer, *Darwin's Plots,* p. 8.

4. Charles Darwin, *Metaphysics, Materialism, and the Evolution of Mind: Early Writings of Charles Darwin,* ed. Paul H. Barrett (Chicago: University of Chicago Press, 1974), p. 20.

5. See Gillian Beer, "Darwin's Reading and the Fictions of Development," in *The Darwinian Heritage,* pp. 543–588.

6. Nora Barlow, ed., *The Autobiography of Charles Darwin* (New York: W. W. Norton, 1958), pp. 138–139.

7. See Jonathan Arac, *Commissioned Spirits: The Shaping of Social Motion in Dickens, Carlyle, Melville, and Hawthorne* (New Brunswick, N.J.: Rutgers University Press, 1979), pp. 126, 131. Arac offers an interesting discussion of how scientific language pervades the novel.

8. Ann Wilkinson, "*Bleak House:* From Faraday to Judgment Day," *ELH,* 34 (1967): 225–247. I draw on this excellent essay frequently in my discussion of *Bleak House.*

9. See E. Gaskell, "More about Spontaneous Combustion," *Dickensian,* 69 (1973): 23–35.

10. Alexander Welsh, *The City of Dickens* (Oxford: Oxford University Press, 1971), p. 117.

11. Harvey Sucksmith, *The Narrative Art of Charles Dickens* (Oxford: Oxford University Press, 1970), p. 171.

12. E. T. Cook and Alexander Wedderburn, eds., *The Library Edition of the Works of John Ruskin,* 39 vols. (London: George Allen, 1905), VII, 7.

13. K. J. Fielding, ed., *The Speeches of Charles Dickens* (Oxford: Oxford University Press, 1960), p. 403.

14. Ibid., p. 404.

15. Ibid.

16. Charles Dickens, *The Posthumous Papers of the Pickwick Club,* ed. Robert L. Patten (Harmondsworth: Penguin Books, 1972), pp. 73–74, 647.

17. Charles Dickens, *The Haunted Man,* in *The Christmas Books,* 2 vols. (Harmondsworth: Penguin Books, 1971) II, 322.

18. Lavoisier quoted in Gerald Holton, *Introduction to Concepts and Theories in Physical Science* (Princeton: Princeton University Press, 1985), p. 231.

19. Charles Dickens, "Gamekeeper's Natural History," *All the Year Round* (September 10, 1859): 474.

20. Charles Dickens, "The World of Water," reprinted from *Household Words*, in *Home and Social Philosophy; or, Chapters on Everyday Topics* (New York: G. P. Putnam, 1852), p. 245.

21. See *All the Year Round* (September 17, 1859): 490.

22. Stephen Toulmin and June Goodfield (in *The Discovery of Time*) cite Boswell quoting Johnson: "But, sir, it is as possible that the Ourang-Outang does not speak, as that he speaks. However, I shall not dispute the point. I should have thought it not possible to find a Monboddo; yet he exists" (p. 98).

23. Charles Dickens, *The Life and Adventures of Martin Chuzzlewit* (London: Thomas Nelson, n.d.), p. 7.

24. Charles Dickens, "Our Nearest Relation," *All the Year Round* (May 28, 1859): 114–115.

25. Bowler, *Evolution*, pp. 123–124.

26. Charles Dickens, "Nature's Greatness in Small Things," *Household Words* (November 28, 1857): 513.

27. Cf. Bowler, *Evolution*, p. 222.

28. T. H. Huxley, *Man's Place in Nature* (1863; repr., Ann Arbor: University of Michigan Press, 1959), p. 122.

29. Review of *The Origin of Species*, *All the Year Round* (July 7, 1860): 293, 299.

30. "Species," *All the Year Round* (June 2, 1860): 176.

31. Charles Dickens, *Sketches by "Boz"* (London: Macmillan and Co., 1958), p. 128.

32. Dickens, "Nature's Greatness," p. 511.

33. Cf. John Romano, *Dickens and Reality* (New York: Columbia University Press, 1978): "A single-minded stress on the separation of Dickens' fictive world from our own slights the realistic or representational elements in the novels and their persistent claim, like the claim in Tolstoy, that they are set in the real world" (p. 3). Romano's important claims for Dickens as realist locate a crucial distinction of the realist's enterprise: "The realist puts in doubt the very enterprise of art. As a corollary of the affirmation of the real, realism discredits the precondition of its own existence, that form which confines, distorts, de-actualizes reality in the process of assimilating it" (p. 84). But it is precisely here where the claims for Dickens as a realist might be most usefully assimilated to my argument about natural theology and uniformitarianism. Dickens's commitment to realism is combined with something like a natural-theological faith in the order of experience itself; hence, there is no real sign in Dickens of that characteristic Thackerayan irony at the expense of art itself.

34. Charles Dickens, *Oliver Twist* (Harmondsworth: Penguin Books, 1966), pp. 36, 35.

35. Charles Dickens, *Our Mutual Friend* (Harmondsworth: Penguin Books, 1971), p. 893. Page numbers in the text refer to this edition.

36. George Eliot, *Adam Bede* (Harmondsworth: Penguin Books, 1980), p. 198.
37. Ibid., p. 410.
38. Dickens, *Sketches,* pp. 185, 199.
39. Herschel, *Preliminary Discourse,* p. 14.
40. George Eliot, "The Sad Fortunes of the Reverend Amos Barton," in *Scenes of Clerical Life* (Harmondsworth: Penguin Books, 1973), p. 81.
41. Two recent essays discuss this contradiction in the novel. The form of the novel itself seems to run counter to expressed narrative intent. This problem deserves yet fuller treatment. See Susan R. Cohen, "A History and a Metamorphosis: Continuity and Discontinuity in *Silas Marner,*" *Texas Studies in Literature and Language,* 25 (Fall 1983): 410–446; Donald Hawes, "Chance in *Silas Marner,*" *English,* 31 (1982): 213–218.
42. "The effect of Dickens's characteristic method," says Harland Nelson, "is an impression of an all-pervading design in human affairs, unexpectedly encompassing and harmonizing the profusely various elements of the story; not (as in a novel by Collins) an impression of an unbroken chain of events, unobtrusively laid down and given a final shake to bring the whole linked series at once into view" ("Dickens's Plots: 'The Ways of Providence' or the Influence of Collins?" *Victorian Newsletter,* 19, 1961: 11). But the strain to make the harmony is evident in the great novels, and the ways of providence or of natural theology are often challenged by the methods designed to affirm them.
43. Charles Dickens, "Magic and Science," *All the Year Round* (March 23, 1861): 562.
44. Mayr, *The Diversity of Life,* p. 37.
45. Manier's summary of Darwin's views on the wars of nature that help further natural selection demonstrates how inappropriate moral and generalizing application of Darwinian theory was:

> Success in these wars of organic being could *not* be traced to variations which were *favorable* in some *absolute* sense; on the contrary, a variation must be understood to be *successful in relation* to some particular segment of the range of alternative variations, and in the context of the chances of life which happened to be available in the given physical circumstances or conditions. Such expression as "chance offspring" or "round of chances" alluded to the complexity of the predictions used in Darwin's hypothesis, and implied that this complexity could not be reduced in the way that Newton had reduced the complexity of planetary motion by formulating a few generally applicable laws. (*The Young Darwin*, pp. 121–122)

46. Beer, *Darwin's Plots,* p. 53.
47. Mayr, *The Diversity of Life,* p. 283.
48. Beatty, "Speaking of Species," p. 265.
49. Mayr, *The Growth of Biological Thought,* pp. 304–305, 38.
50. Darwin was not totally consistent in his rejection of essentialism. Not only

is such consistency impossible given the nature of our language, but Darwin could himself employ arguments that imply an essentialist reality. This is particularly so when, under the pressure of antievolutionary arguments, he revises the *Origin* in later editions and makes many more concessions on the inheritance of acquired characteristics. Daniel Simberloff emphasizes the importance of the rejection of essentialist typology in the Darwinian revolution and points to various Darwinian ideas that nevertheless revert to essentialism. See "A Succession of Paradigms in Ecology: Essentialism to Materialism and Probabilism," *Synthese,* 43 (1980): 3–29, where he cites Richard Lewontin in arguing that Darwin's belief in the blending (not particularistic) theory of inheritance is "readily traced to [his] attachment to essentialist, typological thought." Darwin was hampered by his lack of knowledge of genetics and of Mendelian theory, which made evident that inheritance was not a blending but a 1:2:1 distribution of genetic materials, some of which would not be visible in the parent. Simberloff also notes Lewontin's description of Darwin's "retreat to idealism or essentialism" in his theory of "pangenesis," for the hypothetical "gemmules" are "egregiously ideal essence-conferring entities" (pp. 6–7). See also R. D. Lewontin, "Darwin and Mendel—the Materialist Revolution," in *The Heritage of Copernicus: Theories "More Pleasing to the Mind,"* ed. J. Neyman (Cambridge, Mass.: MIT Press, 1974), pp. 166–183.

51. Barbara Hardy, *"Martin Chuzzlewit,"* in *Dickens and the Twentieth Century,* ed. John Gross and Gabriel Pearson (Toronto: University of Toronto Press, 1962), pp. 107–120. Hardy points out that the abruptness of Martin's conversion is typical of Dickens: "There is no point in comparing Dickens's conversions here with the slow and often eddying movement traced in George Eliot or Henry James. But I think this change is even more abrupt in exposition, relying heavily on compressed rhetoric, than the fairly abrupt conversions of David Copperfield or Bella Wilfer, though the important difference lies in the context of dramatized moral action. Dombey, Steerforth, Gradgrind, Pip, and other major and minor examples of flawed character—not necessarily changing—are demonstrated in appropriate action, large and small" (p. 114). Hardy is criticizing not the abruptness but the way the abrupt change is dramatized and contextualized. The Dickensian mode of change, belonging to a very different tradition, has its own constraints.

52. Bersani, *A Future for Astyanax,* p. 18.

53. Charles Darwin, *The Expression of the Emotions in Man and Animals* (1872; repr., Chicago: University of Chicago Press, 1965), p. 175.

54. Beer, *Darwin's Plots,* p. 8.

55. Ghiselin, *The Triumph of the Darwinian Method,* p. 61.

6. *Little Dorrit and Three Kinds of Science*

1. For an excellent survey of contributions to thermodynamic theory by Carnot (a book of 1824), Joule, Mayer (major papers in 1842 and 1845), Clausius (a book

of 1850), and Helmholtz (a definitive paper in 1847), see Charles Gillispie's *The Edge of Objectivity,* pp. 352–405.

2. See "The Conservation of Force: A Physical Memoir," in *Selected Writings of Hermann von Helmholtz,* ed. Russell Kahl (Middletown: Wesleyan University Press, 1971).

3. "The Mysteries of a Tea-Kettle," *Household Words,* 2 (November 16, 1850): 176–181; James Prescott Joule, "On Matter, Living Force, and Heat," reprinted in *Science Before Darwin,* ed. Howard Mumford Jones and I. Bernard Cohen. The teakettle essay was one of several written by Percival Leigh, at the suggestion of Dickens, based on lectures that Faraday delivered at the Royal Institution. Dickens clearly wanted his readers to understand the materials of their lives in the terms of the most advanced science available to them. Although there is a quality of "believe-it-or-not" about some of the facts explained, the essay provides wonderfully lucid characterizations of "latent heat," and, more important, comfortably demonstrates the complexity of the ostensibly simple materials of ordinary life. The world of the teakettle is a world of transformations, in which no energy or matter is destroyed. Joule's essay also aspires to a clarity necessary for the intelligent lay audience, and he makes the overall thesis that points to the law of the conservation of energy much more overtly. He does it, moreover, by making physical law compatible with divine: "We might reason, *a priori,* that . . . absolute destruction of living force cannot possibly take place, because it is manifestly absurd to suppose that the powers with which God has endowed matter can be destroyed any more than that they can be created by man's agency; but we are not left with argument alone, decisive as it must be to every unprejudiced mind . . . How comes it to pass that, though in almost all natural phenomena we witness the arrest of motion and apparent destruction of living force, we find that no waste or loss of living forces has actually occurred? . . . Experiment has shown that wherever living force is apparently destroyed or absorbed, heat is produced" (pp. 178–179).

4. See Lynn Barber, *The Heyday of Natural History* (Garden City, N.J.: Doubleday, 1980): "Victorian natural history books were written with the aim of encouraging their readers to see evidence of God's existence and attributes in the natural organisms around them, by means of natural theology. And on this basis it was quite legitimate to pass over any facts which did not immediately illustrate God's goodness or wisdom" (p. 72). Although a popular account, Barber's is sensible and reliable and makes no claim to consideration of the more serious scientific word of the time. *Household Words,* although it often attempted fuller accounts of natural history, frequently presented essays full of theo- and antropocentric readings of nature, and fully confirms Barber's general account of popular versions of natural history.

5. See James R. Moore, *The Post-Darwinian Controversies.* Dickens was ready for Darwin's *Origin* when it arrived; but he was not ready for the way Darwin rejected

the argument from design to make his own case. The kind of Darwinism Dickens might have subscribed to was like Robert Chambers's in his notorious *Vestiges of Creation,* or like Asa Gray's teleological version in "Natural Selection Not Inconsistent with Natural Theology." See Asa Gray, *Darwiniana: Essays and Reviews Pertaining to Darwinism* (New York: D. Appleton and Co., 1884). In a popular exposition of the theory in December 1860, Henry Fawcett argued: "Those who, like Mr. Darwin, endeavour to explain the laws which regulate the succession of life, do not seek to detract one iota from the attributes of a Supreme Intelligence. Religious veneration will not be diminished, if, after life has been once placed upon this planet by the will of the Creator, finite man is able to discover laws so simple that we can understand the agency by which all that lives around us has been generated from those forms in which life first dawned upon it" (*Macmillan's Magazine,* 3, December 1860: 145–146).

6. William Thomson, Lord Kelvin, might well have been an intellectual ringleader in an attempt to "get" Darwin. David L. Hull, in "Darwinism as a Historical Entity" (in *The Darwinian Heritage*), argues that "the only social group of scientists . . . that formed to oppose Darwin had as its sociological exemplar Lord Kelvin. The number of Darwin's critics that had close ties to Lord Kelvin is so high that it is difficult not to suspect some sort of conscious intent on their part" (p. 798). Kelvin's publications on the debate of physics with Darwinism are extensive and damaging. See, for example, "On the Secular Cooling of the Earth," *Philosophical Magazine,* 25 (1863): 1–14; "The Doctrine of Uniformity in Geology Briefly Refuted," *Proceedings, Royal Society of Edinburgh,* 5 (1865): 512; "On Geological Time," *Transactions of the Glasgow Geological Society,* 3 (1871): 1–28; "The Age of the Earth as an Abode Fitted for Life," *Philosophical Magazine,* 47 (1899): 66–90. For a brief interesting discussion of Kelvin's argument, see Hull, *Darwin and His Critics* (Chicago: University of Chicago Press, 1973), pp. 349–350.

7. Herbert Spencer, *First Principles,* 5th ed. (New York: A. L. Burt, n.d.), pp. 444–445.

8. Brush, *Temperature of History,* pp. 57–60.

9. Note how Joule regards his discovery of the law of conservation of energy, and how quickly assimilable it is to natural theology:

> Indeed the phenomena of nature, whether mechanical, chemical, or vital, consist almost entirely in a continual conversion of attraction through space, living force, and heat into one another. Thus it is that order is maintained in the universe—nothing is deranged, nothing is ever lost, but the entire machinery, complicated as it is, works smoothly and harmoniously. And though, as in the awful vision of Ezekiel, "wheel may be in the middle of the wheel," and everything may appear complicated and involved in the apparent confusion and intricacy of an almost endless variety of causes, effects, conversions, and arrangements, yet is the most perfect regularity presented—

the whole being governed by the sovereign will of God. ("On Matter," p. 184)

10. For a *relatively* compressed discussion of this development to higher forms, see Spencer's *First Principles,* pt. 2, chaps. 12–18. Spencer's theory does comprehend the idea of "persistence of force" and of economy of energy, so that dissipated energy turns into integration of matter. The two processes of all organisms are evolution and dissolution. It is not clear to me that he believes in an ultimate entropic dissolution of life.

11. John Tyndall, *Heat: A Mode of Motion,* 5th ed. (1863; repr., London: Longmans, Green and Co., 1875), p. 490.

12. Using the basic argument of Gillespie's *Charles Darwin and the Problem of Creation* for my purposes here, I suggest that *Little Dorrit* is a text at the intersection of two competing ways of knowing, the "creationist" and the positivist. Dickens's narrative seeks signs of the creator but is constrained by the condition of a world that seems irredeemably secular (see Gillespie, p. 10).

13. William Thomson, "On the Age of the Sun's Heat" (1862), in *Popular Lectures and Address,* 3 vols. (London: Macmillan and Co., 1891–1894), I, 349.

14. Ibid., I, 350.

15. Helmholtz's researches that led to his formulation of a theory of conservation of force had beginnings in physiology. He argues that "there is a close connection between both the fundamental questions of engineering and the fundamental questions of physiology with the conservation of force. For getting machines into motion, it is always necessary to have motive power, either in water, fuel, or living animal matter." He points out that those who thought they had in every animal a machine that was a perpetual mover "were not aware that eating could be connected with the production of mechanical power"; see "The Application of the Law of the Conservation of Force," in *Selected Writings of Hermann von Helmholtz,* p. 115. Helmholtz is much less exuberantly engaged with the potential moral implications of treating the human body as matter than Tyndall, whose treatment of the subject in *Heat* follows Helmholtz's lines. Tyndall insists on a final total separation between science and questions that do not deal with matter, and lapses into lyricism frequently, as on page 489:

> Still, though the progress and development of science may seem to be unlimited, there is a region beyond her reach—a line with which she does not even tend to inosculate . . . When we endeavour to pass . . . from the region of physics to that of thought, we meet a problem not only beyond our present powers, but transcending any conceivable expansion of the powers we now possess. We may think over the subject again and again, but it eludes all intellectual presentation. The origin of the material universe is equally inscrutable. Thus, having exhausted science, and reached its very rim, the real mystery of existence still looms around us. And thus it will ever

loom—ever beyond the bourne of man's intellect—giving the poets of successive ages just occasion to declare that

> We are such stuff
> As dreams are made of, and our little life
> Is rounded by a sleep.

16. See Charles Dickens, *Little Dorrit,* ed. Harvey Peter Sucksmith (Oxford: Oxford University Press, 1979), p. xiii.

17. Ibid., p. xviii.

18. We don't need science for this: there are more obvious places to turn, for instance, to Calvinism and bureaucracy, two of the targets of Dickens's anger in *Little Dorrit.* At the time he was writing *Little Dorrit* Dickens was both enraged and depressed by the failures of British institutions. Edgar Johnson has shown how the horrifying official ineptness during the Crimean War, the failures to effect parliamentary and administrative reform, the conditions in England itself, and the developing disasters of his own marriage helped feed his imagination of uncontrollable disorder; see Johnson's *Charles Dickens: His Tragedy and Triumph,* 2 vols. (New York: Simon and Schuster, 1952), esp. chap. 6. A man of action, Dickens was finding England a great circumlocution office, teaching daily how not to do it. Everywhere, even (worryingly) in himself, he found a failure of will to change. As Lionel Trilling argued, in his well known essay on the novel, it is about "the will and society"; "at the time of *Little Dorrit* [Dickens] was at a crisis of the will which is expressed in the characters and forces of the novel" (*"Little Dorrit,"* in *The Opposing Self,* New York: Viking Press, 1955, pp, 57, 63).

19. Barbara Hardy, in response to an earlier version of this chapter, objected to the treatment of images and patterns without countering reference to the full context for them. Arguing that *Little Dorrit* is not about entropy, she demonstrated the pattern of growth toward love and life to be traced through the relationship of Amy and Clennam. I do not, of course, claim that the novel is "about" entropy; I only suggest that it sustains within itself a pattern subversive of the overall narrative direction, and that this pattern shares with the developments in thermodynamics certain qualities which might be usefully juxtaposed in a demonstration of the mutual mythic enterprises of science and literature.

20. Charles Dickens, *Little Dorrit* (Harmondsworth: Penguin Books, 1967), p. 39. Page numbers in the text refer to this edition.

21. Gerald Holton and Stephen Brush, *Introduction to Concepts and Theories in Physical Science* (Princeton: Princeton University Press, 1985), p. 287.

22. It is the movement of time that is central to *Little Dorrit*'s narrative meaning, and, as Beaty says, "Human life in time is like . . . the road"; see Jerome Beaty, "The 'Soothing Songs' of *Little Dorrit*: New Light on Dickens's Darkness," in *Nineteenth-Century Literary Perspectives,* ed. Clyde de L. Ryals (Durham: University of North Carolina Press, 1974), p. 229.

23. Edward Eigner, *The Metaphysical Novel in England and America* (Berkeley: University of California Press, 1978), p. 116.

7. *The Darwinian World of Anthony Trollope*

1. "Our new journal," he later wrote, "became an organ of liberalism, free-thinking, and open inquiry"; see Anthony Trollope, *An Autobiography* (London: Oxford University Press, 1950), p. 164. Page numbers in the text refer to this edition.

2. Ruth apRoberts, connecting the main stream of Victorian realistic fiction with German historicism, locates in Trollope many of the qualities that mark him as of the Darwinian ethos. The title of her essay "Trollope and the *Zeitgeist*," *Nineteenth Century Fiction*, 37 (1982), suggests the interplay between aspects of culture that I am trying to demonstrate in these chapters. What apRoberts appropriately attributes to historicism is part of what fed into the way Darwin ultimately worked out his theory: "Historicism emphasizes process, relativity, unceasing change, development. Trollope's novels exhibit all these; his people develop, age, and die. But perhaps the most notable mark of his process-art is the openness of form of his novels: the continuation of one set of characters into another is the objective correlative of Darwinian process" (p. 271). apRoberts's focus on the individual character, "multiform, infinitely variable, infinite in potential," is equally to the point.

3. In an early essay I argued that Trollope was sensitive in his exploration of the grounds for and the possibilities of women's rejection of male authority. But in *Can You Forgive Her?* his plots deliberately thwart the insights aroused, and finally treat with condescension those gestures at rebellion that, for many chapters, carried considerable weight. See "Can You Forgive Him? Trollope's 'Can You Forgive Her?' and the Myth of Realism," *Victorian Studies*, 18 (1974): 5–30. It is not that Trollope is taking nature as a model for gradualist change, since nature does not play a significant role in his novels, except in fox-hunts, but that he imagines it "natural" within society for change to be gradual. It is "natural" that men are stronger and wiser than women, despite the many strong and wise women in his novels.

4. *An Autobiography*, partly because it *seems* so much the product of an intellectually ordinary mind, is a peculiarly interesting and controversial document that has attracted a very large literature, particularly in recent years. In a combative defense of Trollope's genius, A. L. Rowse comments on the notorious effect of *An Autobiography* on Trollope's reputation. Rowse argues, "There is not the slightest doubt that, the most honest and sincere of writers that ever lived, Trollope had no intention [to mislead]. He really did think that he was no genius; there he was wrong. He thought that he was not in the same class with Thackeray; but he was wrong . . . ("Trollope's *Autobiography*," in *Trollope Centenary Essays*, ed. John

Halperin, New York: St. Martin's Press, 1982, p. 137). The most interesting recent consideration of the *Autobiography* is Walter Kendrick, *The Novel Machine: The Theory and Fiction of Anthony Trollope* (Baltimore: John Hopkins University Press, 1980). Kendrick's argument is that while the *Autobiography* looks like a "banal assertion of clichés," it is "the clearest, most comprehensive statement of the theory of realism that realism itself has ever produced" (p. 3). Kendrick's argument is useful for my purposes in that he shows that for Trollope "the realistic novel . . . is never a static structure to be contemplated or reflected upon. It is always dynamic, a process rather than an object" (p. 4). The aesthetic, contemplative tradition of James, a reaction against Trollopean realism, can be viewed in one respect as an attempt to fix time, to withdraw an object, a scene, a moment itself, from the stream of time. Art on this model resists the revelations of Darwinian "reality," sets itself up as a value against it. Hence, the deep tensions in Darwinian-inspired writers like James himself and Conrad, where the attempt to stop time is constant, and constantly failing.

5. The question of the structure of Trollope's novels has been receiving increasingly aggressive formalist analysis. A particularly interesting example of this has reached me only as I finish this book: Christopher Herbert, *Trollope and Comic Pleasure* (Chicago: Chicago University Press, 1986). Herbert, like Rowse and Kendrick, rejects those critics who take the autobiographical self-deprecations as valid, and he argues impressively against my contention (which I nevertheless maintain) that "technically. . . , Trollope is profoundly uninteresting." Herbert successfully shows, however, that Trollope's novels are designed as comedies, in a strict sense, "based upon the formal patterns of standard stage comedy . . . possessing a traditional thematic structure" (p. 3). "Comedy," he argues, enabled Trollope "to focus on a particular phenomenon of which he was obsessively aware: the appearance of a rapidly deepening split in the field of literature between the popular and the serious" (p. 5). Agreeing with Kendrick that Trollope valued reality over literature, he points out—and this is certainly correct—that his practice and his theory are "sharply at odds" (p. 6) because he studied and used the traditions of comedy in English literature quite extensively. "The realism proclaimed in the title of *The Way We Live Now,*" claims Herbert, given the number and pervasiveness of Trollope's borrowings from earlier comedies, is "in fact of a dissembling or oblique kind at best" (p. 7). Kendrick's argument on this point seems stronger, that Trollope's realism did indeed employ "conventions," but that the conventional nature of Trollope's art is one of the conditions of realism itself.

6. N. John Hall, ed. *The Letters of Anthony Trollope,* 2 vols. (Stanford: Stanford University Press, 1983), I, 447.

7. Trollope, *Letters,* II, 804.

8. Anthony Trollope, *Doctor Thorne* (Boston: Riverside Press, 1959), p. 30. Page numbers in the text refer to this edition. The novel was published in 1858, at about the time Darwin was beginning work on the *Origin.*

9. For the record, here are a few facts about Trollope's connections with science. (My information about Trollope's library is derived from Richard H. Grossman and Andrew Wright, "Anthony Trollope's Libraries," *Nineteenth Century Fiction,* 31, 1976: 48–64.) Although his library included a run of the *Fortnightly Review* through 1873, if his own descriptions are accurate, he surely knew little about its scientific essays. Trollope inherited a great many books from a friend and seems to have sold off those on natural history and science—"a sparse area" among Trollope's holdings (p. 53). John Clark points out, however, that a dialogue in *Barchester Towers* between Charlotte Stanhope and Eleanor Bold alludes to the well-known scientific debate on the question of the "plurality of worlds": "Are you a Whewellite or a Brewsterite, or a t'othermanite, Mrs. Bold?" asks Charlotte (*The Language and Style of Anthony Trollope,* London: Andre Deutsch, 1975, p. 161). Trollope would not have needed to read the books by Brewster and Whewell to work up the conversation, but the allusion does make for an interesting anomaly in the Trollope canon. Clark notes no other potentially informed allusion to naturalists or scientists in any of Trollope's novels.

10. For example, on December 13, 1855, Darwin notes without comment having read "The Warden (a novel)," and on July 15, 1859, he notes reading "Bertram." *The Bertrams,* which includes a sequence in which the protagonist's spiritual faith is compromised by the realities of a pilgrimage to the East, and which lacks a fully satisfactory "happy ending," was published in March (Vorzimmer, "The Darwin Reading Notebooks," pp. 147, 152). Elsewhere he alludes to Trollope unfavorably: "Remember," he writes, "what Trollope says in Can You Forgive Her? about getting into Parliament, as the highest ambition . . . a poor-sighted view" (*Life and Letters,* III, 41).

11. John Stuart Mill classified contemporary civilization as a condition in which "power passes more and more from individuals . . . to masses"; see "Civilization," originally published in 1836 in the *Westminster Review,* reprinted in *Essays on Politics and Culture,* ed. Gertrude Himmelfarb (Garden City, N.J.: Doubleday, 1962), p. 53. His own autobiography shares the rhetorical strategies of Trollope and Darwin.

12. See N. John Hall, "Trollope the Person," in *Trollope Centenary Essays.* Hall's analysis of why Trollope behaved as he did is much to the point of my argument about the autobiographies: "Trollope, more than most men, prepared a face to meet the faces that he met. He assumed a role, put on a mask, acted out a part. It was a strategy to keep people . . . at arm's length, to camouflage the insecurities and fears that lay so close to the surface. It was his way of coming to terms with his early years" (p. 178).

13. Gertrude Himmelfarb, *Darwin and the Darwinian Revolution* (Garden City, N.J.: Doubleday, 1959), p. 140.

14. Charles Darwin, *The Autobiography of Charles Darwin, 1809–1882,* ed. Nora Barlow (New York: W. W. Norton, 1958), p. 140.

15. Henry James, "Anthony Trollope," in *Henry James: Essays on Literature, American Writers, English Writers* (New York: Library of America, 1984), p. 1335.

16. Herbert, *Trollope and Comic Pleasure,* p. 114.

17. Kendrick rightly argues that Trollope's whole enterprise is antifigurative. He points out, however, that Hawthorne's famous metaphor about Trollope's realism demonstrates that Trollope worked metonymically—"where the reader comes face to face with an enclosed chunk of the same earth he stands on" (*The Novel Machine,* p. 127). See also Michael Riffaterre, "Trollope's Metonymies," *Nineteenth Century Fiction,* 37 (1982): 272–292. The "realistic" detail's "true function," says Riffaterre, is to refer to and ultimately symbolize "something other than its 'natural' referent" (p. 272). It does not thereby lose its realistic function, but implicitly suggests the intrusion of the author, guiding the real. Trollope's (like Darwin's) ostensible empiricism, that is to say, cloaks a guiding intention. The irresistibly figurative nature of language entraps Trollope, as it must any "realist," even as he keeps the pressure on the particular and resists the general and the representative.

18. Gould, *Panda's Thumb,* p. 66.

19. Ruth apRoberts, *The Moral Trollope* (Athens: Ohio University Press, 1971), pp. 82–83.

20. See Herbert, *Trollope and Comic Pleasure,* in which he discusses Trollope's use of the traditions of stage comedy, and Robert Tracy, *Trollope's Later Novels* (Berkeley: University of California Press, 1978). Tracy argues that when we recognize in Trollope the techniques of Elizabethan and Jacobean drama (he left 257 carefully annotated plays in his library), we find that "those of his novels that have been most often condemned as shapeless emerge as conspicuous formal achievements" (p. 10). "Far from being straggling, episodic, and innocent of plan, form, and structure, Trollope's later novels represent an attempt to adapt the multiple plot structure of Elizabethan and Jacobean dramatists to the novel" (p. 39). Both Herbert and Tracy are convincing about the way Trollope borrowed from earlier drama for the structures of many of his novels. While this discovery is a very useful one and should help rescue Trollope from the literal readings of his *Autobiography,* the patternings discovered do not constitute particularly original uses of narrative conventions and do not, I believe, undercut the basic antiplot formulas Trollope used. It is possible to claim that, given his own perceived weakness in plotting, Trollope fell back on the rigorous strategies of dramatists to help provide a skeleton on which to build his realist structures.

21. ApRoberts, *The Moral Trollope,* p. 87.

22. Eigner, *The Metaphysical Novel,* p. 17.

23. As Walter Kendrick has put it, "Life, for the realist, is always said to come first, and writing follows" (*The Novel Machine,* p. 5).

24. See Shirley Robin Letwin, *The Gentleman in Trollope: Individuality and Moral Conduct* (Cambridge, Mass.: Harvard University Press, 1982). Letwin

insists on the complexity of the idea of the "gentleman" in Trollope and posits it as an issue not so much of "class" as of morality. Nevertheless, association with an ideal of class remains strong in Trollope's ideal of the gentleman. This is evident even in Letwin's convincing conclusion, in which she describes the gentleman in this way: "But whatever disagreement he encounters, however uncongenial he may find his neighbours or his fortune, he will always be thoroughly at home in the human world because he can enjoy absurdities and has no ambition to overleap mortality" (p. 269). ApRoberts is certainly right about the moral implications of Trollope's need for truth telling in fiction, however, when she says that for Trollope's scientific experimental method, "a failure in realism is . . . an unethical manipulation of data" (*The Moral Trollope,* p. 43).

25. Anthony Trollope, "Novel Reading," *The Nineteenth Century* (January 1879): 43.

26. Donald Fleming, "Charles Darwin, the Anaesthetic Man," *Victorian Studies,* 4 (1961): 232.

27. James Kincaid, *The Novels of Anthony Trollope* (Oxford: Clarendon Press, 1977), p. 40.

28. E. S. Dallas, unsigned notice, in *Trollope: The Critical Heritage,* ed. Donald Smalley (New York: Barnes and Noble, 1969), p. 104.

29. Anthony Trollope, *Thackeray* (New York: Harper and Brothers, n.d.), p. 120.

30. Herbert, *Trollope and Comic Pleasure,* p. 114.

31. Anthony Trollope, *Framley Parsonage* (London: T. Nelson and Sons, n.d.), p. 275.

32. Anthony Trollope, *The Small House at Allington* (London: J. M. Dent and Sons, 1950), pp. 243, 258. Page numbers in the text refer to this edition.

33. Anthony Trollope, *The Claverings* (New York: Dover Publications, 1977), p. 79. Page numbers in the text refer to this edition.

34. Alexander Welsh, *George Eliot and Blackmail* (Cambridge, Mass.: Harvard University Press, 1986).

8. The Perils of Observation

1. John Locke, *An Essay Concerning Human Understanding,* ed. A. S. Pringle-Pattison (Oxford: Clarendon Press, 1924), pp. 42, 43. While we know that for Locke "experience" did not imply the rather purer sensationalism of, say, Hume or of later, nineteenth-century scientific empiricists, there can be no question about the importance of his turn to experience as opposed to authority for the source of knowledge.

2. Herschel, *Preliminary Discourse,* p. 58.

3. Maurice Mandlebaum, *History, Man, and Reason: A Study in Nineteenth Century Thought* (Baltimore: Johns Hopkins University Press, 1971), p. 350.

4. T. H. Huxley, "On the Advisableness of Improving Natural Knowledge," in *Methods and Results* (London: Macmillan and Co., 1893), p. 40.

5. "Advertisement to *The Germ,*" reprinted in *Strangeness and Beauty,* ed. Eric Warner and Graham Hough (Cambridge: Cambridge University Press, 1983), p. 105.

6. Michel Foucault, *Discipline and Punish: The Birth of the Prison* (New York: Vintage Books, 1979). See the discussion of examinations, pp. 184–194, and the following chapter, "Panopticism," pp. 195–228.

7. Ibid., p. 192.

8. Ibid., p. 193.

9. John Herschel, "Review of Whewell's *History* and *Philosophy of the Inductive Sciences,*" *Quarterly Review,* 69 (1840): 179.

10. Baden Powell, *The Unity of Worlds and of Nature: Three Essays* (London, 1856), p. 77.

11. Stocking, *Victorian Anthropology*, p. 22.

12. See Stocking's discussion of this episode in Victorian anthropological history in *Victorian Anthropology,* pp. 248–257.

13. Charles Pickering, *The Races of Man, and Their Geographical Distribution,* to which is prefixed "An Analytical Synopsis of the Natural History of Man," by John Charles Hall (London: H. G. Bohn, 1854), pp. lxxii, xiii.

14. Darwin, *The Expression of the Emotions,* pp. 14, 17.

15. Ibid., p. 36.

16. W. K. Clifford, "Philosophy of the Pure Sciences," *Lectures and Essays,* ed. Leslie Stephen and Frederick Pollock (London: Macmillan, 1901), I, 335–336.

17. Karl Pearson, *The Grammar of Science* (1892; repr., London: Everyman, 1937), pp. 60–61.

18. Ibid., p. 61.

19. Welsh, *Eliot and Blackmail,* p. 337.

20. Ibid., p. 339.

21. See Frank Kermode, *The Romantic Image* (New York: Vintage Books, 1957).

22. George Eliot, *Middlemarch* (Harmondsworth: Penguin Books, 1965), p. 226

23. Kermode, *Romantic Image,* p. 9.

24. Welsh, *Eliot and Blackmail,* p. 104.

25. Oscar Wilde, *The Picture of Dorian Gray* (Harmondsworth: Penguin Books, 1983), p. 68.

26. Eliot, *Adam Bede,* p. 341.

27. Welsh: "Whereas Grandcourt preys on Gwendolen's fear of being found out, Deronda tries to cultivate in her a sense of guilt. The distinction is sometimes hard to discern, since Gwendolen's 'fear' is prevalent throughout; but the fear that is natural to her is fear of shame or exposure, and her sense of guilt has to be

enlarged" *Eliot and Blackmail,* p. 293). Welsh links these alternative intentions with the Freudian project which seeks, through close observation of the patient, to find those points of shame and guilt that are the source of neurosis, and by exposure to overcome them.

28. W. M. Thackeray, *Pendennis,* 2 vols. (London: Everyman, 1959), I, 251.

29. See, for example, W. B. Carpenter's review of Whewell's *History of the Inductive Sciences,* where he describes the special difficulty of the physiologists, who will be successful in discovering the laws of organisms "in the inverse proportion to the derangement of the train of vital actions which shall have been created by the violence he is compelled to inflict" on them in his observations (pp. 333–334).

30. Emma Hardy, *The Early Life of Thomas Hardy* (London: Macmillan and Co., 1928), p. 198.

31. Alexander Welsh, "Theories of Science and Romance, 1870–1920," *Victorian Studies,* 17 (1973): 135–154.

32. Thomas Hardy, *The Woodlanders* (Harmondsworth: Penguin Books, 1981), pp. 41, 43. Page numbers in the text refer to this edition.

33. Allon White, *The Uses of Obscurity: The Fiction of Early Modernism* (London: Routledge and Kegan Paul, 1981), see esp. pp. 30–36, 71. Meredith's fiction, with its extraordinary evasiveness and circuitous approach to major scenes, plays out in a different way the preoccupation with "observation" and with its dangers. The crisis of *The Egoist,* for example, comes with Crossjay's overhearing Sir Willoughby's proposal to Laetitia. Although this kind of scene has roots that reach as far back as restoration drama, the novel's concentration on Sir Willoughby's terror of being observed has modern reverberations. On the whole, I would argue, the nineteenth-century English novel's preoccupation with propriety and class is importantly linked to novelistic experiments with perception and observation. The scientists' valuing of observation and the society's valuing of gossip converge. This idea is strongly borne out by Welsh in his study of Eliot.

34. Thomas Hardy, *A Pair of Blue Eyes* (London: Macmillan, 1975), p. 267.

35. Thomas Hardy, *Tess of the D'Urbervilles* (Harmondsworth: Penguin Books, 1978), p. 200.

36. G. H. Lewes, *Problems of Life and Mind,* 5 vols. (London: Kegan Paul, Trench, Trübner, 1874–1879). For Lewes's definition of "feeling" as "a generalised expression for what all mental states have in common," including "ideation," see "Third Series," "Problem II," chap. 1.

37. Thomas Hardy, *Two on a Tower* (London: Macmillan and Co., 1975), p. 86.

38. Ibid., p. 146.

39. Arthur Balfour, *Defence of Philosophic Doubt: Being an Essay on the Foundations of Belief* (London: Macmillan and Co., 1879), p. 288.

40. Arthur Balfour, "Reflections Suggested by the New Theory of Matter," Presidential Address to the British Academy for the Advancement of Science (London: Longmans, Green, 1904), pp. 18–19.

41. Wilde, *Dorian Gray,* p. 124.
42. Joseph Conrad, Preface to *The Nigger of the "Narcissus"* (New York: Doubleday, 1924), p. xiv.
43. Joseph Conrad, "Heart of Darkness," in *Youth* (New York: Doubleday, 1924), p. 48.

9. From Scott to Darwin to Conrad

1. See Hillis Miller's interesting discussion of the willed and self-oriented basis of Trollope's protagonists' values in *The Form of Victorian Fiction.* In a very different way, Miller is arguing the newly secular basis of Victorian narrative, and his analysis curiously echoes Darwinian argument: "The collective consciousness is like each individual consciousness in that it exists as the spontaneous will to sustain itself as continuous with itself. Like the individual self, it is sustained by nothing outside mankind. It is for this reason that Trollope puts such a high value on the unbroken historical continuity of English society . . . The fact that culture is a collective game, the shared will to go on living by certain rules of action and judgment, is not seen by those living with the culture and accepting its rules as absolutely valid" (p. 135).
2. See Robert Young, "The Impact of Darwin on Conventional Thought," in *Darwin's Metaphor,* p. 20.
3. Frederick Engels, *The Dialectics of Nature,* Preface and notes by J. S. Haldane (New York: International Publishers, 1940), p. 19.
4. T. H. Huxley, *Evolution and Ethics* (London: Macmillan and Co., 1911), pp. 26–27.
5. Ibid., p. 45.
6. See Georg Lukács, *The Historical Novel,* trans. Hannah Mitchell and Stanley Mitchell (Boston: Beacon Press, 1962). Scott, whom Lukács describes as a continuer "of the great realistic social novel of the eighteenth century" (p. 31), is great "in his capacity to give living human embodiment to historical-social types. The typically human terms in which great historical trends become tangible had never before been so superbly, straightforwardly and pregnantly portrayed. And above all, never before had this kind of portrayal been consciously set at the centre of the representation of reality" (p. 35).
7. Walter Scott, *Old Mortality* (Harmondsworth: Penguin Books, 1975), p. 94. Page numbers in the text refer to this edition.
8. Joseph Conrad, *Under Western Eyes* (New York: Doubleday, 1963), p. 12. Page numbers in the text refer to this edition.
9. George Lukács, *Studies in European Realism,* intro. Alfred Kazin (New York: Grosset and Dunlap, 1964). Lukćas opposed impressionism and all contemporary narrative entrapped by the atypical subjectivity of its protagonists. Such writing "describes . . . quite individual, non-recurring traits," and breaks down the

conception of the "complete human personality" (p. 8). One might note, by the way, that there are conflicting scientisms at work in this argument. Lukaćs aspires to a science in which all details are seen in a larger, general context. His rejection of impressionism is a rejection of the extreme of empiricism I invoked in the last chapter with discussion of Pearson. The problem for such empiricism was to transform the unique into the general, a possibility denied by Pater's sort of art and theory, for example. Empiricist fragmentation becomes, in art, bourgeois mystification. The question of the unique phenomenon has remained since Darwin an extremely difficult one at all levels of discourse. See George Lukaćs, *Realism in our Time: Literature and the Class Struggle* (New York: Harper and Row, 1962), where he makes his most sustained case against the subjectivity and abstractness of modernism as against the healthy balance of subjective and objective of the great realists, like Balzac.

10. See the discussion in John Hedley Brooke, "The Relation between Darwin's Science and His Religion," p. 60. Brooke points to other interesting discussions of the problem, in particular Dov Ospovat, "God and Natural Selection: The Darwinian Idea of Design," *Journal of the History of Biology,* 13 (1980): 184–189; Sylvan Schweber, "The Young Darwin," *Journal of the History of Biology,* 12 (1979): 187–188; and Edward Manier, *The Young Darwin and His Cultural Circle,* p. 121. In the Duke of Argyll's *The Reign of Law* (1866), he argues that what appears to be chance is the convergence of different and unrelated causal streams:

> The notion . . . that the uniformity or invariableness of the Laws of Nature cannot be reconciled with their subordination to the exercise of Will, is a notion contrary to our own experience. It is a confusion of thought arising out of the ambiguity of language. For let it be observed that, of the senses in which the word Law is used, there is only one in which it is true that laws are immutable or invariable; and that is the sense in which Law is used to designate an individual Force . . . There are no phenomena visible to Man of which it is true to say that they are governed by an invariable Force. That which does govern them is always some variable combination of invariable forces. But this makes all the difference in reasoning on the relation of Will to Law. (pp. 97–98)

11. Beer, *Darwin's Plots,* p. 64. See also her slightly different version of the argument in "Darwin's Reading and the Fictions of Development," in *The Darwinian Heritage,* esp. pp. 575–577.

12. Shuttleworth, *Eliot and Nineteenth-Century Science,* p. 83.

13. Jane Millgate, in *Walter Scott: The Making of the Novelist* (Toronto: University of Toronto Press, 1984), argues that Morton's survival is contingent on his "power of choice," that he "insists . . . on acting like a free man" (p. 127). While the latter is true, and while it is interesting to see Claverhouse and Burley as incapable of freedom, being caught up in the inevitable movements of history, I

believe that Morton's power of survival is dependent on Scott's rendering him passive in each situation in which his life is on the line. He is pushed around by history and saved by contingencies while he insists on his freedom. He takes up the stance so much admired by the Victorians as the courageous spokesman for liberal values, but his career tends to demonstrate the inadequacy of those values—as is so often the case with Victorian protagonists, who are frequently weak and passive figures.

14. Ioan Williams, ed., *Sir Walter Scott on Novelists and Fiction* (New York: Barnes and Noble, 1968). See Scott's review of *Emma,* p. 230.

15. William Thackeray, *The History of Henry Esmond, Esq.* (New York: Holt, Rinehart, and Winston, 1962), p. xxxviii. Page numbers in the text refer to this edition.

16. Barbara Hardy, *The Exposure of Luxury* (Pittsburgh: University of Pittsburgh Press, 1972). Hardy concedes that "Thackeray was a conservative in social attitudes, despite his interest in Liberal politics . . . Certainly much that he said and did seems to show an acceptance, if not a total liking, for society as he found it." She argues, however, that the tale, rather than the teller is to be trusted, and that Thackeray's tales "expose" "a cruel, cold, mad world" (pp. 15–16). This seems to me undeniably true, but the structures and strategies of those tales carefully defend the narrator from what is exposed and affirm compromise rather than revolution.

17. Charles Dickens, *A Tale of Two Cities* (New York: Random House, 1950), p. 256.

18. Ibid., p. 225.

19. Ibid., p. 251.

20. For an important discussion of this function of the servant in literature, although with particular reference to nineteenth-century fiction, see Bruce Robbins, *The Servant's Hand* (New York: Columbia University Press, 1986).

21. Redmond O'Hanlon, in *Joseph Conrad and Charles Darwin,* argues for the centrality of Darwin to Conrad's imagination, but insists that it is the late Darwin, actually Lamarckian, and impressed with the possibility of degeneration rather than progress, who is most profoundly operative in Conrad's work.

22. See Allan Hunter, *Joseph Conrad and the Ethics of Darwinism* (London: Croom Helm, 1983), pp. 220–239.

23. J. Hillis Miller, *Poets of Reality* (Cambridge, Mass: Harvard University Press, 1965), p. 20.

24. Allan Hunter reads the deeply melancholy letters to Cunningham Grahame as ironic and satirical and not occasions to "lapse into fashionable melancholy." Conrad is "involved in far more complex discussions of man's nature than these easy visions of despair" (*Conrad and the Ethics of Darwinism,* pp. 10, 12). To be sure, Conrad's consideration of these large issues was complex and rich, but the "fashionable melancholy" becomes in his own letters something other than fashionable, and the question of the materialist reading of nature, the determinist

insistence on scientific law and cause and effect, are starting points for his imagination.

25. *Joseph Conrad's Letters to Cunningham Grahame* (Cambridge: Cambridge University Press, 1969), p. 56.

26. William Graham, *The Creed of Science* (London: Kegan Paul, 1881), p. 148.

27. *Life and Letters,* I, 315.

In his *Autobiography,* Darwin wrote similarly and more extensively, on the subject: "Then arises the doubt—can the mind of man, which has, as I fully believe, been developed from a mind as low as that possessed by the lowest animal, be trusted when it draws such grand conclusions? May not these be the result of the connection between cause and effect which strikes us as a necessary one, but probably depends merely on inherited experience? Nor must we overlook the probability of the constant inculcation in a belief in God on the minds of children producing so strong and perhaps an inherited effect on their brains not yet fully developed, that it would be as difficult for them to throw off their belief in God, as for a monkey to throw off its instinctive fear and hatred of a snake" (p. 93).

Index